Springer Series on Biofilms
Volume 6

Series Editor: J. William Costerton

Pittsburgh, USA

For further volumes:
http://www.springer.com/series/7142

Steven L. Percival • Derek C. Knottenbelt •
Christine A. Cochrane
Editors

Biofilms and Veterinary Medicine

Springer

Editors
Prof. Steven L. Percival
Medical School, Department of Pathology
West Virginia University
Morgantown, WV 26506
United States
biofilms@hotmail.co.uk

Dr. Christine A. Cochrane
Faculty of Health and Life Sciences
Institute of Ageing and Chronic Disease
University of Liverpool
Leahurst
Neston. CH64 7TE
United Kingdom
C.A.Cochrane@liverpool.ac.uk

Prof. Derek C. Knottenbelt
University of Liverpool
School of Veterinary Science
Equine Hospital, Leahurst
Chester High Road
CH64 7TE Neston
United Kingdom
knotty@liverpool.ac.uk

Series Editor
J. William Costerton
Director of Microbiology Research
Department of Orthopedic Surgery
Allegheny General Hospital and
Director of Biofilm Research
Allegheny-Singer Research Institute
320 East North Ave.
Pittsburgh PA, 15212
USA

ISBN 978-3-642-21288-8 e-ISBN 978-3-642-21289-5
DOI 10.1007/978-3-642-21289-5
Springer Heidelberg Dordrecht London New York

Library of Congress Control Number: 2011935431

Printed on acid-free paper

Springer is part of Springer Science+Business Media (www.springer.com)

Steven L. Percival would like to dedicate this book to a number of people who have provided support, guidance and reassurance throughout the 20 years he has been studying and researching microbiology and biofilms. These include Carol, Alex, Tom, Mum, Dad, Nigel, Emma, Jorge, Sam, Elisha and Kara.

Christine A. Cochrane would like to dedicate this book to Paul (for your constant support and encouragement), our sons Paul and Mark and grandchildren Jennifer and Callum. Finally, to my Mother Betty who never doubted my ability to become a scientific researcher.

Preface

> Imagination should give wings to our thoughts but we always need decisive experimental proof, and when the moment comes to draw conclusions and to interpret the gathered observations, imagination must be checked and documented by the factual results of the experiment. (Pasteur L)

Louis Pasteur (1822–1985) was an amazing, persevering, perceptive and determined scientist who today is widely regarded as the father of the "Germ Theory" and bacteriology. He is revered for possessing the most important qualities of a scientist: he had an unrivalled ability to scrutinise data on almost any subject and then to develop profound and often fundamental questions from them. He had an uncanny ability to identify the solutions to problems based on analytical scrutiny of data – even without any of the sophisticated statistical tools we have today. He possessed an almost unique and certainly an enviable reputation for patience and drive to research under strictly controlled conditions regardless of the contemporary scepticism that accompanied the current dogma. These are also the characteristics of the scientists who have taken on the needs associated with biofilm research in the modern era. While Pasteur was not the first to propose that disease was the result of pathogenic microorganisms, he developed the principals and theories and conducted the experiments that clearly indicated their relevance.

Since the advent of the antibiotic age man has sought to find "chemical" strategies to overcome pathogens in particular. These have also involved the "devious" manipulation of the immune system through the development of vaccines and hyperimmune sera. The immune stimulating approach to disease is "super-efficient" in that the various cascades of the cellular and humoral immune systems are mobilised specifically at target organisms and in this process, there are a few adverse side effects. The organisms have difficulty overcoming the amazing versatility and target accuracy of the immune system. This results in prevention of disease in the case of vaccine production and/or the limitation of a disease to the extent that the pathogen causes mild or sub-clinical changes. In the event that a pathogen is introduced into a totally susceptible host, there is a race between the immune system and its attempts to both overcome and to eliminate the pathogen and the pathogen's own ability to trigger inflammatory, cytopathic, toxic or other damaging processes. Viruses, bacteria, fungi, yeasts, protozoa and parasites are all capable of causing disease and the survival of any species is surely a testament to the "innate" and "acquired" immune systems that through evolution have developed strategies to at least limit the damage and in many cases to prevent any effect whatever. However, when an infection challenges a naive host, there is a significant delay in the mobilisation of the immune systems resources. During this time, disease can develop and so the objective of modern medicine (including the veterinary and related science and biological professions) is to try to limit the effects of the infection without harming the host animal. Having been exposed to a disease the immune system will react in a co-ordinated fashion to ensure that the disease is as short and as mild as possible and so antimicrobial drugs in these situations would become largely unnecessary.

It is widely accepted that antimicrobial drugs (whether antiviral, antibacterial, antifungal, or antiprotozoal or antiparasitic) are inherently flawed as a long-term strategy for controlling and treating disease because of the evolutionary pressure that will inevitably result in resistance. The concept that "there is an antibiotic that will work if the dose is high enough" is definitely counter to all principles of antimicrobial therapy, and yet it is one of the commonest approaches. It is born out of frustration and lack of understanding as to why bacteria can survive against all the odds. It is surely far better to understand the reasons for failures of efficacy and to address these specifically than it ever is to simply add more and more antibacterial drugs! Whilst there is no doubt at all that antimicrobial strategies have reduced the incidence of disease and reduced the duration of illness associated with infections, the rate of new-molecule development has not kept pace with the ability of the microorganisms to resist them. In many circumstances, failure of efficacy is simply blamed on "resistance" but it is clear from biofilm research that there is much more to "resistance" than meets the eye. The spectrum of drugs used in veterinary species is relatively narrow – a few antibiotics (largely those that are not used in human's medicine!) are used widely. Veterinarians have taken their responsibility for rational use seriously and it is unfair to blame the veterinary profession for the development of antibiotic resistance. There are certainly specific circumstances when antibiotics and the hosts' own immune and reparative processes fail to control infections and one of the most interesting of these is the development of biofilms that protect and "shield" the organisms from potentially damaging environmental and host defences. Biofilms are the most common mode of bacterial growth in nature and are highly resistant to antibiotics.

Biofilms are implicated in many common medical problems including urinary tract infections, catheter infections, middle-ear infections, dental plaque, gingivitis, and some less common but more lethal processes, such as endocarditis, infections in cystic fibrosis. However, biofilms have only recently been given their true importance in the overall process of disease pathogenesis. Bacterial biofilms are one of the fundamental reasons for incipient wound healing failure in that they may impair natural cutaneous wound healing and reduce topical antimicrobial efficiency in infected skin wounds. Their existence explains many of the enigmas of microbial infection and a better grasp of the process may well serve to establish a different approach to infection control and management. Biofilms and their associated complications have been found to be involved in up to 80% of all infections. A large number of studies have been performed targeted at the bacterial biofilms and many of these are referred to in this book, which is the first of its kind. These clinical observations emphasise the importance of biofilm formation to both superficial and systemic infections and the inability of current antimicrobial therapy to "cure" the resulting diseases even when the in vitro tests suggest that they should be fully effective.

In veterinary medicine, the concept of biofilms and their role in the pathogenesis of disease has lagged seriously behind that in human medicine. This is the more extraordinary when one considers that much of the research has been carried out using veterinary species in experimental situations. The clinical features of biofilms in human medicine are certainly mimicked in the veterinary species but there is an inherent, and highly regrettable indifference to the failure of antimicrobial therapy in many veterinary disease situations and this is probably at its most retrograde in veterinary wound management.

Leahurst, UK
March 2011

Derek C. Knottenbelt

Contents

Contributors

Lucy A. Brunton Faculty of Biological Sciences, University of Leeds, Leeds, LS2 9JT, UK

Peter D. Clegg Department of Musculoskeletal Biology, University of Liverpool, Leahurst Campus, Neston, Cheshire, CH64 5SD, UK

Christine A. Cochrane Faculty of Health and Life Sciences, Institute of Ageing and Chronic Disease, University of Liverpool, Leahurst, Neston. CH64 7TE, United Kingdom

Helena Cruz Division of Microbiology and Risk Assessment, National Food Institute, Technical University of Denmark, Mørkhøj Bygade 19, Building H, 2860 Søborg, Denmark

Ana B. García Division of Microbiology and Risk Assessment, National Food Institute, Technical University of Denmark, Mørkhøj Bygade 19, Building H, 2860 Søborg, Denmark

Alice J. Gardner Institute of Biological, Environmental and Rural Sciences, Aberystwyth University, Ceredigion, SY23 3AL, UK

John Heritage Faculty of Biological Sciences, University of Leeds, Leeds, LS2 9JT, UK

Samuel J. Hooper Oral Microbiology Group, Tissue Engineering and Reparative Dentistry, School of Dentistry, Cardiff University, Heath Park, Cardiff, CF14 4XY, UK

Jerry S. Knapp Faculty of Biological Sciences, University of Leeds, Leeds, LS2 9JT, UK

Derek C. Knottenbelt Faculty of Health and Life Sciences, Department of musculoskeletal Biology, University of Liverpool Leahurst, Neston, South Wirral CH64 7TE, UK

Tomoari Kuriyama Department of Oral and Maxillofacial Surgery, Graduate School of Medical Science, Kanazawa University, Kanazawa 920-8640, Japan

Michael A.O. Lewis School of Dentistry, Cardiff University, Heath Park, Cardiff, CF14 4XY, UK

Thomas W. Maddox Faculty of Health and Life Sciences, School of Veterinary Science, University of Liverpool, Leahurst Campus, Neston CH64 7TE, UK

Sladjana Malic Oral Microbiology Group, Tissue Engineering and Reparative Dentistry, School of Dentistry, Cardiff University, Heath Park, Cardiff CF14 4XY, UK

Marielle B. Melchior MBM Veterinary Consultancy, 8111 RA, Heeten, The Netherlands

Steven L. Percival Department of Pathology, Medical School, West Virginia University, Morgantown, WV, USA

Marcello P. Riggio Infection and Immunity Research Group, Dental School, College of Medical Veterinary and Life Sciences, University of Glasgow, 378 Sauchiehall Street, Glasgow G2 3JZ, UK

Sónia da Silva Centre for Biological Engineering, IBB-Institute for Biotechenology and Bioengineering, Universidade do Minho, Campus de Gualtar, 4710-057 Braga, Portugal

Samantha J. Westgate Faculty of Health and Life Sciences, Department of musculoskeletal Biology, University of Liverpool, Leahurst, Neston, South Wirral CH64 7TE, UK

David W. Williams Oral Microbiology Group, Tissue Engineering and Reparative Dentistry, School of Dentistry, Cardiff University, Heath Park, Cardiff, CF14 4XY, UK

Introduction to Microbiology, Zoonoses and Antibiotics

Steven L. Percival, Jerry S. Knapp, David W. Williams, John Heritage, and Lucy A. Brunton

Abstract Microorganisms are biological entities (organisms) which are so small they cannot be visualised without the aid of some type of microscope. There are six groups that make up the microorganisms – archaea, bacteria, protozoa, fungi, algae and viruses. Despite their small size, it is clear that microorganisms have a profound influence on human and animal life and indeed on all aspects of the biosphere. Prokaryotes come in a variety of shapes and sizes. Probably the most frequently encountered are cocci (coccus – singular) (round or oval cells), bacilli (bacillus singular) (rod-shaped) and vibrios (curved). Algae are photosynthetic eukaryotes with the cells containing chloroplasts. Algae are autotrophic primary producers and do not cause infections; they are thus of limited importance in the veterinary field. The fungi are an important and diverse group of eukaryotes; although formerly considered to be plants, they are now known to be more closely related to animal cells. Protozoa, otherwise known as protists, are also a very varied group. Protozoa are nearly all chemoheterotrophs ranging from free-living cells to obligate parasites. Viruses are infectious particles which lack a cellular structure. Since viruses do not possess the mechanisms needed to produce energy and the ribosomes required to synthesise proteins, they are incapable of independent metabolism, replication or movement. As a result, viruses are completely dependent on the host cells, which they effectively hijack, to produce new virus particles. For survival microbes require sources of energy, carbon and several other elements including nitrogen, oxygen, phosphorus, sulphur, potassium, sodium, calcium,

S.L. Percival (✉)
Department of Pathology, Medical School, West Virginia University, Morgantown, WV, USA
e-mail: Biofilms@hotmail.co.uk

J.S. Knapp • J. Heritage • L.A. Brunton
Faculty of Biological Sciences, University of Leeds, Leeds LS2 9JT, UK

D.W. Williams
Oral Microbiology Group, Tissue Engineering and Reparative Dentistry, School of Dentistry, Cardiff University, Heath Park, Cardiff CF14 4XY, UK

S.L. Percival et al. (eds.), *Biofilms and Veterinary Medicine*,
Springer Series on Biofilms 6, DOI 10.1007/978-3-642-21289-5_1,

magnesium and iron. Trace elements are also needed but in relatively small amounts. All these elements are required for the maintenance of life processes and the synthesis of new biomass. Animals are host to large numbers of microbes, many of which contribute to the health of their host. However, the majority of these microbes have the ability to cause disease. Many of these microbes may only be able to infect a single species, but others are able to cross the species barrier to infect other species, including humans. Diseases that can be passed between vertebrate animals and humans are known as zoonotic diseases, or zoonoses.

1 What Are Microorganisms?

Microorganisms are biological entities (organisms) which are so small; they cannot be visualised without the aid of some type of microscope. This is an unusual definition since the overall grouping of microorganisms includes some members (viruses) that cannot truly be considered to be organisms since they have no cellular structure and are the ultimate parasites, incapable of independent reproduction. In many ways, the term microbe might be preferable to microorganisms. There are six groups that make up the microorganisms – archaea, bacteria, protozoa, fungi, algae and viruses. It is notable that despite the prefix "micro", there are some members of these groups which are in fact macroscopic and visible to the naked eye. So, for example, while many fungi (e.g. yeasts) are microscopic for their entire life cycle, others have large macroscopic fruiting bodies (mushrooms and toadstools); it is paradoxical that the largest living organism (certainly in terms of its area) is probably a fungus, *Armillaria ostoyae*, which is known to grow to over 1,500 acres in size. Similarly many algae are relatively large (e.g. some of the seaweeds such as kelp), and although most protozoa are microscopic, some are visible to the naked eye (e.g. *Amoeba proteus* which can reach 0.75–1 mm in length).

The existence of microorganisms was first demonstrated by Anthonie van Leeuwenhoek in the later part of the seventeenth century. However, the systematic study of microorganisms only really started with the work of Louis Pasteur and Robert Koch from the 1860s onwards. Despite their small size, it is clear that microorganisms have a profound influence on human and animal life and indeed on all aspects of the biosphere.

Within the group of microorganisms, there exists great diversity not only of microbial size but also of structure, nutrition, ecology and genetics. The term microorganism has no real taxonomic significance as the overall grouping contains many entities with no taxonomic relationships. In many ways, what unifies the microorganisms is that they are studied by a range of similar laboratory techniques encompassed by the science of microbiology; these include microscopic observation, artificial cultivation (often in pure culture) and a range of biochemical and genetic/molecular biological techniques. As mentioned above, in many ways the term "microbe" is preferable to microorganisms as it makes fewer assumptions about taxonomy, structure and "lifestyle".

It is now thought that living organisms can be divided into three distinct domains based on cell structure. These domains are the archaea, the bacteria and the eukarya, the latter group including all plants and animals as well as the eukaryotic microorganisms. The relationship between the domains can be confirmed by a detailed analysis of the structure and sequences of the RNA molecules in the small subunit of the ribosomes – 16S for the prokaryotes and 18S for eukaryotes. The genes for these RNA molecules are present in all organisms, always have the same function, are large enough to be able to show divergence in sequence and appear to have changed slowly over time.

Microorganisms can also be divided into three sections.

1. Non-cellular obligate parasites – the viruses. These entities are infectious particles containing a nucleic acid genome and some proteins and occasionally lipids, but they lack the complete cellular "machinery" to generate energy and to synthesise proteins and nucleic acids. Viruses are completely reliant on the cells of the host to reproduce new particles and so cannot on their own be considered living organisms. There appear to be viruses which parasitize all types of living cellular organisms.
2. Prokaryotic microbes – the archaea and bacteria. Prokaryotes are true cellular organisms, but have a relatively simple basic cell structure lacking internal membrane-bound nuclei and organelles. Prokaryotes have 70S ribosomes consisting of 30S and 50S subunits. Most species are unicellular and in general they do not exhibit differentiated cell types – although there are some exceptions such as production of resting stage spores by *Bacillus* and *Clostridium* species and the production of heterocysts (cells specialised for di-nitrogen fixation) by some cyanobacteria such as *Nostoc*. Archaea and bacteria are similar in many respects and until recently archaea were all classed as bacteria. However, it is now clear that there are significant differences between the groups in relation to the presence and structure of cell walls, the composition of cell membranes, and also in relation to the molecular biology of their genomes and the processes used for protein synthesis. Many archaea are extremophiles (able to live in very hot, saline or acidic conditions) or have unusual metabolic pathways, such as the ability to generate energy during the production of methane.
3. Eukaryotic microorganisms – this group includes the algae, the protozoa and the fungi, along with other lesser known groups. Eukaryotes have more complex cells with a true membrane-bound nucleus and membrane-bound organelles (normally including mitochondria). The classification of eukaryotic microorganisms has changed greatly in recent years and taxonomy based on gene and protein sequences has revealed previously unsuspected evolutionary relationships. Eukaryotes are a diverse group; all possess the more complex eukaryotic cell structure, and while some are simple unicellular organisms, others are more complex and multicellular and display considerable differentiation and complexity of life cycles. Eukaryotic cells are typically more complex than prokaryotes. They have a membrane-bound nucleus, which contains more than one linear chromosome, and generally have membrane-bound structures or organelles, such as mitochondria, hydrogenosomes and chloroplasts.

The flagella (singular *flagellum*) (or cilia) of eukaryotic cells are more complex in structure than those of prokaryotes. Eukaryote cytoplasm also has larger 80S ribosomes than prokaryotes. It is now accepted that eukaryotes evolved partly due to the acquisition of symbiotic bacterial cells, which were incorporated into ancestral cell types and over time became permanent. This is highlighted by the fact that mitochondria and chloroplasts have their own genomes and prokaryote-type ribosomes.

2 Bacteria and Archaea

2.1 Cell Shape and Arrangement of Bacteria and Archaea

Prokaryotes come in a variety of shapes and sizes. Probably the most frequently encountered are cocci (coccus – singular) (round or oval cells), bacilli (bacillus – singular) (rod-shaped) and vibrios (curved). Very short rod-shaped bacilli are referred to as coccobacilli. A spirillum is a helical or spiral-shaped cell, while long slender spiral cells with flagella within the turns of the helix are called spirochaetes. A significant number of bacteria and archaea exist as long filaments. In recent years, more bacteria have been discovered with unusual shapes, some are rather irregular and lobed; indeed, there is even a flat square archaea. Certain types of prokaryotes are pleiomorphic with cells varying in shape and size at different stages of growth.

While many bacteria and archaea grow as single cells, others form distinctive arrangements and are often found as chains (filaments), pairs or fours, or cubical packets of eight (these are called sarcinae). Other bacteria, such as staphylococci, form irregular clumps giving the classical appearance of a "bunch of grapes". These characteristic arrangements are caused by the failure of daughter cells to fully separate after cell division and reflect the symmetry, or otherwise, of successive rounds of cell division.

2.2 Cell Size

Prokaryotes are always small, but do show wide variation in size. Typical dimensions are in the order of 0.75 to 2 μm in diameter for cocci, and for bacilli 0.75 to 2 μm in width to about 3 to 8 μm in length. There are, however, many exceptions; some species of *Mycoplasma* have cells of about 0.2 to 0.3 μm. These *Mycoplasma* cells have very small genomes and are often parasitic, with some being able to be grown in complex artificial media. At the other extreme, the largest bacterium is the recently discovered *Thiomargarita namibiensis* which is a spherical marine chemolithotrophic bacterium that is theoretically visible to the naked eye with a diameter of 750 μm. In the case of this species, it has to be said that a large part of the cell's volume is occupied by a vacuole and it is not all cytoplasm.

2.3 Genetic Material

Prokaryotes, unlike eukaryotes, do not have a true membrane-bound nucleus, but have a single circular chromosome. In addition, prokaryotes may carry a range of plasmids. Plasmids are small DNA elements, usually circular, which generally encode a range of non-essential, though often useful, characteristics such as antibiotic resistance or pathogenicity determinants.

2.4 Cell Membranes

Bacterial and archaeal cytoplasm is enclosed within a cytoplasmic membrane which controls the integrity of the cell and is situated within the cell wall, if the cell has one. The cell membrane is a phospholipid bilayer containing a range of embedded functional proteins. It controls the entry and exit of materials and has a crucial role in energy generation though the establishment of transmembrane gradients. Though similar, there are a range of differences between bacterial and archaeal cell membranes.

2.5 Cell Walls

Prokaryote cells are generally contained within an outer rigid cell wall which protects the cell from osmotic stress. However, there are some exceptions, with some cells having cell walls that lack rigidity. In a few genera, such as *Mycoplasma* and *Chlamydia* in the bacteria and *Thermoplasma* in the archaea, there is even a complete absence of a cell wall. While there are various types of bacterial cell walls, all contain a layer of the polymer peptidoglycan, otherwise known as murein, this provides the rigidity and strength. Peptidoglycan consists of many polysaccharide chains cross linked by short peptide bridges and it is uniquely found in bacteria.

Within the bacteria there are two major groups distinguished by their cell wall structure and differentiated by the classical Gram stain. Gram-positive bacteria retain a crystal violet-iodine complex and appear purple, whereas Gram-negative cells lose this complex and display the pink counter stain. Gram-positive bacteria contain thick, multiple layers of peptidoglycan along with other polymers including teichoic and teichuronic acids. Gram-negative cell walls have a single layer of peptidoglycan, but outside this they have an outer membrane layer, composed of lipopolysaccharide and protein. The archaeal cell walls do not contain peptidoglycan, but have a variety of different substances including pseudomurein, which is structurally similar to peptidoglycan. Other materials found in archaeal cell walls include polysaccharides, glycoproteins and proteins.

The cell wall and cytoplasmic membrane together are often referred to as the cell envelope.

2.6 External Structures

Outside the cell wall, some bacteria and archaea have an additional layer called a capsule or a slime layer. Capsules and slime layers are normally made of polysaccharide, although some are made of protein. Capsules are much more defined and tightly attached, whereas slime layers are more diffuse, covering a greater area. Capsulated bacteria tend to resist engulfment by phagocytic cells and therefore help pathogens evade the immune response. Even when the bacterium is engulfed, a capsule can protect the microorganism from intracellular killing. These layers also assist attachment of microbes to surfaces and are important in formation of biofilms.

Other external structures found in some bacteria are fimbriae and pili. These filamentous protein structures have a range of functions. Fimbriae are involved in attachment to cells and other surfaces and may have a role in pathogenic processes. They also aid attachment of bacteria to each other and are involved in formation of pellicles and biofilms. Pili have a range of functions including attachment, genetic exchange and movement.

Some prokaryotes are capable of movement and motile archaea and most motile bacteria move by means of flagella. These proteinaceous appendages protrude through the cell wall and move with a rotary motion. They may be attached at the poles of the cell or all around it (polar and peritrichous flagellation, respectively). A range of motile bacteria do not possess flagella and can only move in contact with surfaces, some of these move with the aid of retractile pili (twitching motility), but others with "gliding motility" have no obvious locomotor organelles and mechanisms for their movement are still a matter of debate.

2.7 Algae

Algae are photosynthetic eukaryotes with the cells containing chloroplasts. Algae are autotrophic primary producers and do not cause infections; they are thus of limited importance in the veterinary field.

2.8 Fungi

The fungi are an important and diverse group of eukaryotes; although formerly considered to be plants, they are now known to be more closely related to animal cells. Fungi are chemoheterotrophs and classically comprise a mass of protoplasm contained within a filamentous structure called a hypha. Fungal cell walls, in some ways, resemble plant cell walls but are chemically different, generally consisting largely of chitin. Most fungi grow as a network of branching hyphae known as a mycelium which grows centrifugally and can spread out to cover a very large area. Fungi generally reproduce by production of spores, which may be a sexual or asexual process. However, some fungi, known as yeasts, grow as single cells rather than hyphae and generally reproduce by budding rather than binary fission. The yeasts are not a taxonomic grouping, with the term "yeast" describing a type of morphology. Other fungi include moulds, mushrooms and toadstools.

Most fungi are aerobes, although there are some strict anaerobes. Indeed, a number of anaerobic fungi play an important role in digestion of herbage in the rumen and caecum of herbivores. Most fungi are saprophytic absorbing dissolved nutrients, or secreting enzymes which decay, and solubilise macromolecules which can then be used as nutrients. In this regard, fungi play a major role in the decomposition of dead tissues particularly plant matter. In fact, fungi are the agents *par excellence* of decomposition of woody materials. However, a number of fungi are pathogenic, particularly towards plants, but there are also a wide range of human and animal pathogens. Some cause superficial infections of the skin (e.g. *Trichophyton* spp.) or in the mouth and vagina (e.g. *Candida* spp.). Other species cause systemic infections (e.g. of the lungs, *Cryptococcus*, *Pneumocystis* and *Coccidioides*). *Pneumocystis* is an interesting organism as it was once thought to be a protozoan, but genetic analysis now shows it to be a fungus, although it has little resemblance to one. In recent years, it has been demonstrated that some chytrid fungi are severe pathogens of frogs and are causing widespread death of frogs worldwide.

Fungi also have a range of industrial uses that make them economically very important. Best known of these is the use of yeast in the production of alcoholic beverages and bread. Fungi are also a source of some of the major antibiotics such as penicillin (*Penicillium chrysogenum*) and cephalosporins (*Cephalosporium acremonium*) and a range of valuable fine chemicals.

2.9 Protozoa

Protozoa, otherwise known as protists, are also a very varied group. Protozoa are nearly all chemoheterotrophs ranging from free-living cells to obligate parasites. *Euglena* are exceptions in that they have chloroplasts and thus are able to photosynthesize, although some can also ingest and digest bacteria. Of the heterotrophs, some use dissolved nutrients, while others are predatory. Some protozoa are free-living, and some are involved in symbioses (e.g. some anaerobic ciliates are symbionts in the rumen, while cellulose-degrading flagellates are important symbionts in the guts of wood-eating termites). Examples of parasitic protozoa and the diseases they cause are given in Table 1, and some of these diseases are zoonotic.

Table 1 Examples of some parasitic protozoa and the diseases

Organism	Disease caused
Entamoeba histolytica	Amoeboid dysentery
Giardia lamblia	Gasteroenteritis
Cryptosporidium parvum	Gasteroenteritis
Trypanosoma brucei	African sleeping sickness
Plasmodium spp.	Malaria
Toxoplasma sp.	Toxoplasmosis
Eimeria	Coccidiosis
Trichomonas vaginalis	Vaginitis

2.10 Viruses

Viruses are infectious particles which lack a cellular structure. Since viruses do not possess the mechanisms needed to produce energy and the ribosomes required to synthesise proteins, they are incapable of independent metabolism, replication or movement. As a result, viruses are completely dependent on the host cells, which they effectively hijack, to produce new virus particles.

Outside the host cell, viruses exist as tiny particles called virions, but when they invade host cells their components are separated and become interspersed within the host cytoplasm. The infecting virus remains devoid of a defined structure until the point when new virus components are constructed and assembled by the host cell. These new viruses can then be released by the host cell and will serve to infect other host cells.

Virions contain a nucleic acid genome and some proteins which may have structural roles or enzymic activity (which can be crucial to the viral life cycle – e.g. reverse transcriptase in retroviruses). The viral genome may be composed of DNA or RNA, which may be either single- or double-stranded; the former may be either positively or negatively stranded. Some viruses may use both DNA and RNA at different stages of their life cycle, but the virion will only contain one of these at a given point in time.

Viral nucleic acid is contained within a protein coat called a capsid; these structures vary greatly in shape and size but are always very small – generally below the resolution of light microscopy (ca. 20–300 nm), although giant mimiviruses exist and are ~750 nm in diameter. In addition to the capsid, some virions are enveloped, surrounded by a lipid bilayer membrane containing proteins or glycoproteins. Not surprisingly given their small size and very limited capabilities, the genomes of viruses are very small ranging from 3.5 to 150×10^3 base pairs, although some are a little smaller and the largest genome is about 1.2×10^6 base pairs (mimivirus). Viral genomes are sufficient enough to code for necessary proteins to instigate viral replication or persistence in a host cell.

Classification of viruses is principally according to the type of host organism (animal, plant, bacterium, etc.) they infect, the type of nucleic acid in their genome (RNA/DNA, single or double strand, retrovirus) and whether they are enveloped or not.

By following the old adage "big fleas have little fleas upon their backs to bite them, little fleas have lesser fleas and so ad infinitum", viruses are known to parasitize all forms of life including animals, plants, fungi, protozoa and bacteria (such viruses are known as bacteriophages) and even other viruses. In the case of the latter, these are called virophages and an example of this is the Sputnik virus that exploits the replication processes of a mimivirus.

As parasites, viruses damage the host cells generally causing diseases, although some are well tolerated by their normal hosts within which they coexist, causing little apparent damage. Some such viruses can, on occasion, spread to new hosts which have not evolved tolerance often with disastrous results; this ability to "jump the species barrier" can give viruses zoonotic potential.

Although generally adverse in their effects, viruses can have a positive role in causing genetic variation and this has been exploited through the use of viruses as biological control agents for insect pests. Similarly, bacteriophage use in a limited way has been proposed, especially in the former USSR, to treat human bacterial infections, although to date their value is currently limited.

2.11 Microbial Nutrition and "Lifestyle"

For survival microbes require sources of energy, carbon and several other elements including nitrogen, oxygen, phosphorus, sulphur, potassium, sodium, calcium, magnesium and iron. Trace elements are also needed but in relatively small amounts. All these elements are required for the maintenance of life processes and the synthesis of new biomass. The type of energy source utilised varies with phototrophs using the energy of light, and chemotrophs the energy generated from the oxidation of reduced compounds or elements. The types of energy sources and nutrition used are described by a range of terms defined in Table 2.

Table 2 Types of energy source and nutrition used

Autotroph	Obtains all of its carbon from carbon dioxide
Heterotroph	Obtaining its carbon from organic compounds
Phototroph	Able to use the energy of light via photosynthesis
Chemotroph	Obtains energy from the oxidation of reduced organic compounds or elements
Lithotroph	Obtains energy from oxidation of inorganic compounds or elements
Chemoheterotroph	Obtains both energy and carbon from the organic compounds
Photoautotroph	Obtains energy from light and carbon from CO_2
Photoheterotroph	Obtains energy from light and carbon from organic compounds
Chemolithoautotroph	Obtains energy from oxidation of inorganic compounds (e.g. hydrogen, ammonia or sulphide) and carbon from CO_2
Mixotroph	Uses inorganic energy sources but organic carbon sources
Parasite	Obtain some of their nutrients from a host organism in a relationship which damages the host. Some organisms are obligate parasites while others may be capable of independent life

Many microorganisms employ only one type of nutrition, but some are able to use different modes of metabolism according to the environment they find themselves in and the relative availability of sources of carbon and energy. So, for example, a cyanobacterium may be a photoautotroph when light is available or chemoheterotrophic in the dark and when suitable organic chemicals are present.

Between them, microorganisms cover all the different nutritional types and lifestyles available in the biosphere. Microorganisms vary from the extreme autotrophs capable of growth on inorganic chemicals alone, to parasites like viruses, which obtain all their requirements from their host. Some heterotrophic microorganisms only obtain their full nutrient and/or environmental requirements by living within another organism. These organisms are either parasites or obligate symbionts. In the case of symbionts, both the host and the microbe derive mutual benefit, and with parasites the host organism is damaged or even killed by the relationship. Some parasitic bacteria e.g. rickettsias or chlamydias are obligate parasites and cannot be grown outside their hosts, but still have certain, if incomplete, metabolic capabilities. Others such as some mycobacteria and mycoplasmas may be parasitic but can be cultured on artificial media if suitable nutrients and conditions are provided. These microorganisms are described as facultative parasites.

The majority of bacteria are chemoheterotrophs, although there are also many phototrophs and chemolithotrophs too. Chemoheterotrophs require carbon sources for both energy generation and synthesis of biomass. In some cases, the same compounds are used for energy generation and biosynthesis, but this is not always the case and organic substrates which serve as precursors for biosynthesis of macromolecules cannot be degraded to generate energy.

Many bacteria can use a wide range of individual organic carbon and energy sources (in excess of a hundred) and such bacteria are referred to as being nutritionally versatile. These bacteria, typified by species of *Pseudomonas*, *Rhodococcu*s, *Acinetobacter* and many *Mycobacterium* species, have very important roles in the degradation of chemical pollutants. Other bacteria may be very limited in the types of carbon sources they use and this is restricted to one or a very few energy substrates.

The nutritional groupings of microorganisms depend mainly on the carbon and energy sources used, although obviously other elements are important. Some elements are available in many forms and those utilised by particular microbes depend on the organism's ability to synthesise particular metabolites and allow cellular uptake. Some organisms can synthesise all their required nutrition from just inorganic ions, whereas others need preformed organic molecules. For example, bacteria of the genus Pseudomonas can grow in the presence of a single organic carbon source together with inorganic sources of nitrogen, phosphorus, sulphur, magnesium potassium and iron salts. From these nutrients, *Pseudomonas* can synthesise all amino acids, nucleotides, vitamins, and carbohydrates required to produce new biomass.

Other bacteria are unable to synthesise the carbon skeletons needed to generate many amino acids, vitamins, etc., and so a wide range of preformed organic molecules are required to support their growth in addition to the organism's main carbon and energy sources. These compounds are referred to as "growth factors" and like most minerals tend to be required in small amounts compared to the main carbon source. The lactic acid bacteria (including *Lactobacillus*, *Streptococcus*, and *Leuconostoc*) are well known for their complex growth factor requirements and some members of this group have greater nutritional requirements than humans! Growth factors may be obtained by some free-living microbes from their environments; for example, lactic acid bacteria may grow in nutrient-rich materials such as milk or fermenting plant products. Alternatively these bacteria may grow as harmless commensals within the bodies of animals (for example in the mouth, gasterointestinal tract or vagina). However, some microbes may infect plants or animals and so obtain their nutritional requirements by parasitizing their host organism.

Organisms requiring many growth factors typically lack the genes coding the required enzymes for their synthesis and generally these organisms have small genomes. There is a metabolic cost to organisms in making growth factors themselves (in terms of the energy required for synthesis and the carriage of a large genetic component) so if their typical environment can provide these growth factors it may be advantageous to use preformed molecules rather than to make their own. It is thought that this is because they have lost metabolic capabilities during their evolution.

Most microorganisms (including all archaea, bacteria, algae, fungi and some protozoa) are osmotrophs, which take in their nutrients by diffusion across the cytoplasmic membrane. Some (though not all) protozoans are, however, phagotrophs which engulf particles of food into their cells by phagocytosis. The food particles may include live prey (bacteria and other microbes) as well as decaying organic matter. Many ciliate and amoeboid protozoa capture live prey and can be regarded predatory; there are even a few species of bacteria which predate other bacteria (e.g. *Bdellovibrio* sp.).

2.12 Microbial Growth

Many microorganisms can be grown in the laboratory in artificial growth media. Such media may be simple or complicated to prepare, depending on the nutritional requirements of the organism. We generally refer to growth media as either being defined or complex. Defined media contain only specific pure chemical components in appropriate quantities, although they may contain many components if a lot of growth factors and trace element are needed. Complex media may contain some defined pure chemicals, but always contain some components of indeterminate composition which provide most of the growth factors and trace elements needed. Examples of such indeterminate components include meat extracts, protein hydrolysates and yeast extracts. Complex media are very simple to prepare and may contain few components, but their inexact composition makes it difficult to standardise microbial growth and so variable results may be obtained using different sources of the material.

Many microorganisms are currently non-culturable in the laboratory. One of the best known examples of this is *Treponema pallidum*, the causative organism of syphilis. There are, however, many other examples of non-culturable microbes from a wide range of environments. The reasons for an inability to culture microorganisms include the absence of a required nutrient from the medium or the provision of incorrect environmental conditions (e.g. concentration of dissolved oxygen, carbon dioxide or hydrogen, as well as temperature and pressures levels). In addition, some microbes are labile and easily damaged during transfer from their natural habitat into the laboratory. Although the existence of non-culturable bacteria has been known for many years, a large number of new non-culturable microbes are discovered every year as the results of exploration of new and often extreme environments (Hegarty et al. 2001). It is now thought that there are many more microbes which we are unable to culture in the laboratory than ones we can. For some of these microbes, the growth conditions may well eventually be defined to effect their in vitro growth.

Until recently, these non-cultivable organisms were largely studied by observation, but with the development of modern molecular biological methods, including PCR, genome sequencing and transcriptomics, it is now possible to learn a lot about the biochemistry and physiology of such organisms.

Viruses do not grow on their own in artificial media in the laboratory, but some can be propagated using tissue culture where cells of suitable host organisms (mammals, insect, plants, etc.) are grown in artificial media under well-defined environmental conditions. Similarly, bacteriophages can be grown in bacterial cell cultures. It has also proved possible to culture a number of obligately parasitic bacteria (like *Chlamydia trachomatis*) in tissue cultures of mammalian cells.

2.13 Oxygen

Microbes vary in their response to oxygen. Strict or obligate aerobes cannot grow in the absence of oxygen, which is required for energy generation via respiration. Anaerobes grow without oxygen, usually employing fermentation processes. Some microbes, however, are inhibited or even killed by exposure to oxygen (due to the inability to cope with toxic oxygen radicals) and are therefore strict/obligate anaerobes, an example being *Clostridium tetani*, the cause of tetanus. Other microbes can grow either with or without oxygen and these are called facultative anaerobes. Generally, facultative anaerobes grow better with oxygen, as respiration is more efficient for energy generation than fermentation and this is the case with *Escherichia coli*. Lactic acid bacteria, however, cannot respire aerobically and so growth is no better and sometimes a little worse in the presence of oxygen; these are referred to as aerotolerant anaerobes. Another group, the microaerophiles, require oxygen but at low concentrations (often 2–10% saturation) and *Campylobacter* species are typical examples. For some organisms which normally respire aerobically, nitrate can substitute for oxygen as an electron acceptor allowing anaerobic growth.

Certain microorganisms have the capacity for very rapid growth, but their growth rate is of course constrained by the conditions pertaining in their environment. Particularly important are the temperature, pH and the types and concentrations of nutrients available. The rate of most chemical reactions increases with temperature and this is true within limits for microbial growth rates. Such limits being that all microbes have optimum, minimum and maximum growth temperatures. Growth limits are imposed by the heat sensitivity of microbial proteins and cell membranes, which tends to vary with the ambient temperatures pertaining in their normal environments. Microbes can be grouped according to their normal optimum growth temperature as being psychrophiles (<20°C), mesophiles (20–40°C) or thermophiles (>40°C). Some thermophiles have extremely high temperature optima, often >100°C, and are referred to as hyperthermophiles; these are all archaea and live in geothermal regions. Psychrotrophs grow at very low temperatures (0–5°C), but have their optimum temperature in the mesophilic range; these microorganisms can cause problems with the preservation of foods in cold storage.

Most microbes have an optimum pH for growth in the neutral range i.e. pH 6–8, but some (acidophiles) grow optimally at acid pHs below 5, while others grow best at pH values above 9 (alkaliphiles).

Many fungi grow very well at pHs below 5 and are certainly acid tolerant if not philic. Fungi often produce organic acids especially when growing on carbohydrates and thus reduce the local pH. Extreme acidophiles can grow in environments with pH values <2; bacteria of this type can be found growing in the mammalian stomach or by oxidising sulphur or metal sulphides.

To grow a microbe requires a full range of nutrients at appropriate and balanced concentrations. In practice, normally one or more nutrients are in limited supply compared to the others, but where growth is unlimited and under optimum environmental conditions, it can be very fast. Most bacteria multiply by means of binary fission and their growth can easily be modelled mathematically. Unlimited growth proceeds according to a geometric progression e.g. 1–2–4–8–16–32, etc., the number doubling every generation. A plot of cell number against time gives a curve of ever increasing gradient. This can be described thus:

If an initial population of N_0 bacteria is left to grow for z generations, the number at time t (N) is given by the expression

$$N = N_0 2^z.$$

The mean generation time, T, is the average time taken for the population to double in number (or mass). The number of generations in a time period of t is as follows:

$$z = t/T.$$

If we take logarithms of the expression $N = N_0 2^z$, then we get:

$$\log_{10} N = \log_{10} N\, 2^z = \log_{10} N_0 + z \log_{10} 2.$$

By substituting and rearranging this, we get:

$$\log_{10}N = \log_{10}N_0 + t\,\log_{10}2/T.$$

This latter equation is that for a straight line: $\log_{10}2 = 0.301$; if $\log_{10}N$ is plotted against t (time) then the intercept is $\log_{10}N_0$ and the gradient is $0.301/T$.

Growth of this type, with cells dividing at the same, constant rate is known as exponential or logarithmic growth and is demonstrated by a straight line relationship between $\log_{10}N$ and time. The same relationship holds for the change of biomass (X) with time.

Bacteria do not always grow at a constant rate. Growth in laboratory culture usually displays a lag phase, without obvious growth, as cells adapt to new conditions. This lag phase is followed by an acceleration phase during which growth rate steadily increases to the maximum which is achieved during the exponential phase. When nutrients are limited or conditions become toxic, growth enters the deceleration phase, followed by the stationary phase in which there is no net growth. After this, the population number may decrease in the decline phase. The relative duration of these phases will depend on the environment and nutrition.

Bacterial growth can also be modelled in terms of the proportional increase in biomass or numbers per unit time.

$dN/dt \propto N$; therefore $dN/dt = \mu N$, where μ is a proportionality constant, which describes the proportional increase in numbers per unit time; by re-arranging we get $dN/N = \mu\, dt$; if this equation is integrated it gives $\log_e N/N_0 = \mu t$.

This derivation allows calculation of the specific growth rate constant (μ), with units of time^{-1}, usually hour^{-1}. Mean generation time (also called doubling time) is calculated in units of time, usually hours. Specific growth rate and mean generation time can be inter-related by the equation $\mu = 0.693/T$.

The mean generation time and growth rate constant can only be calculated during exponential growth as this is the only phase when these parameters are constant. Growth rate/generation time depends (inter alia) on the concentration of the limiting nutrient, a relationship which can be described mathematically.

Growth rate during the exponential phase is the fastest a bacterium is capable of *under the conditions imposed*, but the organism may be able to grow much faster if conditions are more optimal. The growth rate/generation time is a measure of how well an organism is adapted to the growth conditions. Some microbes are capable of extremely fast growth – the shortest generation times being around 15–20 min and generation times of an hour are common – it should be stressed that these are in artificial culture under ideal conditions. Theoretically such growth rates would allow a bacterium to produce progeny equal to the weight of the earth in a few days. This clearly cannot happen as in practice conditions rarely remain optimal for very long and the organisms run out of nutrients or create toxic conditions which limit their growth. *Escherichia coli* can grow with a generation time of 20–30 min and one individual cell can grow on solid agar media to produce a visible colony overnight. However, many bacteria are intrinsically slow growing and even with optimal conditions cannot grow fast; generation times for *Mycobacterium tuberculosis* would be in the order of 24 h or more – it would take several weeks for single cells to produce visible colonies. For many microbes, generation times will be much longer (i.e. growth is slower) and this is especially so in the natural environment in which generation times may be in the order of days, weeks or even longer due to sub-optimal conditions.

In laboratory culture on solid (agar) media, bacteria grow in colonies, the shape and appearance of which can be distinctive. In liquid culture, bacteria often grow individually as well-separated cells in fairly homogenous suspension; this is not always the case and some species grow as aggregates, while many adhere to sold surfaces and grow as a biofilm. A biofilm is a more or less continuous layer of cells often with layers of cells packed on top of each other. In nature, bacteria and archaea generally grow as micro-colonies often attached to solid surfaces. Sometimes the surfaces are actually an organic material that the microbes are utilising as a nutrient. In the veterinary field, bacteria grow in biofilms on the surface of teeth, and biofilms often form on implanted surgical devices.

2.14 Interactions Between Microbes and Animals

There are several different ways in which microbes interact with animals. Symbiosis is a commonly used term which denotes that organisms "live together". Often in a fashion which is generally advantageous to both parties and sometimes this is referred to as mutualism. Some authors reserve the term symbiosis for beneficial relationships; others use it for any "living together" existence. The exact relationship varies considerably as does its importance to the partners. The degree to which a relationship is important and advantageous often depends on the exact circumstances, the environment and the nutrition available. Some symbioses are casual but advantageous; others are obligate under most conditions.

Commensalism is a relationship which is advantageous for one partner, but may have little or no advantage for the other, but in general causes no harm. The skin of animals and their guts often provides a habitat for a wide range of microbes that generally do not harm the host organism but are beneficial. Microorganisms that exist in this manner are generally referred to as the commensal microbiota or microflora. In some cases, the relationship is stronger and is considered symbiotic.

Parasitism can be a considered a type of symbiosis in which one partner benefits greatly at the expense of the other. Parasitic relationships cause a huge range of diseases for humans and animals. Some parasites are obligate and have not been shown to grow in the absence of a host organism – examples include *Mycobacterium leprae* (cause of leprosy), *T. pallidum* and many of the protozoal parasites such as *Plasmodium* species which cause malaria. However, many bacterial pathogens are not obligate parasites and live as saprophytes, their "lifestyle" depending on where they happen to be.

Symbiosis has had an important role in evolution. The mitochondria found in most eukaryotes and responsible for energy generation have evolved from the obligate and permanent relationship between a bacterium and a host "ancestral eukaryote"; mitochondria have their own prokaryote-type genome distinct from the genome in the cells nucleus. The same it true for chloroplasts (evolved from symbiont cyanobacteria).

Microbes are obviously a cause of animal disease, but in the veterinary context they can be very important, indeed vital, to many animals as symbionts. Probably, the best known and important example is the role of symbiotic microbes in the digestion of plant-based food in the guts of many herbivorous animals. Much of the herbage eaten by many animals as their staple diet cannot be digested by the animals' digestive system. This is because the ability to produce enzymes capable of digesting the lignocellulose of plant cell walls appears to be restricted to microbes, and lignocellulose makes up a very high proportion of the animal diet. Ruminant animals are the most effective users of herbage due to the presence of huge microbial populations in their rumens. The rumen contains a range of cellulolytic microorganisms including bacteria, ciliate protozoa and anaerobic fungi. This is a "foregut fermentation" and is particularly effective as all the products of the ruminal fermentation including bacterial cells are available for digestion and absorption by the animal's own digestive system. Foregut fermentations are not restricted to ruminants but are quite widespread, being found in camels, kangaroos, sloths, some leaf-eating monkeys (*Colubus*) and in at least one bird, the Hoatzin.

Other herbivores often have a "hindgut fermentation" with large microbial populations occurring in an enlarged large intestine or caecum; this fermentation is seen in rabbits, horses, elephants, rhinoceroses and some birds and reptiles. It is less efficient than foregut fermentation since, although products from the microbial fermentations are available to the host animal, the nutrients contained in microbial cells are not. Some hindgut fermenters, rabbits for example, overcome this to some extent by eating their own faeces! Even non-herbivores benefit from the presence in the gut of a healthy population of bacteria which help to protect the animal from pathogens and also to provide essential nutrients such as vitamins (e.g. vitamin K and B_{12}). Some gnotobiotic (germ-free) animals may require additional nutrients to overcome the lack of microbes in their guts.

Many insects (e.g. aphids) are dependent on the presence of intracellular bacterial symbionts that have important roles in their nutrition.

The full range of involvement of microbes in causation of animal diseases cannot be covered in this introduction, but it is important specifically to mention here the role of animals in causation of human diseases.

3 Zoonoses

As already discussed, animals are host to large numbers of microbes, many of which contribute to the health of their host. However, the majority of these microbes have the ability to cause disease. Many of these microbes may only be able to infect a single species, but others are able to cross the species barrier to infect other species, including humans. Diseases that can be passed between vertebrate animals and humans are known as zoonotic diseases, or zoonoses. Zoonoses can be caused by numerous microbes including bacteria, viruses, fungi and protozoa and occur in all animals. Zoonoses are thought to be the biggest contributor to human disease worldwide with it having been estimated that 61% of all human diseases are zoonotic in origin. The high incidence of zoonotic disease in the human population was likely to have been initiated with the domestication of animals leading to closer contact between animals and humans (Christou 2011). In recent centuries, factors such as increases in trade, travel and urbanisation may have all contributed to the increase in the prevalence of zoonotic diseases. Some zoonoses are specific to a few species, but there are others that can infect a wide range of wild and domestic animals. Examples of these are the bacterial zoonoses listeriosis, pasteurellosis, salmonellosis, tularaemia and yersiniosis, and the protozoal zoonosis cryptosporidiosis.

3.1 Routes of Transmission

Microbial diseases can be spread between hosts in many ways, and many diseases produce symptoms that can help facilitate the spread of microbes; for example, sneezing aids the spread of influenza virus particles. Like all microbial diseases, zoonoses can also be spread in many ways. However, there are two primary routes of transmission, which are commonly associated with zoonoses, and these are food-borne transmission and occupational transmission.

Food-borne transmission is typically associated with the zoonotic diseases found in food-producing animals such as cattle, pigs and poultry. Some common examples of food-borne zoonoses include salmonellosis and campylobacteriosis in poultry and pigs, and *E. coli* O157:H7 in cattle (Ferens and Hovde 2011). This route of transmission is facilitated by poor hygiene and inadequate food preparation. The slaughter methods in place at abattoirs as well as procedures such as carcass scalding are designed to reduce the amount of microbial contamination on meat carcasses. However, in some cases, contamination will occur. For example, due to the anatomy of poultry, it is extremely difficult to eviscerate the birds without causing some faecal contamination of the carcasses. Thus, it is important that meat is prepared and cooked adequately to remove any microbial contamination. Failure to do this may result in food-related illnesses such as gastroenteritis.

Occupational transmission occurs when zoonoses are spread between animals and the humans that work with them. Those at risk of occupational transmission are anyone who has close contact with animals, such as farm workers, veterinarians and abattoir workers. Examples of occupational zoonotic microorganisms include *Streptococcus suis* in pigs and *Bacillus anthracis* in sheep. Occupational transmission can be managed by strict monitoring of the health of animals and workers and by using personal protective measures when in contact with diseased animals such as gloves, overalls and face masks. However, many diseases do not always present symptoms in animals so it is important that good hygiene is observed and protective measures are in place for the day-to-day handling of animals.

3.2 *Zoonoses in Farm Animals*

Some of the most prevalent zoonoses are associated with farm animals. As mentioned above, some well-known zoonoses in farm animals are associated with the consumption of infected meat. Bacterial zoonoses transmitted in this manner include salmonellosis and campylobacteriosis in poultry and pork, yersiniosis, which can occur in many domestic animals, and *E. coli* O157:H7 in cattle and other ruminants. Protozoal zoonoses such as toxoplasmosis can also be transmitted through the consumption of infected meat, and pork in particular. Another zoonosis associated with the consumption of meat which has not yet been discussed is bovine spongiform encephalopathy (BSE), which occurs in cattle (known as "mad cow disease") (Wells 2003). This disease differs from the other zoonoses mentioned here in that it is not caused by a microorganism. The causative agents of BSE are prions that are proteinaceous agents. Infectious prions initiate encephalopathies by altering the folding structure of proteins in the neural tissue. Consumption of beef from cows with BSE can lead to "new variant" Creutzfeldt–Jakob disease (CJD) in humans, an incurable neurodegenerative disease.

Many of the zoonoses associated with farm animals are not associated with the consumption of infected meat or animal products. Instead, they are often transmitted to humans through close contact and inhalation of dust or aerosols. Some of the most common of these, particularly in Europe, include leptospirosis caused by *Leptospira* spp. found in the urine of rats and cattle, orf caused by poxvirus in sheep and goats, and ringworm caused by fungi of the *Tinea* spp. and found in horses, cats and dogs. A number of bacterial zoonoses are found in farm animals. Brucellosis, a febrile disease, is caused by bacteria of the *Brucella* genus and is found in many animals including cattle, sheep, goats and pigs. Anthrax is caused by *B. anthracis* and is found in many animals, particularly sheep, cattle and horses. Low-level exposure to the spores of *B. anthracis* can lead to cutaneous anthrax lesions in persons working with animal hides and wool. This organism has also been cultivated in vitro for use as a biological weapon. Bovine tuberculosis caused by *Mycobacterium bovis* can be contracted from cattle, badgers and deer and, along with brucellosis and anthrax, is a notifiable disease in the UK. Other bacterial zoonoses associated with farm animals include *S. suis* in pigs, ovine chlamydiosis, which causes enzootic abortion in ewes, psittacosis in poultry caused by *Chlamydia psittaci*, and Q-fever in sheep and cattle as well as other mammals, caused by *Coxiella burnetti*. Most of these zoonoses present with "flu-like" symptoms in humans, but can lead to more complicated illnesses. A protozoal zoonosis associated with farm animals is cryptosporidiosis caused by protozoa of the *Cryptosporidium* genus, which are found in numerous animals worldwide. Cryptosporidia is excreted in the faeces of animals and often contaminates water supplies. Viral zoonoses associated with farm animals include orf in sheep and goats as mentioned above, hepatitis E in pigs and influenza in pigs and poultry. Influenza is an extremely important zoonotic disease and is discussed further under emerging zoonoses.

3.3 Zoonoses in Pets

Human contact with animals is not restricted to food consumption or occupation. Many of us live with animals as pets and are constantly exposed to the microbial populations they host. In fact, man's best friend may be anything but! A number of zoonotic diseases have been associated with dogs, including bacterial zoonoses such as campylobacteriosis, leptospirosis, Glanders or farcy, caused by the bacteria *Burkholderia mallei* and *E. coli* O157:H7 (Weber 2005). Ringworm is a common fungal zoonosis of dogs, as are the protozoal zoonoses toxocariasis, leishmaniasis and Hydatid disease caused by the protozoan *Echinococcus vogeli*. Cats are no less a source of zoonotic diseases, also being host to campylobacteriosis, Glanders, ringworm, toxocariasis and toxoplasmosis. Cats are the definitive host for *Toxoplasma* spp. and are implicated in the infection of pigs with this protozoan.

Horses are another animal which many humans have prolonged social contact with. One of the most common zoonoses associated with horses is ringworm. Horses are also known to be a source of anthrax, Glanders and *E. coli* O157:H7. They can also serve as a secondary reservoir for the Hendra virus of fruit bats (known in horses as equine morbillivirus), which if transmitted to humans can cause a fatal encephalopathy.

The introduction of more exotic pets into our homes has exposed us to some exotic strains of common bacterial zoonoses. In recent years, there have been reports of humans having been infected with *Salmonella chameleon* and *Salmonella arizonae* from pet reptiles.

3.4 Emerging Zoonoses

The term "emerging zoonoses" can have a number of meanings. Diseases that were not previously considered zoonotic, but have the potential to be so, may be considered emerging. In addition, well-known zoonoses that are coming to the fore due to a number of contributing factors are often described as "emerging". A good example of this is influenza (Taubenberger and Kash 2010). Influenza is a virus which has a highly unstable genome and is thus continually mutating. It has long been known that influenza has the ability to cross the species barrier, and when it does so it can be highly virulent with devastating effects. The influenza pandemic which followed World War II killed around 21 million people and it is thought that the strain which was responsible for this pandemic may have originated from pigs. Influenza made a return to the public eye at the beginning of the twenty-first century with outbreaks of avian influenza H5N1 in birds, and of influenza H1N1 in humans. Pigs are often implicated in influenza outbreaks as they carry receptors for both avian and human influenza virus particles. It is possible that pigs could serve as a "mixing vessel" in which the virulent H5N1 avian influenza might be able to recombine with a human strain giving it the ability to infect humans. Therefore, there is much anticipation of the global zoonotic threat that influenza might pose in the future.

A potentially emerging zoonotic bacterium that was not previously considered as such is *Clostridium difficile. Clostridium difficile* is a bacterium which causes disease associated with the administration of antibiotics. It is therefore usually considered a hospital-acquired infection. However, it is also found in a number of animals including pigs, cattle and horses. Recent studies have shown an increase in human infections in the Netherlands with the type of *C. difficile*, which infects pigs, known as PCR ribotype 078, although there is as yet no evidence to confirm transmission of this organism from pigs to humans.

Another emerging zoonotic bacterium which is again considered to cause hospital-acquired infection is methicillin-resistant *Staphylococcus aureus* (MRSA). *Staphylococcus aureus* and MRSA are well-known pathogens of humans, but recently MRSA has been isolated from a variety of animals including pigs, rabbits and birds (Springer et al. 2009). Some studies have identified MRSA infections in humans who have had close contact with infected animals, and research has shown that veterinarians may be more likely to be carriers of MRSA than the general population. The emergence of MRSA in animals brings a new dimension to the study of zoonotic diseases. Antimicrobial resistance can only add to the arsenal of zoonoses, contributing to the future challenges of tackling these diseases.

Microbes certainly often have the potential to cause diseases and it is important to consider how this disease may be controlled or cured by antimicrobial chemotherapy – the use of chemical agents to kill microbes or prevent their growth – some of these antimicrobials are, of course, produced by microbes.

4 Antibiotics

Antibiotics are the products of microbes that, in dilute solution, inhibit or kill other organisms. Antimicrobial agents include antibiotics and synthetic compounds that have the same effect. Naturally occurring antibiotics may be modified to give semi-synthetic derivatives, which often differ from their parent compound in their antimicrobial activity or their pharmacological properties.

Often the term "antibiotic" is applied very loosely and includes antibacterial agents as well, although this is strictly incorrect. An antibiotic is a chemical produced by a microorganism that, in small amounts, inhibits or kills the growth of other microorganisms. Since sulphonamides and later drugs such as trimethoprim and the quinolones are man made, they cannot be accurately described as antibiotics, although commonly this is how they are categorised. The real problem, however, comes with the semi-synthetic compounds such as ampicillin, which is an artificial derivative of penicillin. As it is a derivative of an antibiotic rather than being the product of a microorganism, it cannot truly be described as an antibiotic, but it is not entirely artificial. To save bother, all but the most pedantic of microbiologists frequently use the term "antibiotic" to cover true antibiotics, their semi-synthetic derivatives and the artificial antimicrobial compounds.

Antimicrobial agents may have one of two effects: they may inhibit microbes or they may kill them. Agents that inhibit microbes are said to be STATIC; those that kill microbes are said to be CIDAL. The effects of static and cidal antimicrobials can easily be seen by looking at the effects of drugs on growing populations of bacteria. Static drugs prevent growth until the drug is removed; cidal drugs cause the death of cells in a culture (Fig. 1).

Static antibiotics combat many infections because they "buy time" for the host defences to eliminate the causative infectious agent, although there are conditions where only the use of cidal drugs will effect a cure. The terms "-cidal" and "-static" are used to describe the action of disinfectants as well as antibiotics. For example, chloramphenicol is bacteriostatic and gentamicin is bactericidal; phenol is germicidal, whereas mercury ions are bacteriostatic. Germicidal is distinct from bactericidal since it may represent an agent that is active against many different microorganisms and not just bacteria.

In Fig. 1, the down arrows indicate the point at which drug is added to the culture; the up arrow indicates where the static drug is removed from the culture, permitting growth to resume.

The efficacy of an antimicrobial agent may be assessed by measuring its ***M***inimum ***I***nhibitory ***C***oncentration, or MIC for short. A serial dilution of the drug to be investigated is made in a suitable broth medium and a standard inoculum of the test organism is added to each tube. Two controls are included, broth with no antibiotic, to show that the organism grows under the test conditions that are being used and a sterile broth to which no drug or inoculum is added and thus should remain sterile. The cultures are incubated at a suitable temperature and for an appropriate time.

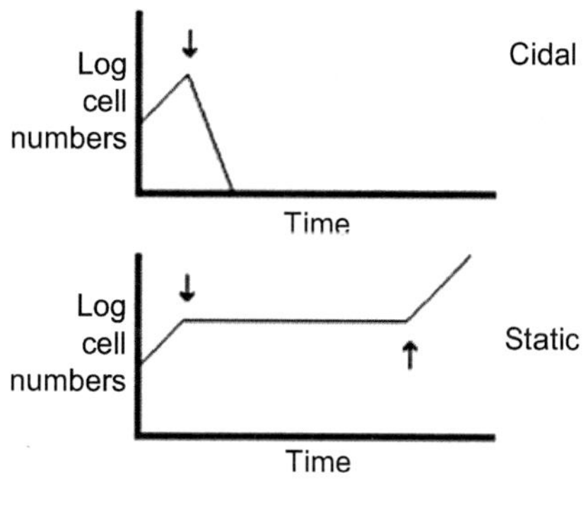

Fig. 1 The effects of cidal and static antimicrobials on the growth of a microbial batch culture

The MIC is the lowest concentration of the agent that prevents growth of the inoculum. The broth will remain clear in that tube but will appear cloudy in subsequent dilutions where the test organism is able to grow. Subculturing from all the clear tubes onto drug-free medium will permit estimation of the ***M***inimum ***B***actericidal ***C***oncentration, or MBC, of an antibacterial antibiotic. This is defined as the minimum concentration required to kill the cells in the culture, as shown by the inability of the subculture to yield colonies on subculture. With static drugs, the MIC and the MBC will differ vastly; with cidal drugs, the MBC of the drug is very close to its MIC.

Using serial dilutions is a relatively crude measure of drug efficacy, and one tube difference is often seen in the MBC and MIC of cidal drugs. This is because the real MIC will lie between the last tube that is seen to prevent growth and the first tube to permit growth. In practice, MICs on a number of isolates are often determined on agar plates containing a serial dilution of the drug to be tested. This only permits estimation of the MIC, but it has the advantage that the process can be semi-automated with a replica plating machine and up to 50 isolates may be tested on a single set of plates.

Antimicrobial agents may, when used in combination, affect each other. The effect may simply be additive. In some cases, the activity of one drug enhances that of a second drug and this is referred to as synergy. Alternatively, drugs may interfere with each other, which is antagonism. Where drugs are indifferent to one another, a straight line determining where the drug activity stops is seen when increasing concentrations of each drug are tested in a checkerboard pattern. When drugs are antagonistic, growth occurs at higher concentrations of each drug, and when drugs are synergistic, growth is seen at lower concentrations. Antagonistic combinations are problematic for combined drug therapy. An example is the combined use of penicillin and erythromycin. Erythromycin is a static drug that prevents bacterial growth; penicillin is a cidal drug that requires cells to be actively growing for the drug to be effective. By combining these two agents, the effectiveness of penicillin is reduced. Synergistic combinations are often exploited in the chemotherapy of challenging infections. In the treatment of endocarditis, it is essential to use cidal therapy since vegetations seal off the pathogen from host defences. A combination of penicillin and gentamicin is often used in such cases.

4.1 The History of Antimicrobial Chemotherapy

Rational antimicrobial chemotherapy was one of the greatest triumphs of the twentieth century. Before the First World War, a lab technician in the Botany Laboratories at Cambridge University used to collect cultures of *Penicillium* fungi to make a salve to cure "the gatherings". In Victorian times, Joseph Lister pioneered studies on antibiotics. He used his own urine as a growth medium and noted the effect fungi had on bacteria that he cultured. He persisted with irrigation of wounds with fungal extracts for a number of years, but could not obtain sufficient antibiotic to make the therapy reliably successful.

The first antimicrobials were described as "Magic Bullets" by their discoverer, Paul Ehrlich. Together with his student Sahachiro Hata, he pioneered the use of arsenical compounds in the successful treatment of syphilis and fever induced by malaria. These replaced mercury which was an ineffective treatment. To highlight the ineffective nature of mercury on treatment of syphilis, it used to be said that "a night with Venus would lead to a lifetime on Mercury". From the arsenical compounds, sulphonamides, the first successful broad-spectrum antimicrobial compounds, were developed by Gerhard Domagk.

In the 1920s, Alexander Fleming was attributed with the discovery of penicillin. He claimed that the fungus responsible for the antibiotic contaminated his plate during the time of inoculation. This is unlikely since fungi take much longer to grow than do bacteria and more likely was that the agar already had fungal growth on it. The inhibition of bacterial growth by the fungus, as observed by Fleming, is because the fungus had had time to produce sufficient penicillin to prevent bacterial growth. The agent, Penicillin G, was the first true antibiotic and was first used therapeutically by a team led by Howard Florey and Ernst Chain in Oxford in 1941. At that time, it was more valuable than gold, but due to the united Allied efforts under wartime conditions, problems of its mass production were quickly solved. The first patient treated was a reserve policeman with an overwhelming staphylococcal infection. He died when the supply of penicillin ran out, so the next person to receive penicillin was a small boy with a streptococcal infection: he did not need such large amounts of the drug and he survived. When penicillin was in short supply in the early years of the Second World War, penicillin was used to treat soldiers with gonorrhoea rather than those injured in battle. This decision was taken at the highest level, but makes good sense in that a soldier with gonorrhoea can be returned to duty after a single dose of the drug, whereas one suffering wound trauma would need time for injuries to heal before returning to front-line duties.

4.2 *Selective Toxicity*

Bacteria are good targets for the activity of antimicrobial substances. Aspects of their structure and metabolism are significantly different from that of humans. These differences can be exploited to damage the bacterium without harming its host.

Antibiotics may act upon bacterial reactions that are not found in human cells, or, if they are found, are inaccessible to antibiotics. This provides the basis for the selective toxicity of antibiotics, affecting the bacteria but not the human host. Fungi and protozoa have a metabolism that is much closer to that of humans than that of bacteria, while viruses depend almost exclusively upon human metabolism for their replication. Where there are less obvious unique targets for antibiotics, difficulties can arise both in the production of effective agents and in reducing associated toxic effects.

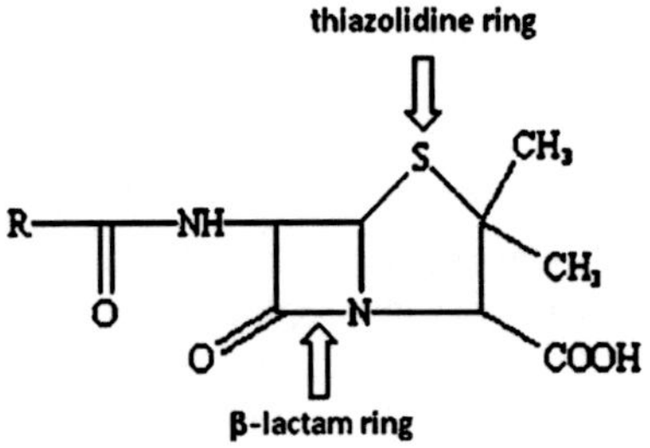

Fig. 2 The structure of the penicillin nucleus indicating the position of the four-membered β-lactam ring and the five-membered thiazolidine ring (variation in the side chain *R*- gives penicillins with different activity)

Selective toxicity is not absolute and a minority of individuals may experience life-threatening allergic reactions to penicillin, suffering anaphylaxis or, much more rarely, Stevens Johnson syndrome, in which the epidermis separates from the dermis. Penicillin allergy is due to the presence of the thiazolidine ring of penicillins. It is the β-lactam ring rather than the thiazolidine ring that is responsible for the antibiotic activity (Fig. 2).

4.3 Antiviral Agents

Viruses are obligate intracellular parasites. Many of the drugs that affect the replication and spread of viruses are too toxic for human use. There are, however, agents to treat specific virus infections. The development of these agents has been enhanced considerably by the emergence of AIDS and research into HIV and other emerging virus infections. Research into compounds that will act as antiviral agents focuses on the replication cycle of viruses; key points in the replication cycle include:

- Entry of the virus into a new host cell
- Uncoating of the virus
- For retroviruses, reverse transcription of the virus genome
- Integration of the virus genome into the host genome, if appropriate
- Protease inhibition
- Virus replication and assembly
- Virus release

Translation of viral RNA is also targeted in a novel manner, using compounds to inhibit transcription factors, antisense RNA, and also using ribozymes. Antisense RNA is a short sequence of RNA that comprises the reverse complement of an mRNA molecule. The antisense RNA binds to its specific mRNA to form a double-stranded structure that prevents translation of that mRNA. Ribozymes are RNA molecules in which the tertiary structure confers enzymatic activity, for example causing hydrolysis of other RNA molecules.

Amantadine is used in the treatment and prevention of Influenza A infections; rimantadine, a racemate of optical isomers, is also used for this purpose. Both antivirals work by targeting the penetration and uncoating process. Acyclovir is a nucleoside analogue used to treat infections caused by the herpes viruses, especially herpes simplex virus and varicella zoster virus. Acyclovir was the first successful antiviral agent to be released and inhibits the activity of the virus thymidine kinase. Gancyclovir is a derivative of acyclovir that is more active against cytomegalovirus. Ribavirin is another nucleoside analogue that has a broad spectrum of activity. It is used to treat infections caused by respiratory syncytial virus, in particular, bronchiolitis, in infants.

Zidovudine (AZT) was the first of a family of reverse transcriptase inhibitors. It is used to slow the progress of the human immunodeficiency virus, HIV, when the patient has developed full-blown AIDS. Lamivudine is another inhibitor of reverse transcriptase. Viruses may produce "polyproteins" that are hydrolysed by specific proteases to release mature virus proteins. These proteases are the target for antiviral drugs known as protease inhibitors; an example is nelfinavir used in the management of HIV infection. Combination therapy (use of a mixture of agents with different mechanisms of action) for people with AIDS is proving successful in the management of individuals with the disease and also in delaying the emergence of drug resistance of HIV. The approach is often referred to as HAART – highly active antiretroviral therapy. To delay emergence of resistance, drugs with different modes of action are used. This approach is not without problems; administration may involve complex timetables for dosing and the drugs are associated with significant side effects, making patient compliance difficult in some cases.

Oseltamivir (Tamiflu) and zanamivir (Relenza) are neuraminidase inhibitors that prevent release of influenza virus from infected cells. Heavy use of oseltamivir during the H1N1 influenza pandemic has led to the emergence of resistant strains of this virus.

4.4 Antifungal Agents

Fungi are eukaryotic and share many similar metabolic processes with humans. There are relatively few antifungal drugs that can be safely used to treat fungal infections. Given the success of agents that target the bacterial envelope, it is perhaps surprising that it took until 2006 to develop agents that attacked the fungal cell wall. The echinocandins are a family of drugs that inhibit the synthesis of glucan in the fungal cell wall. It is thought that they act by inhibiting the 1,3-β glucan synthase, which is an enzyme responsible for polymerisation of the sugars in the wall. Examples include anidulafungin, caspofungin and micafungin.

Polyenes act by binding with sterols in eukaryotic membranes causing their destabilisation. Common polyenes include nystatin used to treat oral and genital *Candida* infections and amphotericin B, used for the treatment of systemic mycoses. Amphotericin B is potentially toxic to humans and is often given with a second antifungal or in a lipid preparation. It penetrates poorly into cerebrospinal fluid when it is used to treat meningitis caused by a fungus and it may be delivered directly into the ventricles of the brain. For life-threatening fungal infections, amphotericin

B remains the drug of choice. It is often administered together with 5-flucytosine since combining these two drugs allows a lower dose of amphotericin B to be used.

Flucytosine (5-FC) is a synthetic pyrimidine that is metabolised in fungi to 5-flurouracil and is primarily active against pathogenic yeasts. As an analogue of nucleotides that occur naturally, it acts by interfering with the synthesis of nucleic acids. It may be administered by the oral route as well as by intravenous infusion, either alone or in combination with amphotericin B, where it is used to treat, for example, systemic candidiasis and cryptococcal meningitis. When used as a single agent, there is a significant risk of selecting resistant mutants of the pathogen being treated, a major drawback in the use of 5-flucytosine, necessitating susceptibility testing during therapy.

The azoles are a large group of synthetic compounds that inhibit ergosterol synthesis, leading to membrane leakage. There are two families of azoles that are used clinically: the imidazoles including clotrimazole, miconazole, econazole and ketoconazole and the triazoles, which include fluconazole, itraconazole, voriconazole and posaconazole. These days, the imidazoles are considered too toxic to be used to treat systemic infections, but are still used topically to treat superficial fungal infections. Triazoles have a wide variety of applications, although fluconazole is only active against yeasts and dimorphic fungi; other triazoles are active against moulds and yeasts. Yeast species are developing resistance to fluconazole and some species, for example *Candida krusei*, are intrinsically resistant to this agent.

Terbinafine is a synthetic antifungal agent introduced into the UK in 1991 and is used to treat skin and nail infections where it inhibits ergosterol biosynthesis. Griseofulvin is a naturally occurring compound and so is a true antibiotic with antifungal properties. It binds to the proteins involved in microtubule formation and prevents separation of chromosomes at mitosis. Why griseofulvin does not affect human cells is not known. It is used in the treatment of ringworm and other fungal infections of the skin or nails.

4.5 Antiprotozoal Agents

Protozoa are unicellular eukaryotes. This makes selective toxicity more difficult to achieve than in the treatment of bacterial infections. Nevertheless, protozoa have targets that can be exploited in antiprotozoal therapy. Examples of antiprotozoal drugs include chloroquine, mefloquine and pyrimethamine. Chloroquine and mefloquine are thought to interfere with the parasite's metabolism of haem, disrupting its nutrition. Pyrimethamine targets the parasite's dihydrofolate reductase, an enzyme essential for the synthesis of folic acid. For some protozoal infections and also in the treatment of infections caused by the fungus *Pneumocystis jiroveci*, a combination of sulphamethoxazole and trimethoprim may be used. These block steps in folic acid synthesis. This combination therapy was used for many years to treat bacterial infections, but in the UK, the combination is no longer available for antibacterial therapy, because of the level of resistance to sulphonamides in clinically important bacteria and also because of the increased risk of side effects for the patient. The combination is still used elsewhere in the world.

Metronidazole was developed as an antiprotozoal drug. It induces strand breaks in the DNA of sensitive organisms and also disrupts membrane integrity. It is used in the treatment of *Trichomonas vaginalis* and *Entamoeba histolytica*. Careful clinical trials showed that it is highly effective in the treatment of infections caused by anaerobic bacteria.

4.6 Antibacterial Antibiotics

The prokaryotic cell offers a number of targets that are suitable for selective toxicity caused by antibacterial agents. These include:

- The cell envelope
- Membrane integrity
- DNA metabolism and packaging
- RNA polymerase
- Protein synthesis

4.7 Agents Affecting the Cell Envelope

The bacterial cell envelope offers a number of targets for antibacterial agents. The bacterial cell wall contains peptidoglycan, a unique biopolymer that contains both L- and D-amino acids. As such it should make an ideal target for agents that show "selective toxicity". In fact, antibiotics that interfere with the integrity of cell walls may be toxic to humans too. An example is bacitracin which in diagnostic laboratories may be used to differentiate *Streptococcus pyogenes* from other β-haemolytic streptococci. However, because of its toxicity it is only used rarely and, when it is used, is applied topically.

Clinically useful inhibitors of cell wall synthesis include:

- Cycloserine
- Fosfomycin
- Glycopeptides
- β-Lactams

4.8 Cycloserine and Fosfomycin

Cycloserine acts inside the bacterial cell and inhibits the racemase enzyme that converts L-alanine into D-alanine (Calvo and Martínez-Martínez 2009). It is rarely used except in the treatment of infections caused by *M. tuberculosis* where other treatments have failed. Fosfomycin, also known as phosphomycin and phosphonomycin, is another drug that acts inside the cell to prevent synthesis of the peptidoglycan precursor. The fact that resistance to this drug readily evolves means that it is not used commonly, but is gaining popularity in the treatment of urinary tract infections.

4.9 Glycopeptides

Glycopeptides such as vancomycin are chemically large and can only act against Gram-positive bacteria, where they prevent incorporation of peptidoglycan monomers into the growing cell wall. The glycopeptide molecules are too large to penetrate the outer membrane of Gram-negative bacteria and thus cannot reach the peptidoglycan target. Vancomycin is the mainstay treatment of infections caused by MRSA and is also used in the treatment of severe pseudomembranous colitis caused by *C. difficile*, when metronidazole has failed.

4.10 The β-Lactams

All β-lactam antibiotics have a four-membered ring structure known as the β-lactam ring. These antibiotics are the most widely prescribed of the antibacterial antibiotics. Benzyl penicillin (penicillin G) was the first true antibiotic in clinical practice. This agent is acid labile so must be delivered parenterally. The first oral penicillin was phenoxymethyl penicillin (penicillin V). Ampicillin was the first penicillin with activity against members of the family *Enterobacteriaceae*. Like benzyl penicillin, this can only be delivered parenterally, the oral equivalent being amoxicillin. Meticillin was the first anti-staphylococcal penicillin and has a side chain that prevents hydrolysis by staphylococcal β-lactamase. Meticillin has largely been superseded by drugs such as cloxacillin and flucloxacillin. The β-lactam ring forms a structural analogue of D-alanyl-D-alanine, found as the terminal residues on the peptidoglycan monomer. The β-lactams antibiotics act by interfering with penicillin-binding proteins which are the enzymes involved in the synthesis and maintenance of peptidoglycan.

Other β-lactam antibiotics include:

- The cephalosporins – a family of broad-spectrum antibiotics that are among the most frequently prescribed antibacterial agents
- Monobactams – drugs active primarily against Gram-negative bacteria and are little used in clinical practice
- Carbapenems, another family of broad-spectrum agents used in the empirical treatment of serious infections, particularly before culture results have been obtained

The cephalosporins have a six-membered thiazine ring next to the β-lactam ring. Monobactams need to only have the β-lactam ring, other rings being coincidental and not necessary for antibacterial activity. In carbapenems, the sulphur atom found in the thiazolidine ring of the penicillins has been replaced by a carbon atom. Clavulanic acid is a β-lactam agent that has very little antibacterial activity. It does, however, act as a competitive inhibitor of many bacterial β-lactamases, enzymes that can hydrolyse the β-lactam ring, destroying the antibacterial activity of these drugs. When used in combination with a β-lactam drug, it will overcome resistance due to β-lactamase, restoring the efficacy of the antibiotic. The first successful combination was amoxicillin and clavulanic acid, marketed as "Augmentin". Now, several combinations of β-lactam and β-lactamase inhibitors are available.

4.11 Agents Affecting Membrane Integrity

There are antibiotics that target the integrity of the bacterial membrane. Some are toxic and are only used when other options are unavailable. The polymyxins are a family of antibiotics with a cyclic peptide structure, principally active against Gram-negative bacteria. They are nephrotoxic and neurotoxic and this has severely limited their clinical use. These antibiotics are, however, increasingly used to treat infections caused by pathogens that are resistant to other therapies.

The gramicidins are used topically to treat infections caused by Gram-positive bacteria, but their use is also limited because of toxicity problems. Metronidazole is a broad-spectrum antimicrobial agent that has very wide applications. It is used in the treatment of certain protozoal infections as well as being the mainstay of treatment of anaerobic bacterial infections. Although its primary target is to inhibit DNA gyrase, it is thought that it also causes damage to membranes as a secondary target.

4.12 Agents Affecting DNA

The sulphonamides were the first antimicrobial agents that successfully targeted DNA metabolism. Humans require an exogenous source of folic acid, a vitamin with a role in nucleic acid synthesis, while bacteria make their own folic acid. This provides target therefore for selective toxicity. Folic acid is involved in the synthesis of purines and many bacteria make folic acid from para-aminobenzoic acid. This pathway serves as a target for the action of sulphonamides and trimethoprim. Sulphonamides specifically inhibit dihydropteroate synthetase, the enzyme that catalyses the first step in folic acid synthesis, while dihydrofolate reductase is the target for trimethoprim. Trimethoprim was first introduced to potentiate the activity of sulphonamides. In in vitro tests, sulphonamides and trimethoprim are synergistic, but this synergy is not seen when the combination is used to treat bacterial infections. Although used in combination for many years, trimethoprim is now available as a single agent and it is widely used in the treatment of urinary tract infections.

Despite the perceived selective toxicity, problems of host toxicity do occur and this, together with high levels of resistance, means that sulphonamides are rarely used these days.

In addition to agents that target DNA metabolism, there are drugs that directly affect DNA. The bacterial chromosome, when linearised, is many times longer than the cell that contains it. Bacterial DNA thus requires packaging to fit into the cell in which it is housed. This is achieved by enzymes such as DNA gyrase, which is responsible for supercoiling DNA. This and similar enzymes introduce twists into the DNA molecule causing it to package in a manner similar to elastic bands that have been twisted into supercoils. Nalidixic acid was the first quinolone antibiotic that directly targeted DNA, but toxicity problems have limited its use. In the mid-1980s, ciprofloxacin, the first of the fluoroquinolones, was introduced into clinical practice. This is a broad-spectrum oral agent that is generally well tolerated. It has a wide range of applications and achieved notoriety when it was used to treat victims of the postal anthrax attack in the USA in 2001. The fluoroquinolones belong to a very successful family and new agents are still being developed. Examples of newer fluoroquinolones include moxifloxacin and levofloxacin. Metronidazole is thought to act in a manner similar to the quinolone drugs.

4.13 Agents Affecting RNA Synthesis

The bacterial DNA-dependent RNA polymerase is inhibited by rifampicin, but this drug has little effect on eukaryotic cells. Rifampicin is active against the mitochondrial RNA polymerase, but its penetration into mitochondria is so poor that it displays very little activity in intact eukaryotic cells. Rifampicin prevents production of messenger RNA, ultimately limiting protein synthesis. Clinically, rifampicin is used in treating tuberculosis and for prophylaxis against meningococcal meningitis. In such cases, it is offered to close contacts of people with the disease. It has also been used to treat infection caused by vaccinia virus. The synthetic antibacterial nitrofuran compounds also act by preventing messenger RNA production.

4.14 Agents Affecting Protein Synthesis

Although similar in structure to the ribosomes found in mitochondria and chloroplasts, bacterial ribosomes are a good target for selective toxicity with several families of drug targeting these structures. The organelle ribosomes are not easily targeted by antibacterial antibiotics because they are protected by the cell plasma membrane and the organelle membranes, making drug penetration difficult.

Antibiotics that inhibit bacterial protein synthesis include:

- Aminoglycosides
- Tetracyclines

- Chloramphenicol
- Macrolides
- Lincosamides
- Fusidic acid
- Mupirocin
- Linezolid
- Streptogramins

4.15 Aminoglycosides

With their broad spectrum of activity and bactericidal action, the aminoglycosides are a clinically important group of antibiotics. The family includes streptomycin, gentamicin, tobramycin, kanamycin, amikacin and netilmicin. Of these, gentamicin is the most commonly used for the treatment of serious infections. The older members of the family are not pure compounds, but rather they comprise mixtures of similar compounds. In contrast, amikacin and netilmicin are pure compounds. These were developed in part with the hope of reducing the toxicity associated with aminoglycosides.

All the aminoglycosides are associated with serious side effects including nephrotoxicity and damage to the VIIIth cranial nerve leading to deafness and balance problems. The therapeutic window is very narrow and use of the aminoglycosides requires careful monitoring to ensure that adequate therapeutic levels are maintained, without the accumulation of toxic levels.

Aminoglycosides have a variety of effects within the bacterial cell, but principally they inhibit protein synthesis. This is achieved by binding to the 30S ribosomal subunit, thus preventing the formation of an initiation complex with messenger RNA. Aminoglycosides also cause misreading of the messenger RNA message, leading to the production of nonsense peptides. Another important function of the aminoglycosides is that they increase membrane leakage. These multiple effects mean that the aminoglycosides are rapidly bactericidal.

4.16 Tetracyclines

The tetracyclines are a family of antibiotics that have a four ring structure, giving this family its name. They inhibit binding of the aminoacyl tRNA to the 30S ribosomal subunit in a wide array of bacteria, giving the family a broad spectrum of activity. The activity of tetracyclines is bacteriostatic and can therefore be alleviated upon removal of the drug. The clinical use of tetracyclines is generally confined to adults as they can affect bone development and can cause staining of teeth in children. Tigecycline is a new, broad-spectrum tetracycline that is cidal in its activity. It is used in the treatment of meticillin-resistant *S. aureus* (MRSA).

4.17 Chloramphenicol

The broad-spectrum bacteriostatic agent chloramphenicol is toxic to humans. It has been recognised as a cause of aplastic anaemia and so its use is restricted to life-threatening infections where no alternative therapy is available, although it is used widely in the "Third World" because it is cheap to produce and is relatively stable. Chloramphenicol blocks the formation of the peptide bond in nascent peptides by inhibiting peptidyl transferase activity following binding to the 50S ribosomal subunit. In eukaryotic cells, it is a potent inhibitor of mitochondrial protein synthesis.

4.18 The Macrolides and Lincosamides

The macrolides are a group of antibiotics that have a large, lactone ring structure. These may be 14- or 16-membered rings. The most widely used macrolides in human medicine are erythromycin and clarithromycin and these are relatively non-toxic antibiotics, mostly active against Gram-positive bacteria. In adults, use of erythromycin is associated with diarrhoea as the drug decreases transit time through the intestine dramatically. This makes compliance in adults problematic. Fortunately, newer macrolides, such as azithromycin and clarithromycin, are less often associated with diarrhoea.

Despite the spectrum of activity being primarily against Gram-positive bacteria, erythromycin is, nevertheless, the treatment of choice for Legionnaire's disease caused by the Gram-negative bacillus *Legionella pneumophila* and it may also be active against isolates of *Haemophilus influenzae*, another Gram-negative bacillus. Erythromycin binds to the 50S ribosomal subunit and inhibits either peptidyl transferase activity or translocation of the growing peptide. The lincosamide antibiotic, lincomycin and its semi-synthetic derivative clindamycin have similar modes of action to the macrolides, although they have a broader spectrum of activity. Clindamycin is not widely used because of its association with pseudomembranous colitis, severe diarrhoea associated with antibiotic use and caused by *C. difficile*.

4.19 Fusidic Acid

The steroid antibiotic fusidic acid is used to treat Gram-positive infections where it acts by preventing translocation of peptidyl tRNA. Resistant mutants may be selected easily and this may occur during therapy. Therefore, fusidic acid is usually administered in combination with another antibiotic which reduces the risk of selecting resistant mutants.

4.20 *Oxazolidones*

Linezolid is the first of a new class of bacterial protein synthesis inhibitors, the oxazolidone antibiotics. Linezolid prevents the initiation of protein synthesis by interfering with the interaction between mRNA and the two ribosomal subunits necessary for the initiation of translation of the messenger RNA into the nascent peptide chain. The antibiotic is active against Gram-positive cocci, including meticillin-resistant *S. aureus* and vancomycin-resistant enterococci (VRE).

4.21 *Mupirocin*

Mupirocin is an analogue of isoleucine and was formerly known as pseudomonic acid. Mupirocin inhibits the iso-leucyl-transfer RNA synthetase, thereby preventing the incorporation of isoleucine into growing polypeptide chains. It is not toxic to humans, but can only be used topically for skin infections. This is because humans rapidly metabolise the drug to an inactive form so in systemic use it is destroyed before it can exert its antibiotic effective. This agent is widely used to clear colonisation with MRSA.

4.22 *Streptogramins*

There are two classes of streptogramin referred to as Type A and Type B. Type A streptogramins have a large, polyunsaturated non-peptide ring. Streptogramins related to streptogramin B are cyclic peptides. Both inhibit bacterial protein synthesis, but they differ in their mode of action. Type A streptogramins distort the ribosome, preventing binding of tRNA, while Type B streptogramins act by blocking translocation of the nascent peptide. Because of the challenges in controlling MRSA, a combination of dalfopristin and quinupristin has recently been introduced. Dalfopristin is a Type A streptogramin and quinupristin is a Type B streptogramin. In combination, these drugs show a synergistic effect.

4.23 Antimycobacterial Agents

Treatment of mycobacterial infections poses serious challenges because of the waxy cell walls of these bacteria and the intracellular habitat of mycobacteria that cause serious infections. Streptomycin was the first drug used successfully to treat tuberculosis. It was discovered by Albert Schatz, working in Selman Waksman's laboratory, and acts by the inhibition of protein synthesis. Rifampicin is also used as an antimycobacterial drug and its mode of action is by interfering with the DNA-dependent RNA polymerase of bacterial cells. The action of rifampicin prevents production of messenger RNA and thus ultimately stops protein synthesis.

The antimycobacterial drug isoniazid inhibits the formation of very long chain fatty acids such as those found in the cell walls of mycobacteria. Isoniazid is used in the treatment of tuberculosis and other mycobacterial infections. Ethambutol is a first-line antimycobacterial drug that inhibits cell wall synthesis, although its mode of action remains to be fully elucidated. Pyrazinamide is another first-line antimycobacterial drug that inhibits mycobacterial metabolism. Again, its mode of action remains to be determined. Pathogenic mycobacteria grow very slowly, and *M. leprae* can only grow in humans and animal models. Consequently, treatment of mycobacterial infection extends over months or years. As a consequence, resistant mutants are easily selected if single agents are used for therapy. Emergence of resistance can be slowed significantly if drugs are combined. Local resistance patterns will help to dictate which combinations should be used. Another issue in the management of mycobacterial infection is patient compliance. Daily observed therapy, "DOT", has been introduced in many areas to ensure that patients take their drugs regularly.

5 Antibiotic Resistance

Antibiotic resistance emerged as a clinical problem almost as soon as antibiotics were introduced into clinical practice and the genes that encode certain types of resistance have been identified in bacteria isolated before the era of antimicrobial chemotherapy (Tanwir and Khiyani 2011). There are several mechanisms by which bacteria resist antibiotics, some of which are intrinsic to the organism, others that result from mutation of the genome of a previously susceptible bacterium, and others again that result from the acquisition of one or more mobile genetic elements that encode resistance (Martinez and Silley 2010). In the laboratory, it is relatively easy to select for resistance to some antimicrobial agents. Indeed, the spontaneous mutation rate from a susceptible to a resistant phenotype may be so rapid that it compromises the therapeutic use of some antimicrobials as single agents. Such resistance results from an alteration in the target structure that prevents interaction with the antibiotic. Mycobacteria that cause human infection are intracellular for much of their life cycles and so cannot easily exchange genetic material with other bacteria. For these pathogens, resistance occurs from mutation in the genome. Another case where resistance of endogenous genes occurs through point mutation is resistance to the quinolone family of antimicrobial agents. In these instances, almost all resistance that causes clinical problems result from alterations in DNA gyrase or topoisomerase, the targets for these synthetic drugs. In most resistant clinical isolates, however, resistance is either intrinsic to the organism or results from the expression of mobile genetic elements that encode resistance.

Bacteria use several mechanisms to evade the action of antibiotics:

- Absence of a target
- Inaccessible target
- Modification of the target
- Bypassing the antibiotic target
- Modification of the antibiotic

A single bacterium may express more than one mechanism of resistance to a particular antibiotic.

5.1 Absence of a Target

The genus *Chlamydia* includes *C. trachomatis*, the bacterium responsible for non-specific urethritis and a major cause of infertility. Worldwide this organism is also responsible for more cases of blindness than any other single agent. These bacteria are not susceptible to the action of cell wall inhibitors such as penicillins because they do not have a peptidoglycan cell wall. Interestingly, chlamydia does contain penicillin-binding proteins, and these are the enzymes that are required to make peptidoglycan on the surface of other bacteria. This leaves us with the so-called chlamydia paradox. Other bacteria that lack peptidoglycan include the rickettsias and mycoplasmas.

5.2 Inaccessibility of the Target and Efflux Pumps

Penicillin G was the first of the β-lactam family of antibiotics to be introduced into clinical practice, although its use now is relatively limited. Its side chain does not permit it to pass through the outer membrane of enteric Gram-negative bacteria such as *E. coli*. The outer membrane encloses peptidoglycan in Gram-negative bacteria and hence these bacteria are intrinsically resistant to penicillin G. The intrinsic resistance of some Gram-negative bacteria to penicillin G is related to the structure of the side chain of penicillin G and its interaction with the Gram-negative outer membrane. Altering the nature of the side chain on the penicillin nucleus modifies the spectrum of activity of the resultant antibiotic. Ampicillin is another antibiotic in the penicillin family, but its side chain allows it to pass through the outer membrane of Gram-negative bacteria such as *E. coli*.

The outer membrane of Gram-negative bacteria acts as a permeability barrier to large antibiotics such as the glycopeptides. Although vancomycin prevents cell wall formation in Gram-positive bacteria, it cannot penetrate the outer membrane of Gram-negative bacteria. Likewise, macrolide antibiotics such as erythromycin are unable to penetrate the outer membranes of many Gram-negative bacteria. In cell-free protein synthesis systems derived from *E. coli*, however, erythromycin inhibits the process as efficiently as it does in cell-free systems from Gram-positive bacteria.

Inaccessibility of a target due to structures such as the outer membrane of Gram-negative bacteria is a passive process. Inaccessibility of a target may also be an active process. There is a family of resistance genes that code for resistance to the tetracycline antibiotics by actively pumping them out of the bacterial cell. These are the "efflux pumps" and the process requires metabolic energy to move the drug up a concentration gradient.

5.3 Modification of the Drug Target

When bacterial mutants are generated in the laboratory, the basis of their antimicrobial resistance is frequently modification of the drug target. Early work on streptomycin resistance yielded numerous mutants in which the structure of the ribosome was altered to a degree where the drug is no longer able to be active. In the case of streptomycin resistance, a second class of mutants exists where bacteria have modifications to their ribosomes so that they cannot function unless streptomycin is present in the growth medium. Such organisms are streptomycin-dependent, being unable to grow in the absence of this drug.

In clinical settings, the consequences of chromosomal mutants may be even more problematic. As mentioned above, an important example of target modification causing clinical resistance is the family of mutations affecting DNA gyrase or topoisomerase enzymes. These mutations confer quinolone resistance on their hosts. Again as mentioned above, the pathogenic mycobacteria are very slow growing. In artificial culture, it may take over 24 h for a cell of *M. tuberculosis* to divide. One consequence of this is that antimycobacterial therapy is prolonged. Using old drug regimes, patients have to take anti-tuberculous drugs, some of which are very toxic to humans, for up to 2 years. The prolonged treatment time allows the opportunity to select for chromosomal mutants that are able to resist the action of the antimycobacterial therapy. It is for this reason that treatment of tuberculosis and leprosy now involves giving patients drugs in combination to delay the selection of resistant strains. If the chance of selecting resistance to a single agent "A" is 10^{-6} and resistance to a second agent "B" emerges at a rate of 10^{-8}, one in one million cells exposed to agent "A" will carry a mutation that confers resistance to drug "A". One in one hundred million bacteria exposed to drug "B" will resist its antibacterial action. When a population of bacteria is exposed to a mixture of both drugs "A" and "B", then the chances of selecting a bacterium that is resistant to both drugs is $10^{-6} \times 10^{-8}$. This is equivalent to 10^{-14}. These are much better odds for avoiding the selection of resistant mutants!

5.4 *Bypassing the Target*

Resistance to trimethoprim frequently results from the expression of a dihydrofolate reductase with reduced affinity for trimethoprim. The genes encoding altered enzymes are frequently found on plasmids that are self-transmissible. Such enzymes are produced in cells that also produce a wild-type dihydrofolate reductase, but the presence of the altered enzyme overcomes the block on folic acid synthesis that trimethoprim imposes on the wild-type enzyme. Sulphonamide resistance works by a similar mechanism, with an alternative dihydropteroate synthetase conferring resistance.

5.5 *Modifying the Drug*

The commonest mechanism by which clinical isolates become resistant to antibacterial drugs is through the acquisition of genes that code for enzymes that modify or destroy the target antibiotic. These genes are frequently located on mobile genetic elements, including plasmids and transposons, and through the very high mobility of these elements, can spread to sites with which they have no genetic homology.

In the case of resistance genes that form part of a transposable element, these may move between the chromosome and other replicons in the bacterial cell, including plasmids. Resistance genes on certain plasmids may also become integrated within the bacterial chromosome. Study of the evolution and dissemination of antibiotic resistance genes shows how remarkably fluid the bacterial genome is, particularly in the face of the selective pressure applied by the use of antibiotics in clinical and veterinary practice and in agriculture.

One of the earliest reports of antibiotic resistance was of a bacterium that produced a β-lactamase, an enzyme that hydrolyses the β-lactam bond in β-lactam antibiotics such as penicillin. The β-lactam bond is essential in the activity of β-lactam antibiotics, acting as an analogue of the peptide bond that joins the terminal D-alanine to the peptidoglycan monomer. The β-lactamases are now a large family of enzymes, all of which cleave this bond to render various β-lactam antibiotics inactive. The earliest β-lactamases only had a narrow spectrum of activity, hydrolyzing the β-lactam bond of either penicillins or cephalosporins. Recently, broad-spectrum β-lactamases that have activity against both penicillins and cephalosporins have evolved. There are families of such enzymes that have arisen as the result of point mutations accumulating in the genes that code for penicillinases. Many of these new enzymes are encoded by self-transmissible plasmids and these new resistance determinants can spread with great ease.

Resistance to chloramphenicol is often accomplished with the aid of an enzyme that acetylates the drug, rendering it inactive. Acetylation is also a common method of inactivating aminoglycoside antibiotics. Other modifications that cause aminoglycoside resistance include phosphorylation and adenylation.

5.6 The MRSA Story

Soon after the introduction of benzyl penicillin into clinical practice, strains of *S. aureus* producing β-lactamase were seen causing infections that were resistant to penicillin treatment. Alternative treatments had to be sought and early in the history of the development of semi-synthetic penicillins, compounds were manufactured that were able to resist the activity of staphylococcal β-lactamase. These drugs had side chains that prevented the staphylococcal β-lactamase from binding to the antibiotic and hydrolysing it. Meticillin, a penicillin that is stable in the presence of staphylococcal β-lactamase, was introduced into clinical practice during the 1960s. Infections caused by strains of *S. aureus* that produce β-lactamase could be treated with a β-lactam antibiotic once more. Shortly after the introduction of meticillin, formerly known as "methicillin", into medical practice, resistant strains of *S. aureus* were isolated from hospital units where the drug was in regular use (Pantosti and Venditti 2009).

Meticillin resistance is greater at 30°C than at 37°C. Resistance is due to the temperature-sensitive production of an extra penicillin-binding protein, PBP 2. This cell wall enzyme, which is not susceptible to inhibition by meticillin, is produced in larger quantities at low incubation temperatures. Meticillin-resistant *S. aureus* also produces β-lactamase and these bacteria are generally resistant to a very wide range of antimicrobials. Infections caused by meticillin-resistant *S. aureus* can thus be difficult to treat (Gould et al. 2011; Pletz et al. 2010). In some cases, almost the only drugs available to effectively treat infections caused are the glycopeptides such as vancomycin. Bacterial resistance to this agent was unknown until relatively recently.

Vancomycin resistance, and cross-resistance to other glycopeptides, first appeared in enterococci, probably due, at least in part, to the practice of feeding farm animals avoparcin, another glycopeptide, used for growth promotion. This practice is now banned in the European Union. Enterococci resistant to glycopeptides are not susceptible to nearly all currently available standard antimicrobial therapies. In an experiment of dubious ethical status, the gene encoding vancomycin resistance was transferred in the laboratory from a vancomycin-resistant enterococcus into a meticillin-resistant *S. aureus*. In 1997, the first naturally occurring vancomycin-resistant, meticillin-resistant *S. aureus* appeared in Japan (Hiramatsu et al. 1997). How long will it be before we enter the post-antibiotic era?

References

Calvo J, Martínez-Martínez L (2009) Antimicrobial mechanisms of action. Enferm Infecc Microbiol Clin 27(1):44–52

Christou L (2011) The global burden of bacterial and viral zoonotic infections. Clin Microbiol Infect 17(3):326–330

Ferens WA, Hovde CJ (2011) *Escherichia coli* O157:H7: animal reservoir and sources of human infection. Foodborne Pathog Dis 8(4):465–487

Gould IM, Cauda R, Esposito S, Gudiol F, Mazzei T, Garau J (2011) Management of serious meticillin-resistant *Staphylococcus aureus* infections: what are the limits? Int J Antimicrob Agents 37(3):202–209

Hegarty JP, Pickup R, Percival SL (2001) Detection of viable but non-culturable bacterial pathogens. In: Gilbert PG, Allison D, Walker JT, Brading M (eds) Biofilm community interactions: chance or necessity? Species consortia. Bioline, Cardiff, pp 39–51

Hiramatsu K, Aritaka N, Hanaki H, Kawasaki S, Hosoda Y, Hori S, Fukuchi Y, Kobayashi I (1997) Dissemination in Japanese hospitals of strains of *Staphylococcus aureus* heterogeneously resistant to vancomycin. Lancet 350(9092):1670–1673

Martinez M, Silley P (2010) Antimicrobial drug resistance. Handb Exp Pharmacol 199:227–264

Pantosti A, Venditti M (2009) What is MRSA? Eur Respir J 34(5):1190–1196

Pletz MW, Burkhardt O, Welte T (2010) Nosocomial methicillin-resistant *Staphylococcus aureus* (MRSA) pneumonia: linezolid or vancomycin? Comparison of pharmacology and clinical efficacy. Eur J Med Res 15(12):507–513

Springer B, Orendi U, Much P, Höger G, Ruppitsch W, Krziwanek K, Metz-Gercek S, Mittermayer H (2009) Methicillin-resistant *Staphylococcus aureus*: a new zoonotic agent? Wien Klin Wochenschr 121(3–4):86–90

Tanwir F, Khiyani F (2011) Antibiotic resistance: a global concern. J Coll Physicians Surg Pak 21(3):127–129

Taubenberger JK, Kash JC (2010) Influenza virus evolution, host adaptation, and pandemic formation. Cell Host Microbe 7(6):440–451

Weber CJ (2005) Update on infections you can get from pets. Urol Nurs 25(6):485–487

Wells GA (2003) Pathogenesis of BSE. Vet Res Commun 1:25–28

Introduction to Biofilms

Steven L. Percival, Sladjana Malic, Helena Cruz, and David W. Williams

Abstract In the seventeenth century, a dry-goods merchant named Antonie van Leeuwenhoek first observed "animalcules" swarming on living and dead matter. Leeuwenhoek's curiosity and inventiveness were remarkable; he discovered these "animalcules" in the tartar on his own teeth and even after meticulous cleansing, the remaining opaque deposits isolated between his teeth were still "as thick as if it were batter". These deposits contained a mat of various forms of "animalcules" that we now know were the bacteria of dental plaque. It is reasonable to suggest that this early study of dental plaque was the first documented evidence of the existence of microbial biofilms. Today, we generally define such biofilms as microbial communities adhered to a substratum and encased within an extracellular polymeric substance (EPS) produced by the microbial cells themselves. Biofilms may form on a wide variety of surfaces, including natural aquatic systems living tissues, indwelling medical devices and industrial/potable water system piping. The vast majority of microbes grow as biofilms in aqueous environments. These biofilms can be benign or pathogenic, releasing harmful products and toxins, which become encased within the biofilm matrix. Biofilm formation is a phenomenon that occurs in both natural and man-made environments under diverse conditions, occurring on most moist surfaces, plant roots and nearly every living animal. Biofilms may exist as beneficial epithilic communities in rivers and streams, wastewater treatment plant trickling beds or in the alimentary canal of mammals. Given the prevalence of biofilms in

S.L. Percival (✉)
Department of Pathology, Medical School, West Virginia University, Morgantown, WV, USA
e-mail: Biofilms@hotmail.co.uk

S. Malic • D.W. Williams
Oral Microbiology Group, Tissue Engineering and Reparative Dentistry, School of Dentistry, Cardiff University, Heath Park, Cardiff CF14 4XY, UK

H. Cruz
Division of Microbiology and Risk Assessment, National Food Institute, Technical University of Denmark, Mørkhøj Bygade 19, Building H, 2860 Søborg, Denmark

S.L. Percival et al. (eds.), *Biofilms and Veterinary Medicine*,
Springer Series on Biofilms 6, DOI 10.1007/978-3-642-21289-5_2,

natural environments, it is not surprising that these growth forms are responsible for infection in humans and animals. In humans, biofilms have been linked with numerous conditions and equally in animals equivalent infections may occur.

1 Introduction and Historical Perspectives

In the seventeenth century, a dry-goods merchant named Antonie van Leeuwenhoek first observed "animalcules" swarming on living and dead matter. Leeuwenhoek's curiosity and inventiveness were remarkable; he discovered these "animalcules" in the tartar on his own teeth and even after meticulous cleansing, the remaining opaque deposits isolated between his teeth were still "as thick as if it were batter". These deposits contained a mat of various forms of "animalcules" that we now know were the bacteria of dental plaque. It is reasonable to suggest that this early study of dental plaque was the first documented evidence of the existence of microbial biofilms. Today, we generally define such biofilms as microbial communities adhered to a substratum and encased within an extracellular polymeric substance (EPS) produced by the microbial cells themselves.

After van Leuwenhoek's early work, it was not until 1940 that the so-called "bottle effect" in marine microorganisms was first observed (Heukelekian and Heller 1940). This showed that the growth of bacteria was substantially increased when they were attached to a surface. Further advancements in our knowledge of biofilms were made by Zobell in 1943 when he noted that bacteria on surfaces were greater in number compared with the surrounding seawater. From his studies, Zobell also postulated that the adhesion of bacteria consisted of a two-stage process of reversible and then irreversible adhesion.

Despite the above studies being the first documented ones on biofilms, the extensive physical and chemical analysis of bacterial biofilms did not begin until the late 1960s and early 1970s, when a few investigators recognised the prevalence of bacterial biofilms (Jones et al. 1969; Characklis 1973; Costerton et al. 1978). Jones et al. (1969) used scanning and transmission electron microscopy to examine biofilms on trickling filters in a wastewater treatment plant. From this work it was shown that biofilms were composed of a variety of different microorganisms and revealed that the matrix material or EPS was primarily composed of polysaccharides. The investigation of biofilms around this time was greatly aided by the use of electron microscopy, which provided information, not only on biofilm structure, but also on the presence of EPS. In 1973, Characklis who investigated microbial slimes in industrial water revealed that biofilms were both tenacious and highly resistant to the antimicrobial effects of chlorine (Characklis 1973).

The first true analysis of biofilms per se was not recognised until 1978 (Costerton et al. 1978), when studies showed that many bacteria spent the majority of their existence within surface-attached, sessile communities.

Work on dental plaque and sessile communities in mountain streams enabled Costerton et al. (1978) to hypothesise the mechanisms by which microorganisms adhered to living and non-living materials and derived benefit from this ecologic niche.

Since the 1970s, the study of biofilms in industrial, ecological and medical settings has followed similar paths. Initial biofilm studies generally concentrated on composition, especially of the polymer matrix or "glycocalyx" that was thought to conserve and concentrate the digestive enzymes released by the bacteria, thus increasing the metabolic efficiency of the cells. Research by Costerton and Geesey (1979) indicated that this glycocalyx also acted as an ionic exchange matrix, trapping nutrients that were then transported into cells by highly efficient permeases.

In 1981 (Costerton et al. 1981), glycocalyx was characterised as a hydrated polyanionic polysaccharide matrix produced by polymerases affixed to the lipopolysaccharide component of the bacterial cell wall. In aqueous environments, biofilm production of glycocalyx is prevalent with organic and inorganic nutrients being concentrated at the solid/liquid interface. Additionally, the glycocalyx provides a physical/chemical barrier that offers partial protection against antibacterial agents.

Since biofilms form under diverse conditions, and may be composed of single or multiple species, the structures of various biofilms will necessarily have distinct features. Nevertheless, biophysical, structural and chemical studies have led to a useful basic concept of a "biofilm model" (Costerton et al. 1995). In this model, microorganisms form microcolonies surrounded by copious amounts of expolysaccharide. Between the microcolonies are water-filled channels, and it has been proposed that these promote the influx of nutrients and the efflux of waste products. This biofilm model will be discussed in detail both in this chapter and later on in the book.

In 1998, important advances to our understanding of the development and behaviour of biofilms were made, when studies that combined molecular genetic approaches with confocal laser scanning microscopy (CLSM) emerged.

Bacteria are remarkably adept at surviving "feast and famine", and also adjusting their needs to accommodate highly diverse environments. Scientific inquiry has discovered a number of the microbial characteristics that facilitate the way bacteria adapt to changing environments. The capacity to form and maintain biofilms is key to these adaptations. Traditionally, microbiologists have performed physiological experiments with microorganisms grown in liquid monocultures where the cells are "free swimming" or planktonic. Whilst it may seem that microbiologists are always striving for pure cultures, most of the bacteria in the world live in these polymicrobial ecosystems called biofilms; a complex community of microorganisms that are not "free swimming", but are instead attached to surfaces.

2 Prevalence and Importance of Biofilms in Animals and Humans

Biofilms may form on a wide variety of surfaces, including natural aquatic systems living tissues, indwelling medical devices and industrial/potable water system piping. The vast majority of microbes grow as biofilms in aqueous environments.

These biofilms can be benign or pathogenic, releasing harmful products and toxins, which become encased within the biofilm matrix.

Biofilm formation is a phenomenon that occurs in both natural and man-made environments under diverse conditions, occurring on most moist surfaces, plant roots and nearly every living animal. Biofilms may exist as beneficial epithilic communities in rivers and streams, wastewater treatment plant trickling beds or in the alimentary canal of mammals (Costerton et al. 1981).

Biofilms are not, however, confined to solid/liquid interfaces, and can also be found at solid/air or liquid/liquid interfaces (e.g. airborne pathogens and deteriogens have been shown to be important factors in the biodeterioration of surface coatings; biofilms at liquid/liquid interfaces have been implicated in hydrocarbon degradation, including fuels, oils and industrial coolants).

In humans, an estimated 65% of all hospital infections are of biofilm origin (Costerton et al. 1999; Donlan 2001; Donlan and Costerton 2002; Douglas 2003; Ramage et al. 2006). Once established, biofilm infections are very difficult to eradicate due to their resilience to removal by host defence mechanisms and antimicrobials. Table 1 outlines several examples of clinically significant biofilm infections. Of particular note are biofilm infections in patients with medical devices, *Pseudomonas aeruginosa* infection in cystic fibrosis patients, tooth decay and periodontal disease, and chronic wound infections. As well as increasing patient morbidity and mortality, there is also an economic burden to biofilm infections with

Table 1 Comparison of biofilm and planktonic lifestyles

Characteristic	Lifestyle
Physiological	
Free swimming, often in an aqueous environment	Planktonic
Metabolic products continuously removed from the milieu, stresses of physical, chemical and biological nature	Planktonic
Slow growth	Biofilm
Found ubiquitously	Biofilm
Generally higher tolerance to antibiotics, hydrogen peroxide, phagocytosis	Biofilm
Stress conditions: lower metabolic activity (e.g. decreased nutrient accessibility, limitation of oxygen, increased osmotic pressure, pH variation, problems with accumulation of waste metabolites)	Biofilm
Bacteria develop stress responses	Biofilm
Source of infections	Planktonic and biofilm
Display unique gene expression patterns	Biofilm
Attachment can influence metabolic activities	Biofilm
Express virulence factors	Biofilm and planktonic
Structural	
Attached to a solid substrate	Biofilm
Many cells are dormant, likely to be smaller, not actively engaged in cell division	Biofilm
Flagella function in transport and initiation of cell-to-surface interactions and detachment	Biofilm

those involving medical devices estimated to cost $20 billion in the USA alone. Furthermore, biofilms in air-conditioning and other water retention systems can cause human infection if ingested or inhaled, and examples include those biofilms caused by *Legionella* species, which persist in water storage tanks despite chlorination (Percival and Walker 1999).

3 Why Do Microorganisms Form Biofilms?

There are a number of possible advantages gained when a microorganism is in a biofilm compared to a planktonic existence. These include the increased expression of beneficial genes, phenotypic changes in colony morphology, acquisition of antibiotic resistance genes by plasmid transfer, the production of copious amounts of extracellular polymers (Costerton et al. 1987), enhanced access to nutrients and closer proximity between cells facilitating mutualistic or synergistic associations and protection (Costerton and Lappin-Scott 1989). This list is by no means complete but helps to highlight the possible advantages of survival in biofilms.

Microbiology research has traditionally focused on the in vitro analysis of single species in liquid culture. The study of planktonic microorganisms has undoubtedly been of value in expanding our knowledge on microbial physiology and biochemistry. However, in recent years, the importance of biofilm growth has been recognised (Percival and Bowler 2004; Wilson 2001) with microorganisms in a biofilm shown to differ markedly from their planktonic counterparts in terms of behaviour, structure and physiology (Table 2). These differences have significance both to the pathogenic potential of microorganisms together with their susceptibility to antimicrobials.

Table 2 Clinically significant examples of biofilm infections

Infection	Typical biofilm organism
Dental caries	*Streptococcus* spp.
Periodontitis	*Fusobacterium nucleatum, Porphyromonas gingivalis, Bacteroides forsythus, Prevotella intermedia*
Otitis media	*Streptococcus pneumoniae, Haemophilus influenzae, Moraxella catarrhalis, S. aureus, S. epidermidis, P. aeruginosa*
Cystic fibrosis	*P. aeruginosa, Burkholderia cepacia*
Chronic wounds	Staphylococci, streptococci, enterococci, facultative Gram-negative bacilli, anaerobic bacteria such as *Fusobacterium* spp. and peptostreptoocci
Foreign body/medical device infection	
Urinary catheters	*Proteus mirabilis, Morganella morganii, P. aeruginosa, Klebsiella pneumoniae, Proteus vulgaris*
Native valve endocarditis (NVE)	Streptococci, staphylococci, Gram negative bacteria, fungi

4 Definition of a Biofilm

Biofilms are not easily defined as they vary greatly in structure and composition from one environmental niche to another. Microbial biofilms are extremely complex microbial ecosystems consisting of microorganisms attached to a surface and embedded in an organic polymer matrix of microbial origin. As well as microbial components, non-cellular materials such as mineral crystals, corrosion particles, clay or silt particles, or blood components, may also be found in the biofilm matrix. Biofilms, particularly in water systems, can be highly complex, whilst others such as those on medical devices, may be simpler, and composed of single, coccoid or rod-shaped organisms. Given these differences, it does not seem plausible to suggest that a true "biofilm model" can be defined that is applicable to every ecological, industrial and medical situation. Therefore the definition of a biofilm has to be kept general and thus may be redefined as "microbial cells immobilised in a matrix of extracellular polymers acting as an independent functioning ecosystem, homeostatically regulated" (Percival et al. 2000).

5 The Biofilm Model

Despite ongoing discussions on the so-called "biofilm model" the diversity of biofilms suggests that strict phraseology for a constantly changing dynamic ecosystem is not possible. It is feasible to suggest that biofilms in distinct settings form different structures comprising different microbial consortia dictated by biological and environmental parameters. These biofilms can quickly respond and adapt both phenotypically, genetically and structurally to constantly changing internal and external conditions (Percival et al. 2000). Whilst a so-called "biofilm model" has been proposed and will be discussed in detail throughout the book, it must be borne in mind that applying laboratory-based observations to an otherwise chaotic, totally unpredictable biofilm in the environment may well aid in clouding independent thought of a "true biofilm model", if in fact one exists. From the authors' experiences and research it is clear that environmental/industrial and medical biofilms cannot be truly replicated and reproduced to the conditions that prevail in the field. Therefore one must endeavour to move away from the single "gold standard" and possibly concede to the fact that a "true biofilm model" does not exist and that nature determines the complexities of a biofilm that cannot be reproduced in vitro. However, we need to start somewhere, which is where laboratory-based analysis is warranted. Therefore to appreciate the biofilm and to address the "biofilm model" concept we need to fully understand what a biofilm is and how it forms. This will be discussed in detail below.

The development of a mature biofilm is a multistage process and is dependent on a number of variables including the type of microorganism, the surface to which attachment occurs, environmental factors and expression of biofilm essential genes

(Carpentier and Cerf 1993; Dunne 2002). Aspects of biofilm development are discussed below and elsewhere in this book.

Primary adhesion between bacteria and abiotic surfaces is often non-specific (e.g. mediated by hydrophobic interaction), whereas adhesion to living tissue tends to be through specific molecular (lectin, ligand or adhesin) mechanisms (Carpentier and Cerf 1993; Dunne 2002). Primary adhesion is reversible and in biological systems occurs between a conditioned surface and a planktonic microorganism. The conditioning film is provided by body fluids bathing the surface and its composition may alter the affinity of a microorganism to the surface (Dunne 2002).

Bacteria and host surfaces express multiple adhesins and receptors to facilitate specific adherence. For example, *Staphylococcus epidermidis* and *Staphylococcus aureus* produce a polysaccharide intercellular adhesion (PIA), which is associated with cell-to-cell adhesion and subsequent biofilm formation (Crampton et al. 1999; Heilmann et al. 1996; Mack et al. 1994, 1996).

After microorganisms have successfully adhered to a surface, they form aggregates and produce an extracellular polysaccharide matrix (Costerton et al. 1999), which serves to encase the initial colonisers and biofilm formers. Coaggregation is a specific means of cell-to-cell adherence (Kolenbrander et al. 2006) which can be highly specific and plays a role in the development of multi-species biofilms in many different environments such as dental plaque (Rickard et al. 2002, 2003a, b), urogenital system (Reid et al. 1988), crops of chickens (Vandevoorde et al. 1992) and in aquatic biofilm-forming bacteria (Buswell et al. 1997; Rickard et al. 2000).

The maturation of biofilms involves recruitment of additional microorganisms from the local environment. Biofilms are heterogeneous environments and in addition to aggregates of microbial cells, interstitial voids and channels develop within the matrix, which separate micro-colonies (Sutherland 2001). The purpose of these channels is to enable the delivery of nutrients and gasses as well as the removal of waste products.

6 Stages in the Formation of Biofilms

The process of biofilm formation is complex, but generally recognised as consisting of five stages (Palmer and White 1997):

1. Development of a surface conditioning film
2. Movement of microorganisms into close proximity with the surface
3. Adhesion (reversible and irreversible adhesion of microbes to the conditioned surface)
4. Growth and division of the organisms with the colonisation of the surface, microcolony formation and biofilm formation; phenotype and genotype changes
5. Biofilm cell detachment/dispersal

Each of these processes will be considered in turn.

6.1 Development of Conditioning Films and Substratum Effects

Within the natural environment, microorganisms do not adhere directly to a substratum per se, but actually adhere to a conditioning film, which is known to form on most substrata. The composition of the conditioning film is complex and results in chemical modification of the original surface, thus influencing the rate and extent of microbial adhesion (Mittelman 1996).

The first documented evidence that a "conditioning" film existed was by Loeb and Neihof (1975), although whether a conditioning film is a pre-requisite for bacterial attachment is still debateable with evidence of this still elusive in the scientific literature.

In aquatic or terrestrial environments, the conditioning layer has been shown to consist of complex polysaccharides, glycoproteins and humic compounds (Chamberlain 1992; Marshall et al. 1971; Baier 1980; Rittle et al. 1990). In comparison human host conditioning films may also be complex and determined by the site being conditioned.

An area of "biofilmology" that has been studied extensively is the role of dental plaque in oral disease. In this regard, the enamel of teeth is conditioned by a proteinaceous "pellicle" composed of albumin, glycoproteins, lipids, lysozyme, phosphoproteins and other components of saliva and gingival crevicular fluids (Marsh 1995). Other types of conditioning films have also been reported, particularly on biomaterials used for human use. In these situations the components of blood, tears, urine, saliva, intravascular fluid and respiratory secretions may all contribute to the conditioning film.

The role of the conditioning film in biofilm development is in its ability to modify the physico-chemical properties of the substratum, as well as providing a concentrated nutrient source and important trace elements. It is important to note that conditioning films may actually inhibit rather than promote the adhesion of certain bacteria.

The topography of the surface to which a microbial cell attaches is also fundamental to biofilm formation. Generally, as the roughness of a surface increases, bacterial adhesion will also increase (Characklis et al. 1990a). There are several possible reasons for this, including most specifically the provision of shelter for the microbes from the effects of shear forces. Apart from increasing the available interfacial area, a rough surface enhances mass transfer coefficients and allows cells to "anchor" to micro-irregularities, where they are better protected from possible desorption (Characklis et al. 1990b). Regardless of surface roughness, the attachment of living particles is energetically favourable if the change in the free energy during the process is negative. Despite metallic surfaces being energetically favourable to the attachment of the pioneer colonisers, the chemical composition of surfaces may interfere with adhesion, cellular metabolism and production of exopolymers (Beech and Gaylarde 1989). The surface effect of certain metals on bacterial adhesion was reported by Vieira et al. (1992), who found that *Pseudomonas fluorescens* preferentially fouled aluminium surfaces within a few hours, followed by copper and brass.

In addition to surface roughness, the physicochemical properties of the surface effects bacterial adhesion in that microorganisms attach more rapidly to hydrophobic, non-polar surfaces such as Teflon and other plastics than to hydrophilic materials such as glass or metals (Pringle and Fletcher 1983; Bendinger et al. 1993; Percival and Thomas 2009). However, such results need to be addressed with care, as many studies have proved contradictory, primarily due to non-availability of standardised methods for surface hydrophobicity measurements.

6.2 Transport Mechanisms Involved in Adhesion of Microorganisms to a Surface

The transport of microbial cells and nutrients to a surface is generally achieved by a number of well-established fluid dynamic processes. These include mass transport, thermal (Brownian motion, molecular diffusion) and gravitational effects (differential settling, sedimentation; Characklis 1981).

Within fluid transport pipes, two main flow conditions exist: laminar and turbulent flow. In the blood stream and urinary system, laminar flow is evident and characterised by parallel smooth flow patterns with little or no lateral mixing with the fastest flow in the centre (Fletcher and Marshall 1982; Lappin-Scott et al. 1993). During laminar flow, microorganisms and nutrients maintain a straight path and remain in a stabilised position dictated by flow rate (Lappin-Scott et al. 1993). In contrast, turbulent flow is random and chaotic, ultimately increasing the mixing of bacteria and nutrients (Characklis et al. 1990a, b) and microbial adherence (Percival et al. 1999).

Eddying currents (random and unpredictable flow) are evident in turbulent flow and these cause up- and down-sweep forces, which propel bacteria to within short distances of the surface thereby increasing the chance of adhesion.

Adhesion in static or quiescent environments is aided by a number of factors including Brownian diffusion, gravity and microbial motility (Bryers 1987). Motility of a bacterium is known to increase the chances of adhesion (Fletcher 1977; Marmur and Ruckenstein 1986). This is possibly due to the provision of enough potential energy to overcome any repulsive forces known to operate between the bacterial surface and the substratum in question. To reinforce this, it is generally found that a reduction in motility decreases adhesion (Fletcher 1977). Another mechanism known to be a factor affecting surface colonisation is gravitational cell sedimentation, often only of relevance in flowing systems when co-aggregation occurs (Walt et al. 1985).

6.3 Adhesion

Following surface conditioning and transport of bacteria into an area close to the substratum surface, adhesion of the microorganism usually takes place. Adhesion, as previously mentioned, was first proposed in 1943 (Zobell 1943), as consisting of a two-step sequence involving reversible and irreversible processes. Reversible adhesion is the initial weak attachment of microbial cells to a surface (Rittman 1989) and irreversible adhesion and permanent bonding of the microorganisms to a surface, generally follows. Bacteria influence this process, through expression of specific bacterial adhesins (Whittaker et al. 1996), which bind to receptors on the substratum and in the extracellular polymeric matrix (Marshall et al. 1971).

Bacterial adhesion appears to be related to the distance between the microorganism and the surface (Busscher and Weerkamp 1987). At distances greater than 50 nm, *Van Der Waals* forces occur, whilst at distances of 10–20 nm, both *Van Der Waals* and electrostatic interactions occur. When a microorganism is less than 1.5 nm from a surface *Van Der Waals*, electrostatic and specific interactions occur between the cell and the surface.

The surface of a microbial cell has a major impact on adhesion to a substratum. Cell surface hydrophobicity, the presence of fimbriae and flagella, and particularly the extent and composition of generated EPS, influence both the rate and extent of microbial adhesion. A possible role of proteins for bacterial adhesion has been proposed with treatment of adsorbed cells by proteolytic enzymes found to cause a marked detachment of bacteria (Bashan and Levanony 1988; Danielsson et al. 1977).

Fimbriae contain a high proportion of hydrophobic amino acid residues (Rosenberg and Kjelleberg 1986), which can affect cell surface hydrophobicity and therefore attachment to a surface. It is probable that such fimbriae are able to overcome the initial electrostatic repulsion barrier that exists between the cell and substratum (Corpe 1980; Bullitt and Makowski 1995).

Korber et al. (1989) used motile and non-motile strains of *P. fluorescens* and demonstrated that the former attached in highest numbers and against flow (back growth). In addition, non-motile strains do not recolonize or seed vacant areas on a substratum as evenly as motile strains, resulting in comparatively slower biofilm formation.

6.4 Extracellular Polymeric Substances

If cells reside at a surface for a certain time, irreversible adhesion forms through the production of extracellular cementing substances. As mentioned earlier, this extracellular material, associated with the cell has also been referred to as glycocalyx (Costerton et al. 1978), a slime layer, capsule or a sheath.

The EPS of biofilms may account for 50–90% of the total organic carbon of biofilms (Flemming et al. 2000). Glycoproteins, proteins and nucleic acids are also found within this organic matrix (Humphrey et al. 1979).

EPS varies in chemical and physical properties, but in the case of Gram-negative bacteria, is primarily composed of neutral or polyanionic polysaccharides. Uronic acids (such as D-glucuronic, D-galacturonic, and mannuronic acids) or ketal-linked pyruvates are known to constitute part of the EPS matrix. These give anionic properties to the biofilm allowing cross-linking of divalent cations such as calcium and magnesium (Flemming et al. 2000; Sutherland 2001). Biofilms that primarily contain Gram-positive bacteria produce an EPS that is primarily cationic.

The involvement of extracellular polymers in bacterial attachment has been documented for both fresh and marine water bacteria (Corpe 1970; Marshall et al. 1971). Analysis of bacteria isolated from these environments has shown such polymers are largely composed of acidic polysaccharides (Fletcher 1980). The extent to which polysaccharides are involved in the adhesion process remains open to question. Some evidence suggests that excess polymer production may even prevent adhesion, although trace amounts of polysaccharide might be required initially for adhesion (Brown et al. 1977).

EPS is highly hydrated, and can be both hydrophilic and hydrophobic with varying degrees of solubility. The polysaccharide content of EPS has a marked effect on the biofilm (Sutherland 2001) as the composition and the structure will determine their primary conformation. Bacterial EPS contains backbone structures of 1,3- or 1,4-β-linked hexose residues, which are rigid and generally poorly soluble or insoluble, whereas other EPS molecules are more readily soluble in water.

EPS provides many benefits to a biofilm (Characklis and Cooksey 1983) including the promotion of cohesive forces, increased absorption of nutrients and heavy metals (Bryers 1984; Marshall 1992), the sequestration of microbial products and other microbes, protection of immobilised cells from environmental changes and the provision of a medium for intercellular communication and transfer of genetic material.

Extending polymers on cell surfaces interact with vacant bonding sites on the surface by polymer bridging and as a result, the cell is held near the surface. Possible mechanisms for polymer bridging have been suggested (Characklis and Cooksey 1983) but remain to be fully elucidated. Bacteria can be connected to the substratum via exopolymer–substratum interactions, which are predominately covalent bonds. Research into the ecology of sessile microbial populations often focused on the extracellular polymers produced by the cells (Corpe 1970; Costerton et al. 1981; Uhlinger and White 1983).

In aquatic habitats, microbial exopolymers commonly occur as discrete capsules firmly attached to the cell surface or as slime fibres loosely associated with, or dissociated from, the cells. While it is now believed that many of the capsular polymers serve as holdfasts, anchoring cells to each other and to inert surfaces, the extent to which they facilitate other interactions between sessile bacteria and their environment is less well understood.

EPS influences the physical properties of the biofilm, including diffusivity, thermal conductivity and rheological properties. EPS, irrespective of charge density or its ionic state, has some of the properties of diffusion barriers, molecular sieves and adsorbents, thus influencing physiochemical processes such as diffusion and fluid frictional resistance. The predominantly polyanionic, highly hydrated nature of EPS also means that it can act as an ion exchange matrix, serving to increase local concentrations of ionic species such as heavy metals, ammonium, potassium, etc., whilst opposite effects are generated on anionic groups. EPS has little effect on uncharged molecules including potential nutrients such as sugars (Hamilton and Characklis 1989). However, biofilm bacteria are thought to concentrate and use cationic nutrients such as amines, suggesting that EPS can act as a nutrient trap, especially under oligotrophic conditions (Costerton et al. 1981). Conversely the penetration of charged molecules such as biocides and antibiotics may be at least partly restricted by this phenomenon (Costerton and Lashen 1984).

7 Microcolony and Biofilm Formation

The adsorption of macromolecules and attachment of microbial cells to a substratum are only the initial stages of biofilm development. These are followed by microbial growth, development of microcolonies and recruitment of additional microorganisms. As attachment of microorganisms occurs, the colonising bacteria grow with the production and accumulation of extracellular polymers. The microorganisms eventually become embedded in this hydrated polymeric matrix and immobilised. As a result, the cells are dependent upon substrate flux from the liquid phase and/or exchange of nutrients with neighbouring cells in the biofilm. An important feature is that the microorganisms are immobilised in relatively close proximity to one another. Additional microorganisms may be located within or on top of the biofilm matrix. Specific functional types of microorganisms may, through their activities, create conditions that favour other complementary functional groups of microorganisms. This leads to the establishment of spatially separated, but interactive, functional groups of organisms, which exchange metabolites at group boundaries achieving physiological cooperation (Blenkinsopp and Costerton 1991).

As biofilm communities tend to be complex, both taxonomically and functionally, there is the potential for synergistic interaction among constituent organisms. Homeostatic mechanisms can develop that protect the biofilm microorganisms from outside perturbations and these are extremely important in natural communities exposed to disturbances, such as episodes of pollution. As biofilm heterogeneity increases, chemical and physical microgradients develop including those of pH, oxygen and nutrients.

Generally, when the biofilm reaches a thickness of 10–25 μm, conditions at its base become anaerobic and indicate that the biofilm is approaching a state of maturity, with high species diversity and stability (Hamilton 1987). The thickness of a biofilm is actually very hard to define, as it is not universal and dependent on the local environment. Medical biofilms tend to be thinner than those found within industrial environments. As already mentioned, biofilms are highly hydrated with microorganisms known to only occupy a small component (Costerton et al. 1995). With the use of staining and dehydration techniques used to prepare biofilms for examination for light and electron microscopy, the structure of the biofilm can become distorted leading to errors in determining biofilms thickness and structure.

This, however, has been overcome with the introduction of CLSM which allows biofilms to be studied in their native, hydrated state providing an accurate picture of structure and dimensions (Lawrence and Neu 1999; Palmer and Sternberg 1999). It is now possible to make some general comments about biofilms based on this form of analysis. Oral biofilms have been reported to be up to 1 mm thick, those in CAPD catheters 30 μm thick, whilst a study of biofilms on 50 indwelling bladder catheters found that their thicknesses varied from 3 to 490 μm with layers of cells documented to be 400 cells deep.

Biofilm structure has now been described in the literature for both mixed and pure cultures in many different environments. Most of these biofilms demonstrate heterogeneity, often evident as a patchy biofilm configuration. With the use of CLSM together with microelectrode measurements, researchers have established that the biofilm consists of cell clusters located in an EPS matrix. These clusters have been shown to vary in shape often ranging from cylinders to filaments forming a "mushroom" structure (Lewandowski et al. 1995; Caldwell et al. 1992; Costerton et al. 1994; de Beer et al. 1994; Keevil et al. 1993). Open channels are evident in biofilms (particularly within potable water systems) and are also referred to as voids and pores, and represent further evidence of biofilm spatial and temporal complexity. Voids facilitate mass transfer, which favours higher nutrient concentrations in the void spaces and also allows cellular metabolites and byproducts to be more concentrated under cell clusters. These stack systems, which are evident within oligotrophic environments, have been replicated in simple computer simulations (Wimpenny and Colasanti 1997).

Overall, the development of a biofilm is generally governed by a number of parameters including ambient and system temperatures, which are in turn related to season, day length, climate and wind velocity; hydrodynamic conditions (shear forces, friction drag and mass transfer); nutrient availability (concentration, reactivity, antimicrobial properties); surface roughness, hydrophobicity and electrochemical characteristics of the surface; pH (an approximately neutral pH of the water is optimal for the growth of most biofilm-forming bacteria); the presence of particulate matter (which can become entrapped in the developing biofilm and thus provide additional attachment sites); and the effectiveness of biofilm control measures. Of these, the four major influencing factors are the surface or interface properties, hydrodynamics, nutrients and biofilm consortia (Stoodley et al. 1997). The controversial factor known to affect biofilm structure is the hydrodynamic forces, which operate within flowing conditions. It is now well established that biofilms exposed to high turbulent flow develop a phenomenon known as "streaming". The significance of this is still unclear.

7.1 Microbial Interaction in Biofilms

In biofilms in natural environments, there is evidence of a high level of cellular interaction and competitive behaviour (Connell and Slatyer 1977; Fredrickson

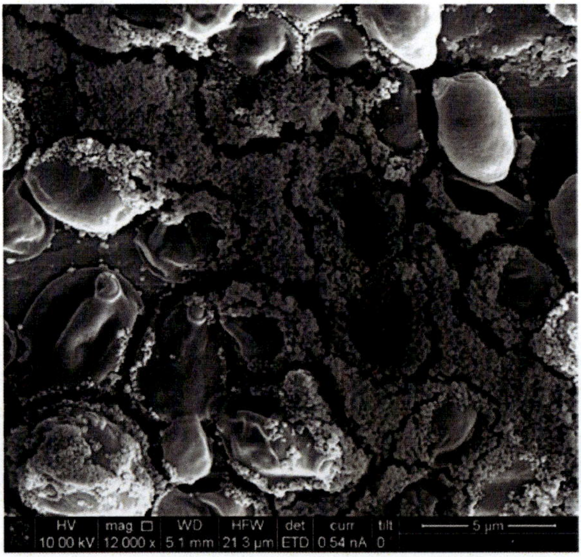

Fig. 1 *Candida albicans* and *Streptococcus mutans* attached to titanium

1977) arising as a consequence of resource availability. As a result of competition strategies by microorganisms, the biofilm system is consistently under a state of flux (Connell and Slatyer 1977; Baier 1984; Wahl 1989) with microbial succession being a common feature. During adhesion, pioneer colonisers have defined requirements, dictated by the conditioning film. The succession of the biofilm community is then governed by a number of physiological and biological events initiated by this initial coloniser (Fletcher and Loeb 1979; Characklis 1981; Baier 1984).

Since microorganisms in a biofilm are in close proximity, they invariably interact and this may be beneficial to one or more of the microorganisms involved, or may be antagonistic (James et al. 1995; Stewart et al. 1997). Figure 1 demonstrates evidence that *Streptococcus mutans* and *Candida albicans* interact together demonstrating a mutualistic relationship. Synergistic mechanisms are considered important in mixed species biofilm development (Hofstad 1992) and a good example of this is the consumption of oxygen by aerobes, which locally increases the redox potential allowing anaerobes to also survive. Other synergistic effects include the sharing of biochemical pathways to fully exploit available nutrients as well combining approaches to protect against host immune defences and antimicrobial agents. In terms of pathogenic potential, polymicrobial interaction may well play a key role for biofilms in delayed wound healing. For example, microorganisms with low invasive capacity may act synergistically with more virulent ones.

Quorum sensing (QS) is microbial cell-to-cell communication through chemical messages (Jones 2005) and is considered key in regulating colonisation and virulence factor expression within the biofilm (Yarwood and Schlievert 2003). QS chemicals are often referred to as microbial pheromones or autoinducers and their concentration often correlates to the population density. Until relatively recently, such inter-bacterial signalling was thought to only occur in a few microorganisms, namely Vibrio fisheri and Vibrio harveyi, Enterococcus faecalis, Myxococcus xanthus and Streptomyces spp. Research into inter-bacterial signals uncovered the use of acyl homoserine lactones (AHLs) as QS molecules in Gram-negative bacteria. These small signal molecules are excreted by cells and accumulate in cultures as a function of cell density, termed the quorum; the accumulated AHLs interact with receptors on the bacterial cell surface and signal to control gene expression. In this way coordinated expression of sets of genes can be achieved in response to local cell density. The dense populations of cells in biofilms led to speculation that AHLs may have specific functions in these communities. AHLs were subsequently detected in natural biofilms growing on submerged stones taken from a river and in biofilms that had formed in vivo on urethral catheters. AHLs in a developing biofilm cause the transformation of individual cells from the planktonic to the biofilm phenotype and coordinate their behaviour in some way so that they build the complex structures of the multicellular communities.

Gram-positive bacteria have not yet been reported to use the autoinducer-based signalling systems; in some, however, an oligopeptide pheromone has been involved in inter-bacterial signalling. Pheromone-based inter-bacterial regulatory systems are thought to regulate bacterial cell density, bioluminescence, conjugative transfer of bacterial plasmids, genetic competence, production of hydrolytic enzymes and secondary metabolism. Several of the systems regulated by QS are involved in the regulation of the virulence determinates of bacterial pathogens. The best understood example of a human pathogen involved in quorum sensing is *P. aeruginosa*. Xie et al. (2000) showed that certain dental plaque bacteria modulate the expression of the genes encoding fimbrial expression (fimA) in *Porphyromonas gingivalis*. *Porphyromonas gingivalis* would not attach to *Streptococcus cristatus* biofilms grown on glass slides whilst *P. gingivalis*, readily attached to *S. gordonii*. *Streptococcus cristatus* cell-free extracts substantially affected expression of fimA in *P. gingivalis*, as determined by a reporter system. *Streptococcus cristatus* is therefore able to modulate *P. gingivalis* fimA expression and prevent its attachment to the biofilm.

8 Detachment and Dispersal of the Biofilm

Characklis et al. (1990a, b) suggested that the detachment of biofilms could be categorised into three areas, namely erosion, sloughing and abrasion. Characklis referred to detachment as an interfacial transfer process, which involved the transfer of cells and other components from the biofilm compartment to the bulk liquid, with the detachment of microbial cells and related biofilm material occurring from the moment of initial attachment.

Many different parameters known to affect biofilm detachment have been examined, including pH, temperature and the presence of organic macromolecules either absorbed on the substratum or dissolved in the liquid phase (McEldowney and Fletcher 1988). The effects these have on bacterial detachment are generally species specific. Surface roughness of the substratum may also be a significant factor in biofilm detachment, with early events in biofilm formation being controlled by hydrodynamic forces (Powell and Slater 1982). As detachment increases with increasing fluid shear stress at the substratum surface, macro and micro roughness may significantly influence the detachment rates of the biofilm due to a sheltering effect from hydrodynamic shear. The detached cells may be transported close to the surface (in a viscous sublayer) resulting in collisions with the surface providing more opportunity for reattachment.

To date, detachment still remains a poorly researched and understood phenomenon, which therefore complicates the formation of satisfactory models. This is surprising considering that the detachment of biofilms from surfaces into surrounding environments does have important implications to public health. In microbiological terms, detachment from surfaces may at first be seen to be a disadvantage to the biofilm. However, it has been found that biofilms with greater detachment rates have larger fractions of active bacteria. It has also been reported that detachment can occur as a result of low nutrient conditions indicating a survival mechanism, which may be genetically determined. Therefore, detachment is not just important for promoting genetic diversity, but also for escaping unfavourable habitats aiding in the development of new niches.

Biofilm dispersal/detachment of biofilm cells are routinely detached from biofilms and have very important implication in medical settings. Raad et al. (1992) determined a relationship between biofilm detachment and catheter-related septicaemia, whilst the detachment of biofilm aggregates from native heart valves has implications in infective endocarditis. It is these clumps of cells from biofilms, which also may contain platelets or erythrocytes, that lead to the production of emboli with serious complications to the host. Furthermore, biofilms in hospital water systems containing potentially pathogenic organisms might also detach as aggregates and, especially for those organisms with a low infective dose, consumption or exposure to water containing these organisms might result in infection. This may be a cause of the increasing incidences of nosocomial infections.

8.1 Resistance of Biofilms to Host Defence Mechanisms

Microorganisms within a biofilm grow in a protected microenvironment largely through production of a biofilm matrix composed of extracellular polysaccharides, proteins and nucleic acids (Davey and O'Toole 2000). The fact that biofilm-based infections are rarely resolved, even in individuals who have a competent innate and adaptive immune response, highlights the high degree of resistance exhibited by biofilms (Stewart and Costerton 2001).

8.2 Tolerance of Biofilms to Antimicrobial Agents

Biofilm structure and the physiological attributes of microorganisms within the biofilm also provide an intrinsic tolerance to antimicrobial agents (antibiotics, disinfectants, germicides or antifungals). Indeed, biofilms can be up to 1,000 times more tolerant to antibiotics than equivalent planktonic cultures (Hoyle and Costerton 1991; Mah and O'Toole 2001). Whilst biofilm cells may employ a variety of mechanisms to resist the action of antimicrobial agents (Mah et al. 2003), studies have also shown that a number of key factors contribute to reduced antimicrobial susceptibility of biofilms (Percival and Bowler 2004).

A reduced ability of the antimicrobials to gain access to the microorganisms located within the matrix of the biofilm is thought to be important in biofilm resistance against certain antimicrobials. This could arise through chemical interaction with extracellular biofilm components or adsorption to anionic polysaccharides. For example, it has been shown that penetration of positively charged aminoglycosides into a biofilm is retarded by binding to negatively charged matrices, such as the alginate produced in *P. aeruginosa* biofilms (Walters et al. 2003). Another study showed that extracellular matrix from coagulase-negative staphylococci reduced the effect of glycopeptide antibiotics, even in planktonic cultures (Fux et al. 2005; König et al. 2001). In addition, Mah et al. (2003) showed that bacteria within a biofilm may actively employ distinct mechanisms including antibiotic sequestration in the periplasm to prevent them from reaching their target sites. The enzyme β-lactamase, produced by several bacterial species can inhibit the activity of β-lactam ring structured antibiotics such as penicillins and cephamycins. β-Lactamase can accumulate in the biofilm matrix due to secretion or cell lysis, and deactivate β-lactam antibiotics at biofilm surface layers more rapidly than they diffuse into the biofilm (Anderl et al. 2003). It is, however, known that the biofilm matrix does not form a completely impenetrable barrier to antimicrobial agents (Percival and Bowler 2004) and other factors are likely to be involved.

8.3 Growth Rate of Biofilm Bacteria and Its Effect on Antimicrobial Tolerance

Antibiotics are generally more effective at killing actively reproducing cells. Hence, reduced activity of microorganisms could render these cells less susceptible. It has been known for sometime that non-dividing bacteria escape the killing effects of antibiotics targeted at growth-specific factors (Davies 2003). Antibiotics, such as ampicillin and penicillin, which inhibit cell wall synthesis, are unable to kill non-dividing cells (Costerton et al. 1999; Stewart and Costerton 2001).

The most likely location of slower growing bacteria in a biofilm is in the lower region of the biofilm (Davies 2003) and these cells are more likely to be metabolically inactive because of reduced access to essential nutrients and gaseous exchange. Biofilm embedded cells may be metabolically dormant and in this state phenotypically equipped to persist in hostile environments (Anwar et al. 1992b; Rhoads et al. 2007). These so-called persister cells represent a small and slow-growing sub-population of the biofilm which have differentiated into an inactive, but highly protected state (Roberts and Stewart 2005; Percival et al. 2011). It has been estimated that these cells constitute 0.1–10% cells in a biofilm and it has been hypothesised after antimicrobial challenge that it is this subpopulation of cells that "re-seed" the biofilm (Harrison et al. 2005; Roberts and Stewart 2005).

8.4 Induction of a Biofilm Phenotype Tolerant to Antimicrobials

Once microorganisms attach to a surface, they may express a general and more virulent biofilm phenotype compared with planktonic counterparts (Mah and O'Toole 2001; Saye 2007). Gilbert et al. (1997) suggested that cells with a biofilm-specific phenotype may be induced. These phenotypes may express active mechanisms such as the expression of bacterial periplasmic glucans to bind to, and physically sequester antibiotics to reduce the efficacy of antibiotics (Gilbert et al. 1997; Maira-Litran et al. 2000; Percival and Bowler 2004).

Altered gene expression by organisms within a biofilm or a general stress response of a biofilm could reduce susceptibility to antimicrobial agents (Brown and Barker 1999). In theory, bacterial cells have a small number of target sites for antibiotics. It could therefore be that biofilm cells use specific genes to phenotypically alter these target sites to protect themselves (Saye 2007). Exact details about the physiological changes that occur during the transition from a planktonic to a biofilm state are not known. It has been suggested that multidrug resistance (MDR) efflux pumps may substantially contribute to the resistance of biofilms (De Kievit et al. 2001; Mah and O'Toole 2001).

8.5 Clinical Importance of Biofilms

Given the prevalence of biofilms in natural environments, it is not surprising that these growth forms are responsible for infection in humans and animals. In humans, biofilms have been linked with numerous conditions and equally in animals equivalent infections may occur.

Native valve endocarditis (NVE) is a condition that results from the interaction of bacteria, the vascular endothelium and pulmonic valves of the heart. The causes of NVE are varied, but it is frequently associated with streptococci, staphylococci, Gram-negative bacteria and also fungi (Braunwald 1997). These microorganisms gain access to the blood and the heart via the oropharynx, gastrointestinal and urinary tract. Once the intact endothelium is damaged, microorganisms adhere to it and non-bacterial thrombotic endocarditis (NBTE) develops at the injury. At the point of injury, the thrombus develops, which is an accumulation of platelets, fibrin and red blood cells (Donlan and Costerton 2002). Treatment is less effective due to a combination of mass transfer limitations and inherent resistance of biofilms (Donlan and Costerton 2002).

Microbial biofilms often develop on, or within indwelling, medical devices, e.g. contact lenses, central venous catheters, mechanical heart valves, pacemakers, peritoneal dialysis catheters, prosthetic joints, urinary catheters and voice prostheses (Donlan 2001; Percival and Kite 2007) and a number of microorganisms can produce biofilms on these surfaces. *Staphylococcus* spp. (*S. aureus*, *S. epidermidis*) are members of the commensal microflora of the skin and can form biofilms on implantable medical devices, such as intravenous catheters, and hip and knee joint prostheses (Bayston 1999; Habash and Reid 1999; Khardori and Yassien 1995).

Implantation of mechanical heart valves causes tissue damage, and circulating platelets and fibrin tend to accumulate where the valve has been attached. Once the tissue is damaged, it facilitates a greater tendency for microorganisms to colonise these locations (Anwar et al. 1992a). The resulting biofilms develop on the heart tissue (Carrel et al. 1998; Illingworth et al. 1999) surrounding the prosthesis or the sewing cuff fabric used to attach the device to the tissue (Donlan 2001; Karchmer and Gibbons 1994).

In humans, urinary catheters are used to remove and monitor urine production from impaired patients. Urinary catheters also facilitate repair of the urethra after surgical procedures and manage urinary retention and incontinence in the elderly and disabled patients (Moore and Lindsay 2001; Stickler 2005). Yeasts such as *Candida parapsilosis* are very adapt at growing on urinary catheters when these are composed of silicone (Fig. 2). Biofilms associated with urinary catheters are particularly important because they cause infections in 10–50% of patients who undergo catheterisation (Mulhall et al. 1988; Stickler 2002). *Proteus mirabilis*,

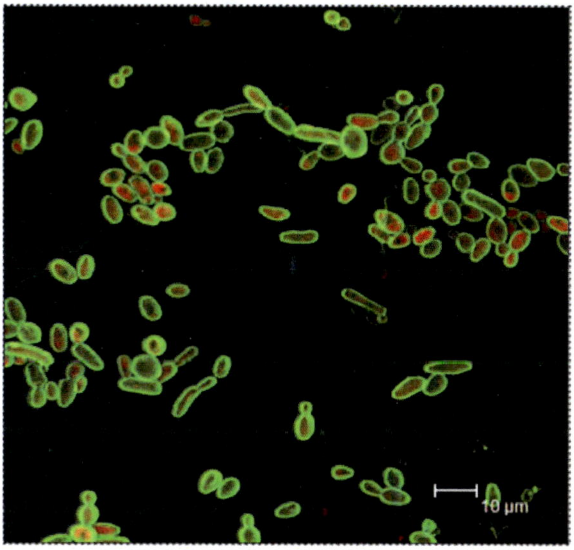

Fig. 2 Evidence of *Candida parapsilosis* on silicone tubing after 2 h exposure to urine

Morganella morganii, *P. aeruginosa*, *K. pneumoniae* and *Proteus vulgaris* are commonly found in urinary catheter biofilms. A number of these bacteria (e.g. *P. mirabilis*) express urease, an enzyme which hydrolyses urea found in the urine, resulting in the production of ammonia. Ammonia causes an increase in the pH of the urine, which in turn allows mineral precipitation including that of calcium and magnesium phosphates, leading to blockage of the catheter and infection (Stickler et al. 1999; Tunney et al. 1999). The problem is compounded because these organisms tend to be antibiotic resistant and as a biofilm forms within the catheter, they are less susceptible to host defences and antimicrobial treatments (Trautner and Darouiche 2004).

Unfortunately, urinary catheters also provide a passageway for bacteria from a heavily contaminated external skin site to a vulnerable body cavity. Polymicrobial communities will eventually develop, but initial infections are usually by single bacterial species (Stickler 2005).

9 Management of Biofilm Infections

Viewing bacteria from the perspective of multicellular behaviour is altering our view of microbiology and of Koch's postulates (Percival et al. 2010). It is evident that 99.9% of organisms prefer attachment, and that bacterial cells have the ability to aggregate into particular three-dimensional assemblages (Davey and O'Toole 2000). Biofilms have been recognised as being important in human disease and the number of biofilm-associated diseases seems to be increasing (Davies 2003). It is important to understand the characteristics of the biofilm mode of growth and the various aspects of biofilm formation. To successfully treat biofilm infections, knowledge of the phenotype of the bacterial population is required. This is important, as antibiotic treatment may not be totally effective if more than one phenotype exists. Some of these cells might remain intact serving to re-infect the host once the antimicrobial treatment has finished (Brooun et al. 2000; Davies 2003; Spoering and Lewis 2001).

A key factor to combating biofilm infections is to understand the physiology of biofilm development. Davies (2003) suggested that chemotherapeutic agents could be developed to promote or prevent transition from one stage of biofilm maturation to the next by targeting unique biofilm regulatory or signalling molecules. Specific agents might be discovered or developed which will interfere with the production of virulence factors, or promote or inhibit the shedding of biofilm bacteria (Davies 2003).

As mentioned before, biofilm resistance depends on aggregation of bacteria into multicellular communities. Therefore, one antimicrobial strategy might be to develop therapies to disrupt the multicellular structure of the biofilm. It could be that host defences might be able to resolve the infection once the multicellularity of the biofilm is reduced, and then the effectiveness of antibiotics might be restored (Stewart and Costerton 2001). Other potential therapies include enzymes that dissolve the matrix polymers of the biofilm, chemical reactions that block biofilm matrix synthesis and analogues of microbial signalling molecules that interfere with cell-to-cell communication, required for normal biofilm formation (Nemoto et al. 2000; Parsek and Greenberg 2000; Yasuda et al. 1993). Already a number of QS inhibitors have been identified such as the inhibitory peptide RNAIII, which inhibits the *agr* system of Gram-positive bacteria (Rhoads et al. 2007). In *P. aeruginosa*, furanones derived from plants have been demonstrated to block AHL pathways (Heurlier et al. 2006).

For in vivo indwelling device-associated infections, effective, preventive and therapeutic strategies still need to be developed. One such therapy could be the production of materials with anti-adhesive surfaces, for example, heparin (Tenke et al. 2004). Tenke et al. (2004) showed that on heparin-coated catheter stents, no biofilm formation was evident between 6 and 8 weeks, whereas uncoated tubes were obstructed within 2–3 weeks. Heparin coating seems one possible solution, but further development of materials resisting bacterial colonisation is needed (Tenke et al. 2004).

Progress has already been made, but the future of biofilm research and management relies upon collaborative efforts to fully explore these complex systems of the microbial world.

10 Conclusion

Infectious disease processes due to bacteria associated with biofilms such as otitis media, periodontitis, cystic fibrosis, native valve endocarditis and chronic prostatitis all appear to be caused by biofilm-associated microorganisms. In addition, indwelling medical devices have been shown to harbour biofilms, which have been implicated in infections.

In hot water and also potable water distribution systems, biofilms have been shown to harbour pathogens, such as *Mycobacterium avium, Legionella pneumophila* and now *Helicobacter pylori* and *Arcobacter* spp. (Percival et al. 2001; Percival and Thomas 2009). How the interaction and growth of pathogenic organisms in a biofilm result in an infectious disease process is presently unknown and warrants extensive research.

To date we can appreciate that biofilms are important in infectious disease processes. The principles by which this is evident is when we consider detachment of cells or biofilm aggregates resulting in the production of emboli, bacteria may exchange resistance plasmids within biofilms, cells in biofilms have dramatically reduced susceptibility to antimicrobial agents, biofilm-associated Gram-negative bacteria may produce endotoxins and biofilms are resistant to host immune system clearance.

Biofilms are highly resistant to most antimicrobial agents and disinfectants. Sessile bacteria within a biofilm are able to acquire resistance through the transfer of resistance plasmids. This acquisition of resistance is particularly important in the healthcare environment for patients with colonised urinary catheters and orthopaedic patients. Many organisms are shown to carry plasmids encoding resistance to multiple antimicrobial agents, particularly in the medical setting.

Microorganisms are capable of growing in both a free form (planktonic) or as biofilms attached to solid surfaces. Biofilms are communities of microorganisms, often adhered to a surface and encased within an extracellular polysaccharide matrix (Kumamoto and Marcelo 2005). Examples of surfaces supporting biofilm growth include inanimate environmental materials, biomaterials interfacing with host tissue and systems or the host tissue itself. The behaviour and phenotype of microbes existing in either planktonic and biofilm states are known to differ significantly and this is perhaps best exemplified by studies on antimicrobial efficacy against the different growth phases (Hill et al. 2003).

In the oral environment, candidal biofilms on prostheses and the oral mucosa have been associated with infection (Kumamoto and Marcelo 2005). Intra-oral biofilms can develop on the tooth enamel, oral mucosa or on introduced oral prostheses and these can all provide a reservoir of potentially pathogenic organisms promoting dental caries and periodontal disease. Furthermore, in the case of oral tissue, certain microorganisms such as yeast of the genus *Candida* have been shown to actually invade the tissue, invoking a pathogenic effect. Similarly, within a chronic wound, it may well be the case that the occurrence of a biofilm results in clinical problems due to the existence of the microbes in a more persistent state. Examples of clinically important biofilms include *P. mirabilis* on urinary tract catheters (Stickler 2005), *Candida* spp. on denture surfaces and bacteria (polymicrobial) within chronic wounds (Hill et al. 2003). Significantly, when biofilms are present on the surface of medical devices, a failure of the device can occur, as encountered in the blocking of urinary catheters and the obstruction of airways within artificial voice box prostheses (Douglas 2003; Van der Mei et al. 2000; Percival et al. 2009). Furthermore, biofilms tend to exhibit heightened resistance to antibiotics, possibly by diffusion limitation or by the presence of biochemically inert microbes within the biofilm (Donlan and Costerton 2002).

References

Anderl JN, Zahller J, Roe F, Stewart PS (2003) Role of nutrient limitation and stationary-phase existence in *Klebsiella pneumoniae* biofilm resistance to ampicillin and ciprofloxacin. Antimicrob Agents Chemother 47:1251–1256

Anwar H, Strap JL, Costerton JW (1992a) Establishment of aging biofilms: possible mechanism of bacterial resistance to antimicrobial therapy. Antimicrob Agents Chemother 36:1347–1351

Anwar H, Strap JL, Chen K, Costerton JW (1992b) Dynamic interactions of biofilms of mucoid *Pseudomonas aeruginosa* with tobramycin and piperacillin. Antimicrob Agents Chemother 36:1208–1214

Baier RE (1980) Substrate influence on adhesion of microorganisms and their resultant new surface properties. In: Bitton G, Marshall KC (eds) Adsorption of microorganisms to surfaces. Wiley, New York, pp 59–104

Baier RE (1984) Initial events in microbial film formation. In: Costlow JD, Tipper RC (eds) Marine biodetermination: an interdisciplinary approach. E & FN Spon, London, pp 57–62
Bashan Y, Levanony H (1988) Active attachment of *Azospirillum brasilense* Cd to quartz sand and to a light-textured soil by protein bridging. J Gen Microbiol 134:2269–2279
Bayston R (1999) Medical problems due to biofilms: clinical impact, aetiology, molecular pathogenesis, treatment and prevention. In: Wilson M, Newman HN (eds) Dental plaque revisited: oral biofilms in health and disease. BioLine, Cardiff, pp 111–124
Beech IB, Gaylarde CC (1989) Adhesion of *Desulfovibrio desulfuricans* and *Pseudomonas fluorescens* to mild steel surfaces. J Appl Bacteriol 67:2017
Bendinger B, Rijnaarts HHM, Altendorf K, Zehnder AJB (1993) Physicochemical cell surface and adhesive properties of coryneform bacteria related to the presence and chain length of mycolic acids. Appl Environ Microbiol 59:3973–3977
Blenkinsopp SA, Costerton JW (1991) Understanding bacterial biofilms. Trends Biotechnol 9:138–143
Braunwald E (1997) Valvular heart disease. In: Braunwald E (ed) Heart disease, vol 2. W.B. Saunders, Philadelphia
Brooun A, Liu S, Lewis K (2000) A dose-response study of antibiotic resistance in *Pseudomonas aeruginosa* biofilms. Antimicrob Agents Chemother 44:640–646
Brown MRW, Barker J (1999) Unexplored reservoirs of pathogenic bacteria: protozoa and biofilms. Trends Microbiol 7:46–50
Brown CM, Ellwood DC, Hunter JR (1977) Growth of bacteria at surfaces: influence of nutrient limitations. FEMS Microbiol Lett 1:163–165
Bryers JD (1984) Biofilm formation and chemostat dynamics: pure and mixed culture considerations. Biotechnol Bioeng 26:948–958
Bryers JD (1987) Biologically active surfaces; processes governing the formation and persistence of biofilms. Biotechnology 3:57–68
Bullitt R, Makowski L (1995) Structural polymorphism of bacterial adhesion pili. Nature 373:164–167
Busscher HJ, Weerkamp A (1987) Specific and non-specific interactions: role in bacterial adhesion to solid substrata. FEMS Microbiol Rev 46:165–173
Buswell CM, Herlihy YM, Marsh PD, Keevil CW, Leach SA (1997) Coaggregation amongst aquatic biofilm bacteria. J Appl Microbiol 83:477–484
Caldwell DE, Korber DR, Lawrence JR (1992) Confocal laser microscopy and digital image analysis in microbial ecology. Adv Microb Ecol 12:1–67
Carpentier B, Cerf O (1993) Biofilms and their consequences, with particular reference to hygiene in the food industry. J Appl Bacteriol 75:499–511
Carrel T, Nguyen T, Kipfer B, Althaus U (1998) Definitive cure of recurrent prosthetic endocarditis using silver-coated St. Jude medical heart valves: a preliminary case report. J Heart Valve Dis 7(5):531–533
Chamberlain AHL (1992) The role of adsorbed layers in bacterial adhesion. In: Melo LF, Bott TR, Fletcher M, Capdeville B (eds) Biofilms-science and technology. Kluwer Academic, Dordrecht, pp 59–67
Characklis WG (1973) Attached microbial growths-II. Frictional resistance due to microbial slimes. Water Res 7:1249–1258
Characklis WG (1981) Fouling biofilm development: a process analysis. Biotechnol Bioeng 23:1923–1960
Characklis WG, Cooksey KE (1983) Biofilms and microbial fouling. Adv Appl Microbiol 29:93–138
Characklis WG, McFeters GA, Marshall KC (1990a) Physiological ecology of biofilm systems. In: Characklis WG, Marshall KC (eds) Biofilms. Wiley, New York, pp 341–393
Characklis WG, Turakhia MH, Zelver N (1990b) Transfer and interfacial transport phenomena. In: Characklis WG, Marshall KC (eds) Biofilms. Wiley, New York, pp 265–340
Connell JH, Slatyer RO (1977) Mechanisms of succession in natural communities and their role in community stability and organization. Am Nat 111:1119–1144

Corpe WA (1970) An acid polysaccharide produced by a primary film forming marine bacterium. Dev Ind Microbiol 11:402–412
Corpe WA (1980) Microbial surface components involved in adsorption of microorganisms onto surfaces. In: Bitton G, Marshall KC (eds) Adsorption of microorganisms to surfaces. Wiley, New York, pp 105–144
Costerton JW, Geesey GG (1979) In: Costerton JW, Colwell RR (eds) Native aquatic bacteria: enumeration, activity, and ecology. ASTM Press, Philadelphia, pp 7–18
Costerton JW, Lappin-Scott HM (1989) Behaviour of bacterial biofilms. Am Soc Microbiol News 55:650–654
Costerton JW, Lashen ES (1984) The influence of biofilm efficacy of biocides on corrosion-causing bacteria. Mater Performance 23:34–37
Costerton JW, Geesey GG, Cheng K-J (1978) How bacteria stick. Sci Am 238:86–95
Costerton JW, Irvin RT, Cheng KJ (1981) The bacterial glycocalyx in nature and disease. Annu Rev Microbiol 35:299–324
Costerton JW, Cheng KJ, Geesey GG, Ladd TIM, Nickel JC, Dasgupta M, Marie TJ (1987) Bacterial biofilms in nature and disease. Annu Rev Microbiol 41:435–464
Costerton JW, Lewandowski Z, de Beer D, Calwell D, Korber D, James G (1994) Biofilms, the customised microniches. J Bacteriol 176:2137–2142
Costerton JW, Lewandowski Z, Caldwell DE, Korber DR, Lappin-Scott HM (1995) Microbial biofilms. Annu Rev Microbiol 49:711–745
Costerton JW, Stewart PS, Greenberg EP (1999) Bacterial biofilms: a common cause of persistent infections. Science 284:1318–1322
Crampton SE, Gerke C, Schnell NF, Nichols WW, Gotz F (1999) The intacellular adhesion (ica) locus is present in *Staphylococcus aureus* and is required for biofilm formation. Infect Immun 67:5427–5433
Danielsson A, Norkrans B, Bjornsson A (1977) On bacterial adhesion – the effect of certain enzymes on adhered cells in a marine *Pseudomonas* sp. Bot Mar 20:13–17
Davey ME, O'Toole A (2000) Microbial biofilms: from ecology to molecular genetics. Microbiol Mol Biol Rev 64:847–867
Davies D (2003) Understanding biofilm resistance to antibacterial agents. Nat Rev Drug Discov 2:114–122
de Beer D, Stoodley P, Roe F, Lewandowski Z (1994) Effects of biofilm structures on oxygen distribution and mass transfer. Biotechnol Bioeng 43:1131–1138
De Kievit TR, Parkins MD, Gillis RJ, Srikumar R, Ceri H, Poole K, Iglewski BH, Storey DG (2001) Multidrug efflux pumps: expression patterns and contribution to antibiotic resistance in *Pseudomonas aeruginosa* biofilms. Antimicrob Agents Chemother 45:1761–1770
Donlan R (2001) Biofilms and device-associated infections. Emerg Infect Dis 7:277–281
Donlan R, Costerton JW (2002) Biofilms: survival mechanisms of clinically relevant microorganisms. Clin Microbiol Rev 15:167–193
Douglas LJ (2003) *Candida* biofilms and their role in infection. Trends Microbiol 11:30–36
Dunne WM Jr (2002) Bacterial adhesion: seen any good biofilms lately? Clin Microbiol Rev 15:155–166
Flemming H-C, Wingender J, Griegbe T, Mayer C (2000) Physico-chemical properties of biofilms. In: Evans LV (ed) Biofilms: recent advances in their study and control. Harwood Academic, Amsterdam, pp 19–34
Fletcher M (1977) The effects of culture concentration and age, time, and temperature on bacterial attachment to polystyrene. Can J Microbiol 23:1–6
Fletcher M (1980) The question of passive versus active attachment mechanisms in non-specific bacterial adhesion. In: Berkeley RCW (ed) Microbial adhesion to surfaces. Horwood, Chichester, pp 67–78
Fletcher M, Loeb GI (1979) The influence of substratum characteristics on the attachment of a marine *Pseudomonas* to solid surfaces. Appl Environ Microbiol 37:67–72

Fletcher M, Marshall KC (1982) Are solid surfaces of ecological significance to aquatic bacteria? Adv Microb Ecol 12:199–236
Fredrickson AG (1977) Behaviour of mixed cultures of microorganisms. Annu Rev Microbiol 33:63–87
Fux CA, Costerton JW, Stewart PS, Stoodley P (2005) Survival strategies of infectious biofilms. Trends Microbiol 13:34–40
Gilbert P, Das J, Foley I (1997) Biofilm susceptibility to antimicrobials. Adv Dent Res 11:160–167
Habash M, Reid G (1999) Microbial biofilms: their development and significance for medical device-related infections. J Clin Pharmacol 39:887–898
Hamilton WA (1987) Biofilm: microbial interaction and metabolic activities. In: Fletcher M, Gray TRG, Jones JG (eds) Ecology of microbial communities. Society for general microbiology symposium 41. Cambridge University Press, Cambridge, pp 361–387
Hamilton WA, Characklis WG (1989) Relative activities of cells in suspension and in biofilms. In: Characklis WG, Wilderer PA (eds) Structure and function of biofilms. Wiley, New York, pp 199–219
Harrison JJ, Turner RJ, Ceri H (2005) Persister cells, the biofilm matrix and tolerance to metal cations in biofilm and planktonic *Pseudomonas aeruginosa*. Environ Microbiol 7:981–994
Heilmann C, Schweitzer O, Gerke C, Vanittanakom N, Mack D, Goetz F (1996) Molecular basis of intercellular adhesion in the biofilm-forming *Staphylococcus epidermidis*. Mol Microbiol 20:1083–1091
Heukelekian H, Heller A (1940) Relation between food concentration and surface for bacterial growth. J Bacteriol 40:547–558
Heurlier K, Denervaud V, Haas D (2006) Impact of quorum sensing on fitness of *Pseudomonas aeruginosa*. Int J Med Microbiol 296:93–102
Hill KE, Davies CE, Wilson MJ, Stephens P, Harding KG, Thomas DW (2003) Molecular analysis of the microflora in chronic venous leg ulceration. J Med Microbiol 52:365–369
Hofstad T (1992) Virulence factors in anaerobic bacteria. Eur J Clin Microbiol Infect Dis 11:1044–1048
Hoyle BD, Costerton JW (1991) Bacterial resistance to antibiotics: the role of biofilms. Prog Drug Res 37:91–105
Humphrey BA, Dickson MR, Marshall KC (1979) Physiochemical and in situ observations on the adhesion of gliding bacteria to surfaces. Arch Microbiol 120:231–238
Illingworth B, Tweden K, Schroeder R, Cameron J (1999) *In vivo* efficacy of silver-coated (Silzone) infection-resistant polyester fabric against a biofilm-producing bacteria. Staphylococcus epidermidis. J Heart Valve Dis 7:524–530
James GA, Beaudette L, Costerton JW (1995) Interspecies bacterial interactions in biofilms. J Ind Microbiol 15:257–262
Jones F (2005) Quorum sensing. Microbiol Today:34–35
Jones HC, Roth IL, Saunders WM III (1969) Electron microscopic study of a slime layer. J Bacteriol 99:316–325
Karchmer A, Gibbons G (1994) Infections of prosthetic heart valves and vascular grafts. In: Bisno AL, Waldovogel FA (eds) Infections associated with indwelling medical devices, 2nd edn. American Society for Microbiology, Washington, pp 213–249
Keevil CW, Dowsett AB, Rogers J (1993) *Legionella* biofilms and their control. Society for Applied Bacteriology technical series: microbiofilms. Society for Applied Bacteriology, Bedford, pp 203–215
Khardori N, Yassien M (1995) Biofilms in device-related infections. J Ind Microbiol Biotechnol 15:141–147
Kolenbrander PE, Palmer RJ, Rickard AH, Jakubovics NS, Chalmers NI, Diaz PI (2006) Bacterial interactions and successions during plaque development. Periodontology 2000(42):47–79
König C, Schwank S, Blaser J (2001) Factors compromising antibiotic activity against biofilms of *Staphylococcus epidermidis*. Eur J Clin Microbiol Infect Dis 20:20–26

Korber DR, Lawrence JR, Sutton B, Caldwell DE (1989) Effect of laminar flow velocity on the kinetics of surface recolonization by Mot+ and Mot- *Pseudomonas fluorescens*. Microb Ecol 18:1–19

Kumamoto CA, Marcelo DV (2005) Alternative *Candida albicans* lifestyles: growth on surfaces. Annu Rev Microbiol 59:113–133

Lappin-Scott HM, Jass J, Costerton JW (1993) Microbial biofilm formation and characterisation. Society for Applied Bacteriology technical series No. 30. Society for Applied Bacteriology, Bedford

Lawrence JR, Neu TR (1999) Confocal laser scanning microscopy for analysis of microbial biofilms. Meth Enzymol 310:131–144

Lewandowski Z, Stoodley P, Roe F (1995) Internal mass transport in heterogeneous biofilms. Recent advances in corrosion/95, paper no. 222. NACE International, Houston

Loeb GI, Neihof RA (1975) Marine conditioning films. Adv Chem Ser 145:319–335

Mack D, Nedelmann M, Krokotsch A, Schwarzkopf A, Heesemann J, Laufs R (1994) Characterization of transposon mutants of biofilm-producing *Staphylococcus epidermidis* impaired in the accumulative phase of biofilm production: genetic identification of a hexosamine-containing polysaccharide intercellular adhesin. Infect Immun 62:3244–3253

Mack D, Fischer W, Krokotsch A, Leopold K, Hartmann R, Egge H, Laufs R (1996) The intercellular adhesin involved in biofilm accumulation of *Staphylococcus epidermidis* is a linear beta-1,6-linked glucosaminoglycan: purification and structural analysis. J Bacteriol 178:175–183

Mah T, O'Toole GA (2001) Mechanisms of biofilm resistance to antimicrobial agents. Trends Microbiol 9:34–39

Mah T, Pitts B, Pellock B, Walker GC, Stewart PS, O'Toole GA (2003) A genetic basis for *Pseudomonas aeruginosa* biofilm antibiotic resistance. Nature 426:306–310

Maira-Litran T, Allison DG, Gilbert P (2000) An evaluation of the potential of the multiple antibiotic resistance operon (mar) and the multidrug efflux pump acrAB to moderate resistance towards ciprofloxacin in *Escherichia coli* biofilms. J Antimicrob Chemother 45:789–795

Marmur A, Ruckenstein E (1986) Gravity and cell adhesion. J Colloid Interface Sci 114:261–266

Marsh PD (1995) Dental plaque. In: Lappin-Scott HM, Costerton JW (eds) Microbial biofilms. Cambridge University Press, Cambridge, pp 282–300

Marshall KC (1992) Biofilms: a overview of bacterial adhesion, activity and control at surfaces. Am Soc Microbiol News 58:202–207

Marshall KC, Stout R, Mitchell R (1971) Mechanism of the initial events in the sorption of marine bacteria to surfaces. J Gen Microbiol 68:337–348

McEldowney S, Fletcher M (1988) Effect of pH, temperature, and growth conditions on the adhesion of a gliding bacterium and three nongliding bacteria to polystyrene. Microb Ecol 16:183–195

Mittelman MW (1996) Adhesion to biomaterials. In: Fletcher M (ed) Bacterial adhesion: molecular and ecological diversity. Wiley-Liss, New York, pp 89–127

Moore PCL, Lindsay JA (2001) Genetic variation among hospital isolates of methicillin-sensitive *Staphylococcus aureus*: evidence for horizontal transfer of virulence genes. J Clin Microbiol 39:2760–2767

Mulhall AB, Chapman RG, Crow RA (1988) Bacteriuria during indwelling urethral catheterization. J Hosp Infect 11:253–262

Nemoto K, Hirota K, Ono T, Murakami K, Nagao D, Miyake Y (2000) Effect of Varidase (streptokinase) on biofilm formed by *Staphylococcus aureus*. Chemotherapy 46:111–115

Palmer RJ, Sternberg C (1999) Modern microscopy in biofilm research: confocal microscopy and other approaches. Curr Opin Biotechnol 10:263–268

Palmer R Jr, White DC (1997) Developmental biology of biofilms: implications for treatment and control. Trends Microbiol 5:435–440

Parsek MR, Greenberg EP (2000) Acyl-homoserine lactone quorum sensing in Gram-negative bacteria: a signaling mechanism involved in associations with higher organisms. Proc Natl Acad Sci USA 97:8789–8793

Percival SL, Bowler PG (2004) Biofilms and their potential role in wound healing. Wounds 16:234–240
Percival SL, Kite P (2007) Catheters and infection control. J Vasc Access 2:69–80
Percival SL, Thomas JG (2009) *Helicobacter pylori* prevalence and transmission and role of biofilms. Water Health 7(3):469–477
Percival SL, Walker JT (1999) Biofilms and public health significance. Biofouling 14:99–115
Percival SL, Knapp JS, Wales DS, Edyvean RGJ (1999) The effect of flow and surface roughness on biofilm formation. J Microbiol Biotechnol 22:152–159
Percival SL, Walker J, Hunter P (2000) Microbiological aspects of biofilms and drinking water. CRC Press, New York
Percival SL, Hegarty JH, McKay G, Reid D (2001) *Helicobacter pylori* in biofilms. In: Gilbert PG, Allison D, Walker JT, Brading M (eds) Biofilm community interactions: chance or necessity. Species consortia. Wiley, New York, pp 59–63
Percival SL, Sabbuba NA, Kite P, Stickler DJ (2009) The effect of EDTA instillations on the rate of development of encrustation and biofilms in Foley catheters. Urol Res 37(4):205–209
Percival SL, Thomas J, Williams D (2010) Biofilms and bacterial imbalances in chronic wounds: anti-Koch. Int Wound J 7(3):169–175
Percival SL, Thomas J, Thomas D, Williams D (2011) Antimicrobial tolerance and role of biofilms and persister cells in wounds. Wound Repair Regen 19(1):1–9
Powell MS, Slater NKH (1982) Removal rate of bacterial cells from glass surfaces by fluid shear. Biotechnol Bioeng 24:2527–2537
Pringle JH, Fletcher M (1983) Influence of substratum wettability on attachment of freshwater bacteria to solid surfaces. Appl Environ Microbiol 45:811–817
Raad II, Sabbagh MF, Rand KH, Sherertz RJ (1992) Quantitative tip culture methods and the diagnosis of central venous catheter-related infections. Diagn Microbiol Infect Dis 15:13–20
Ramage G, Martinez JP, Lopez-Ribot JL (2006) *Candida* biofilms on implanted biomaterials: a clinically significant problem. FEMS Yeast Res 6:979–986
Reid G, McGroarty J, Angotti R, Cook R (1988) *Lactobacillus* inhibitor production against *Escherichia coli* and coaggregation ability with uropathogens. Can J Microbiol 34:344–351
Rhoads DD, Wolcott RW, Cutting KF, Percival SL (2007) Evidence of biofilms in wounds and potential ramifications. In: Gilbert P, Allison D, Brading M, Pratten J, Spratt D, Upton M (eds) Biofilms: coming of age, vol 8. The Biofilm Club, pp. 131–143
Rickard AH, Leach SA, Buswell CM, High NJ, Handley PS (2000) Coaggregation between aquatic bacteria is mediated by specific-growth-phase-dependent lectin-saccharide interactions. Appl Environ Microbiol 66:431–434
Rickard AH, Leach SA, Hall LS, Buswell CM, High NJ, Handley PS (2002) Phylogenetic relationships and coaggregation ability of freshwater biofilm bacteria. Appl Environ Microbiol 68:3644–3650
Rickard AH, Gilbert P, High NJ, Kolenbrander PE, Handley PS (2003a) Bacterial coaggregation: an integral process in the development of multi-species biofilms. Trends Microbiol 11:94–100
Rickard AH, McBain AJ, Ledder RG, Handley PS, Gilbert P (2003b) Coaggregation between freshwater bacteria within biofilm and planktonic communities. FEMS Microbiol Lett 220:133–140
Rittle KH, Helmstetter CE, Meyer AE, Baier RE (1990) *Escherichia coli* retention on solid surfaces as functions of substratum surface energy and cell growth phase. Biofouling 2:121–130
Rittman BE (1989) The effect of shear stress on biofilm loss rate. Biotechnol Bioeng 24:501–506
Roberts ME, Stewart PS (2005) Modelling protection from antimicrobial agents in biofilms through the formation of persister cells. Microbiology 151:75–80
Rosenberg M, Kjelleberg S (1986) Hydrophobic interactions in bacterial adhesion. Adv Microb Ecol 9:353–393
Saye DE (2007) Recurring and antimicrobial-resistant infections: considering the potential role of biofilms in clinical practice. Ostomy Wound Care Manage 53:46–48

Spoering AL, Lewis K (2001) Biofilms and planktonic cells of *Pseudomonas aeruginosa* have similar resistance to killing by antimicrobials. J Bacteriol 183:6746–6751

Stewart PS, Costerton JW (2001) Antibiotic resistance of bacteria in biofilms. Lancet 358:135–138

Stewart PS, Camper AK, Handran SD, Huang CT, Warnecke M (1997) Spatial distribution and coexistence of *Klebsiella pneumoniae* and *Pseudomonas aeruginosa* in biofilms. Microb Ecol 33:2–10

Stickler DJ (2002) Susceptibility of antibiotic-resistant Gram-negative bacteria to biocides: a perspective from the study of catheter biofilms. J Appl Microbiol 92:163S–170S

Stickler D (2005) Urinary catheters: ideal sites for the development of biofilm communities. Microbiol Today:22–25.

Stickler DJ, Morris NS, Winters C (1999) Simple physical model to study formation and physiology of biofilms on urethral catheters. Meth Enzymol 310:494–501

Stoodley P, Boyle JD, Dodds I, Lappin-Scott HM (1997) Consensus model of biofilm structure. In: Biofilms: community interactions and control. Third meeting of the British Biofilm Club, Gregynog Hall, Powys, 26–28 September 1997, pp 1–9

Sutherland IW (2001) The biofilm matrix: an immobilized but dynamic microbial environment. Trends Microbiol 9:222–227

Tenke P, Riedl CR, Jones GL, Williams GJ, Stickler D, Nagy E (2004) Bacterial biofilm formation on urologic devices and heparin coating as preventive strategy. Int J Antimicrob Agents 23:67–74

Trautner BW, Darouiche RO (2004) Role of biofilm in catheter-associated urinary tract infection. Am J Infect Control 32:177–183

Tunney MM, Jones DS, Gorman SP (1999) Biofilm and biofilm-related encrustations of urinary tract devices. In: Doyle RJ (ed) Methods in enzymology. Biofilms, vol 310. Academic, San Diego, pp 558–666

Uhlinger DJ, White DC (1983) Relationship between physiological status and formation of extracellular polysaccharide glycocalyx in *Pseudomonas atlantica.* Appl Environ Microbiol 45:64–70

Van der Mei HC, Free RH, Elving GJ, Van Weissenbruch R, Albers FW, Busscher HJ (2000) Effect of probiotic bacteria on prevalence of yeasts in oropharyngeal biofilms on silicone rubber voice prostheses in vitro. J Med Microbiol 49:713–718

Vandevoorde L, Christiaens H, Verstraete W (1992) Prevalence of coaggregation reactions among chicken lactobacilli. J Appl Bacteriol 72:214–219

Vieira MJ, Oliveira R, Melo L, Pinheiro M, van der Mei H (1992) Adhesion of *Pseudomonas fluorescens* to metallic surfaces. J Dispers Sci Technol 13(4):437–445

Wahl M (1989) Marine epibiosis. 1. Fouling and antifouling: some basic aspects. Mar Ecol Prog Ser 58:175–189

Walt DR, Smulow JB, Turesky SS, Hill RG (1985) The effect of gravity on initial microbial adhesion. J Colloid Interface Sci 107:334–336

Walters MC III, Roe F, Bugnicourt A, Franklin MJ, Stewart PS (2003) Contributions of antibiotic penetration, oxygen limitation, and low metabolic activity to tolerance of *Pseudomonas aeruginosa* biofilms to ciprofloxacin and tobramycin. Antimicrob Agents Chemother 47:317–323

Whittaker CJ, Klier CM, Kolenbrander PE (1996) Mechanisms of adhesion by oral bacteria. Annu Rev Microbiol 50:513–552

Wilson M (2001) Bacterial biofilms and human disease. Sci Prog 84:235–254

Wimpenny JWT, Colasanti R (1997) A unifying hypothesis for the structure of microbial biofilms based on cellular automaton models. FEMS Microbiol Ecol 22:1–6

Xie H, Cook GS, Costerton JW, Bruce G, Rose TM, Lamont RJ (2000) Intergeneric communication in dental plaque biofilms. J Bacteriol 182:7067–7069

Yarwood JM, Schlievert PM (2003) Quorum sensing in *Staphylococcus* infections. J Clin Investig 112:1620–1625

Yasuda H, Ajiki Y, Koga T, Kawada H, Yokota T (1993) Interaction between biofilms formed by *Pseudomonas aeruginosa* and clarithromycin. Antimicrob Agents Chemother 37:1749–1755

Zobell CE (1943) The effect of solid surfaces upon bacterial activity. J Bacteriol 46(1):39–56

Zoonotic Infections: The Role of Biofilms

Ana B. García and Steven L. Percival

Abstract A zoonosis or zoonose is any infectious disease that can be transmitted from non-human animals, both wild and domestic, to humans. Infectious diseases transmitted from humans to non-human animals is sometimes called reverse zoonosis or anthroponosis. Sixty one percent of the pathogens known to affect humans are zoonotic. Biofilm formation is used as a mechanism by zoonotic and environmental pathogens to infect animals and humans. It has been suggested that biofilms are involved in 65–80% of infections treated by doctors in developed countries. Microorganisms can resist extreme temperatures, antibiotic treatments and low levels of nutrients by forming biofilms. Therefore the selection of the right antibiotics to treat human and animal infections caused by biofilms is paramount. It is apparent that more research into biofilm infections in humans and animals, biofilm resistance mechanisms and new strategies for effective treatment need to be developed.

1 Introduction

As mentioned throughout this book, biofilm is a term used to refer to a "vast number of microbial aggregates" as reported by Julian Wimpenny in (2000). However, through the last decade different definitions of biofilms have been reported in the literature as new discoveries in biofilmology are being made. Today biofilms can be defined as a community of microbial organisms which become adherent to each other to form microcolonies. The presence of microcolonies is used as a

A.B. García (✉)
Division of Microbiology and Risk Assessment, National Food Institute, Technical University of Denmark, Mørkhøj Bygade 19, Building H, 2860 Søborg, Denmark
e-mail: agar@food.dtu.dk

S.L. Percival
Department of Pathology, Medical School, West Virginia University, Morgantown, WV, USA
e-mail: biofilms@hotmail.co.uk

S.L. Percival et al. (eds.), *Biofilms and Veterinary Medicine*,
Springer Series on Biofilms 6, DOI 10.1007/978-3-642-21289-5_3,

biomarker for the existence of biofilms. Microcolonies are encased in a matrix of extracellular polymeric substances (EPS) and have been identified in the sputum of cystic fibrosis patients (Bjarnsholt et al. 2009), in chronic wounds (Cochrane et al. 2009) and on catheters (Percival et al. 2005).

Biofilm aggregates have also been found in anthropogenic and natural aquatic environments and are known to harbour pathogenic organisms that may be transmitted to humans (Jennings et al. 2003; Knulst et al. 2003). Consequently, the dissemination of a biofilm known to harbour pathogens is considered to be a public health hazard, particularly in hospital environments. In fact, recent findings have suggested that biofilm dissemination is akin to metastasis of a tumour and as such the detached biofilm fragments might cause serious problems when they detach within the circulatory system of both humans and animals (Stoodley et al. 2001). Detached biofilm fragments once in the circulatory system are free to colonise new niches and as such have been linked to conditions such as endocarditis.

From an evolutionary and natural selection process, the ability of a microorganism to form a biofilm is very important since this phenotypic state will aid in and promote a greater flexibility in microbial adaptation (O'Toole and Kolter 1998; Van Loosdrecht et al. 2002; Costerton et al. 2003).

2 Biofilm Formation and Composition

Bacterial components such as flagella, membrane proteins, pili and fimbriae have been shown to have a role to play in attachment to surface and therefore in biofilm formation (Cloete et al. 1989; Prakash et al. 2003; Lelieveld 2005). The presence of flagella and fimbriae is very important in adhesion as they are able to overcome the repulsive forces bacteria encounter when they first attach to a surface (Corpe 1980; Korber et al. 1989; O'Toole and Kolter 1998; Pratt and Kolter 1998; Giron et al. 2002).

Proteomic studies involving *Pseudomonas aeruginosa* biofilms have enabled identification of the stages bacteria go through during adhesion and biofilm formation, a commonality shared by most if not all prokaryotes. The non-specific binding of bacteria to a surface followed then by multiplication of the adherent microorganisms is the first steps to biofilm formation, as discussed in Chap. 2. Following microbial adhesion the microorganisms multiply, produce EPS culminating in the formation of microcolonies. These microcolonies are maintained and supported at the surface by fluid flow and overtime result in the formation of a "mature" biofilm. The architecture of the "mature" biofilm is affected by the abundance, or limited availability, of nutrients and pH. Biofilms are highly heterogeneous and within in vitro models they have been shown to be complex, being composed of features such as stacks, mushrooms, water channels and streamers. Within a natural biofilm a similar architecture to that found in the in vitro biofilm has been hypothesised to occur. However, within these natural biofilms a more diverse microflora composed of fungi, bacteria, algae and protozoa have been reported which have displayed features significantly different to those observed in pure culture models. A biofilm is able to modulate its internal environment significantly aiming to add stability and protection for the inherent microorganisms present. Whilst the biofilm matrix is in a constant nutritional and biochemical flux a balance between the inherent microbiota of the biofilm is achieved through microbial commensalism, antagonism, mutualism and competition (Lelieveld et al. 2001; Sauer et al. 2002; Prakash et al. 2003; Bartram 2007).

Microbial biofilms which are formed on the surfaces of normal human tissues are purported to offer protection to the host. This phenomenon has been named as the "mucus blanket". It covers the trachea and intestine aiding protection from invasive pathogens (Lambe et al. 1991; Costerton et al. 2003). Biofilms formed by commensal bacteria attached to gut epithelial cells represent a barrier against foodborne pathogens by preventing their attachment (Lee et al. 2000).

Dental plaque on teeth, in both healthy and diseased mouths, is a known biofilm. The microbial composition of these biofilms has been shown to have important implications to health and disease in both animals and humans (Bradshaw et al. 1996; Parsek and Fuqua 2004). In fact, it has been shown that intrauterine infections can be caused by microorganisms initially present in the oral cavity (Fardini et al. 2010). Preterm birth can be caused by infection of the intrauterine environment. Traditionally it has been stated that intrauterine infections predominantly originate from the vaginal tract. However, thanks to technological advances, microbial species that do not belong to the vaginal microflora have been identified in intrauterine infections. Fardini et al. (2010) systemically examined what proportion of the oral microbiome could translocate to the placenta using pregnant mice. Several bacterial species that have been associated with intrauterine infections in humans were identified and the majority of these species were oral commensal organisms. Interestingly, some bacterial species were present with a higher prevalence in the placenta than in oral cavity samples and therefore it was concluded that the placental translocation was species specific.

Many areas of the human body constitute anatomical barriers for preventing the formation of biofilms such as epithelial cells found within the bladder (Uehling 1991). Some organs like the liver need to be maintained in aseptic conditions and as such biofilm development is rare on these organs (Sung et al. 1992).

The formation of biofilms is a dynamic and complex process generating an architecture which aims to protect the inherent microorganisms from chemical or physical removal in some areas and allowing them to colonise new niches (Wimpenny et al. 2000; Allison 2003; Hall-Stoodley et al. 2004; Clutterbuck et al. 2007). Studies have shown that adverse conditions promote biofilm formation and also dispersal (van der Wende and Characklis 1990; Stoodley et al. 2002). It has been suggested that biofilm formation might select in favour of virulent microbial strains (Declerck 2010). It is well documented that some bacteria might help other microorganisms to resist adverse conditions. For example, in biofilms composed of *Pseudomonas fluorescens* their presence has been shown to increase the resistance of *Salmonella typhimurium* to chlorine (Leriche and Carpentier 1995). In addition to this, *Staphylococcus aureus*, a common causative agent of community-acquired and hospital-acquired infections, has been shown to be able to survive and colonise a range of environments due to its metabolic versatility and potential to form a recalcitrant biofilm (O'Neill et al. 2007).

Genetic and environmental factors influence biofilm formation. The genes expressed in bacteria present in biofilms are reported to differ significantly from the genes expressed in the same bacteria in free-living form. In fact, expression of particular genes such as those that regulate flagella, surface-adhesion proteins and the formation of the extracellular matrix seems essential for biofilm formation compared to planktonic phenotypes (Cramton et al. 1999; Whitchurch et al. 2002; Valle et al. 2003; Hall-Stoodley et al. 2004). Advances in molecular technologies such as microarrays and proteomics have been used to investigate gene expression in cells forming biofilms advancing further our understanding of biofilms (Sauer et al. 2002; Tremoulet et al. 2002; Wagner et al. 2003).

3 Exploring Public Health Aspects and Zoonotic Potential of Biofilms

Biofilms have been present since prehistoric times, especially in hydrothermal environments (Hall-Stoodley et al. 2004). They have a role to play in nutrient cycling and from an environmental perspective they can be beneficial in different ways. In addition, biofilms are useful for water treatment, bioremediation and providing colonisation resistance against pathogens on natural mucosal surfaces (Lebeer et al. 2007; McBain 2009). However, they also have adverse effects by causing infections of humans and animals. In fact it has been suggested that biofilms are involved in 65–80% of infections treated by doctors in developed countries (Ghannoum and O'Toole 2004). Zoonotic and environmental pathogens use biofilm formation as a mechanism to infect animals and humans.

Biofilm formation is one of the most important virulence factors for the development of staphylococcal infections. A number of factors are known to induce biofilm formation in staphylococci, including MRSA and MSSA. The presence of glucose is an example, while sodium chloride seems to induce biofilm formation in MSSA isolates. Sodium chloride is known to activate the transcription of *ica* operon and MSSA biofilm formation is dependent on *icaADBC* operon while MRSA biofilm formation is *icaADBC* independent (Fitzpatrick et al. 2005; O'Neill et al. 2007).

The formation of biofilms allows microorganisms to survive hostile environments and to resist conventional treatments. In fact, the formation of biofilms seems to be the preferred method of growth for microbial organisms (Costerton et al. 1999; Watnick and Kolter 2000; Webb et al. 2003; Parsek and Fuqua 2004). By forming biofilms microorganisms can resist the constraints of extreme temperatures, antibiotic treatments and low levels of nutrients (Prakash et al. 2003; Bartram 2007). They are also highly resistant to acid treatments, dehydration exposure to UV light and phagocytosis (Jefferson 2004; Hall-Stoodley and Stoodley 2005). Whilst UV light has been successfully used against pathogenic microorganisms such as *Giardia muris*, *Bacillus subtilis*, *Cryptosporidium parvum* and *Legionella pneumophila* in experimental conditions, it has shown to be ineffective against biofilms in water systems (Zhang et al. 2006).

Human infections associated with biofilm formation frequently involve microbial colonisation of medical devices (Lynch and Robertson 2008). Microorganisms can form biofilms in hospital water and medical equipment used to treat humans and animals. In fact, biofilm formation has been associated with persistent infections in medical tools such as *Mycobacterium avium* infection in an intravascular catheter (Schelonka et al. 1994). Changes induced by biofilm formation in intravascular catheters are known to result in total obstruction of the catheter which has led to the development of novel anti-biofilm agents (Stickler et al. 1993; Percival et al. 2009). Biofilm infections of medical devices seem more commonly associated with urinary catheters (Darouiche 2001; Donlan and Costerton 2002; Kite et al. 2004). Several pathogens of public health importance such as *E. coli*, *S. aureus*, *Clostridium perfringens* and *Candida* have been isolated from biofilms contaminating medical devices. In particular, *S. aureus* infections tend to be more acute producing an acute immune response and significant tissue damage (Lynch and Robertson 2008). Biofilm infections of *Candida albicans* have been associated with high mortality (Kojic and Darouiche 2004: Ramage et al. 2005). Biofilm-forming *S. aureus* and *S. epidermidis* have been isolated from humans (Krepsky et al. 2003), dialysis medical devices (Chaieb et al. 2005), bovine mastitis (Vasudevan et al. 2003) and food processing environments (Moretro et al. 2003).

Biofilm formation has been suggested as a risk factor for chronic bovine intramammary infections caused by *S. aureus*. In fact, chronic mastitis caused by *S. aureus* can be very difficult to treat due to antibiotic resistance (Taponen et al. 2003; Cucarella et al. 2004; Fox et al. 2005). Furthermore, *S. aureus* could be present in milk and dairy products derived from animals with clinical or subclinical mastitis. Dairy cattle suffering from clinical mastitis should be milked last and/or using separate milking equipment if possible. In fact, animals suffering from mastitis should be treated and their milk should not be used for human or animal consumption. However, subclinical mastitis might not be diagnosed posing a risk to public health; the use of pasteurisation and decontamination technologies has been recommended as part of food safety assurance systems applied to the production of milk and dairy products.

Staphylococcus epidermidis present on human skin and *P. aeruginosa*, an environmental microorganism, can cause serious chronic infections in compromised hosts (Costerton et al. 1995). *S. aureus* and *S. epidermidis* are responsible for medical device-associated biofilm infections. *S. aureus* (like *P. aeruginosa*) can regulate virulence factors via two quorum-sensing (QS) systems (Yarwood et al. 2004). The production of adhesins by Staphlococci has been considered as the best-understood mechanism of biofilm development (McKenney et al. 1998). *S. epidermidis* has been recognised as a nosocomial pathogen causing human and medical device-related biofilm infections. Tormo et al. (2005) demonstrated that SarA represents an important regulatory element for *S. epidermidis* virulence factors including biofilm formation. Genetic expression profiling of a *S. epidermidis* biofilm proved that this microorganism exhibited a varied range of genes expressed to increase protection from antibiotics and from the host immune system during biofilm infections (Yao et al. 2005). Regulatory genes such as *agr* and *sarA* are responsible for the production of virulence factors by *S. aureus* and for the expression of specific genes during different stages of infection and biofilm formation (Dunman et al. 2001). In fact, genes can be expressed or repressed during biofilm formation and the use of transcriptome analysis identified 84 genes that were repressed and 48 genes that were induced for *S. aureus* biofilm growth (Beenken et al. 2004). Furthermore, mutations affecting sarA can inhibit *S. aureus* biofilm formation while mutations affecting agr seemed to have a neutral or even increasing effect on *S. aureus* biofilm formation (Beenken et al. 2003). Vectors called staphylococcal cassette chromosome (SCC) contain the mecA gene responsible for methicillin resistance that gets integrated into Staphylococci genetic material to produce the MRSA phenotype (Baba et al. 2002; Boyle-Vavra et al. 2005; Jemili-Ben Jomaa et al. 2006). It has been recognised that mecA genes may spread between humans via *S. epidermidis* (Silva et al. 2001) and 40% of healthcare workers may carry the mecA gene on their hands (Klingenberg et al. 2001). Furthermore, the zoonotic potential of the mecA mobile genetic element responsible for methicillin resistance exhibited by some microorganisms such as Stapylococci has been recognised (Epstein et al. 2009) although more consideration should be given to this zoonotic potential (Guardabassi et al. 2004; Morris et al. 2006; Vengust et al. 2006). In fact, the types of SSCmec found in pets are similar to those found in humans (Malik et al. 2006). MRSA can be embedded in biofilms conferring antimicrobial resistance properties. Furthermore, MRSA may form part of biofilm infections in humans.

Biofilms are also responsible for chronic infections in the urinary tract. Management of urinary infections and the use of a combination of antibiotics (fluoroquinolone and macrolide or fluoroquinolone and fosfomycin) should be considered to treat biofilm infections in the urinary tract (Kumon 2000). Raad et al. (2007) investigated the efficacy of different antibiotics against MRSA present in biofilms. These authors concluded that newer antibiotics such as daptomycin, minocycline and tigecycline should be used in combination with rifampin for antibiotic treatment against MRSA infections. The incidence of MRSA infections is increasing and the emergence of human MRSA infections in hospitals represents a major public health concern (Panlilio et al. 1992; O'Neill et al. 2007). In fact, cross-infection of MRSA between animals and humans has been recognised (Baptiste et al. 2005; Witte et al. 2007) indicating the possibility that mecA genes could spread across species (Boost et al. 2007).

Staphylococcus intermedius has been isolated from pigeons, dogs, dog bites-wound sites (Talan et al. 1989; Futagawa-Saito et al. 2006), from human patients after invasive procedures (Vandenesch et al. 1995) and a human patient with otitis externa (Tanner et al. 2000). *S. intermedius* produces many virulence factors such as coagulase, clumping factor, leukotoxin, enterotoxins and biofilm formation (Raus and Love 1983; Futagawa-Saito et al. 2006). The production of extracellular proteases has been recognised as an important virulence factor in *S. aureus* and

S. intermedius (Karlsson and Arvidson 2002). Meticillin-resistant *S. intermedius* (MRSI) is emerging as a pathogen of concern due to mecA gene acquisitions (Bannoehr et al. 2007). In fact, the zoonotic potential of MRSI and biofilm-forming *S. intermedius* from dogs to humans has also been recognised (Tanner et al. 2000; Futagawa-Saito et al. 2006; van Duijkeren et al. 2008).

Animals might serve as reservoirs for pathogens causing periodontal disease in humans. *Porphyromonas gingivalis* and *Tannerella forsythia* are highly prevalent in humans with periodontitis. Interestingly, *Porphyromonas* and *Tannerella* spp. can be present in the oral cavity of cats. A recent study conducted by Booij-Vrieling et al. (2010) determined the presence and prevalence of *Porphyromonas gulae, P. gingivalis* and *Tannerella forsythia*, and in the oral microflora of cats and their owners by culture and polymerase chain reaction (PCR). This study suggested that the isolates the owners were *P. ingivali* and those isolated from cats were *P. gulae*. However, in one cat/owner couple the *T. forsythia* isolates were identical. Therefore, it was hypothesised that transmission of *T. forsythia* from cats to owners is possible and that cats may be a reservoir for this pathogen (Booij-Vrieling et al. 2010).

Periodontal disease in humans has been associated with chronic obstructive pulmonary disease (COPD). Leuckfeld et al. (2010) found high amounts of Veillonella in the subgingival microflora of COPD subjects. These authors identified subgingival Veillonella isolates by phenotypic and genetic methods in order to assess if Veillonella strain properties correlated with periodontal disease or COPD. The majority of the subgingival Veillonella isolates examined were identified as *Veillonella parvula*. Furthermore, a subgingival and pulmonary isolate obtained from one COPD subject was found to be genetically identical strains of *V. parvula*. Small cocci co-aggregating with larger cocci on the airway epithelium of this patient were observed by electron microscopy. However, no correlation between the subgingival Veillonella strains and the presence of periodontitis or COPD was found. These authors concluded that subgingival *V. parvula* can translocate to the lungs but no particular Veillonella genotype could be associated with periodontal disease or COPD.

L. pneumophila is considered responsible for 90% of human legionellosis cases causing pneumonic and non-pneumonic disease (Yu et al. 2002). *Legionella* grows at high temperatures and hot springs represent natural habitats for this microorganism (Mashiba et al. 1993). However, *L. pneumophila* can transfer from natural habitats to other environments where it can form biofilms and parasitise protozoans as survival mechanisms (Declerck et al. 2005, 2007; Bartram 2007). *Legionella* forms associations with other microorganisms already forming part of biofilms (Watnick and Kolter 2000). *L. pneumophila* can form algal-bacterial biofilms (Hume and Hann 1984). Furthermore, *L. pneumophila* shows necrotrophic growth as another survival mechanism (Temmerman et al. 2006). The survival of *Legionella* forming biofilms also depends on relationships and interference with other microorganisms. In fact, *P. fluorescens* SSD, known as the best bacteriocin producer, had the ability to inhibit *L. pneumophila* present in biofilms (Guerrieri et al. 2008). *L. pneumophila* can survive, replicate and detach from biofilms using infected amoebae such as *Naegleria* spp. which possess an additional flagellate stage. Furthermore, infected amoebae can release very small vesicles containing *L. pneumophila* that get transported over considerable distances posing a risk to human health (Berk et al. 1998; Greub and Raoult 2004; Weissenberger et al. 2007). *Legionella* can spread by aerosols and human infection may result from inhalation. In fact, this organism can colonise hot and cold water systems (at temperatures of 20–50°C), humidifiers, whirlpool spas and other water-containing devices. *L. pneumophila* can survive in air-conditioned systems and has caused fatal casualties in hotels and hospitals (Wright et al. 1989). Biofilms including *Legionella* can be found in poorly maintained buildings and water systems representing a risk for public health (WHO 2005).

Burkholderia pseudomallei (also known as *Pseudomonas pseudomallei*) causes melioidosis, a disease affecting humans mainly in Southeast Asia and Australia. This microorganism can be very persistent in the environment (Stone 2007). Humans can be infected through a break in the skin or through the inhalation of aerosolised *B. pseudomallei* cells. *B. pseudomallei* can be transmitted from infected rodents to humans through biofilms contaminating soil and water causing persisting chronic disease. The mean incubation period for acute melioidosis is 9 days (with a range between 1 and 21 days). Human infection can present a wide range of symptoms; however, the bacteria can hide in the body and some patients may remain symptom free for a very long time (Stone 2007). However, bacteria might detach from biofilms and cause acute infections and bacteraemias. In fact, this seemed to be the cause when hundreds of people stressed by seasonal starvation died of acute melioidosis in northeast Thailand (Vorachit et al. 1995). Glanders is primarily a disease of animals such as horses, mules and donkeys and occasionally cats, dogs and goats and is caused by *Burkholderia mallei*. Glanders disease and melioidosis can cause similar symptoms in humans. These microorganisms are being studied at laboratories worldwide due to their potential use in biological warfare (Wheelis 1998). However, Glanders is rare in humans; it is sporadic and usually an occupational disease affecting people in frequent contact with infected animals such as animal caretakers, abattoir workers, veterinarians and laboratory personnel (Al-Ani and Roberson 2007). In 2000, a human case of Glanders disease was reported in a laboratory worker in the USA (Srinivasan et al. 2001). *B. mallei* is usually sensitive to most common antibiotics (tetracyclines, novobiocin, ciprofloxacin, gentamicin, streptomycin and the sulphonamides) although resistance to chloramphenicol has been reported.

Immune compromised individuals and people with medical devices are more at risk of suffering infections from biofilm formation. However, host defences are usually not effective against biofilms (Khoury et al. 1992). In fact, cells involved in immune defence mechanisms can actually aid in the formation and maintenance of biofilms when emigrating to the injured body area (Walker et al. 2005). *P. aeruginosa* biofilm infection in children suffering from cystic fibrosis has shown to increase mortality rates (Ghannoum and O'Toole 2004).

Some bacteria such as *S. aureus* might even be able to use the immune reaction as a virulence mechanism (Gresham et al. 2000). Staphylococci bacteria growing in biofilms have been associated with resistant infections in humans (Vuong and Otto 2002). In fact, in vitro experiments have shown that bacteria present in biofilms can be 10–100 times more resistant to treatments in comparison with the same strain free floating bacteria (Amorena et al. 1999; Olson et al. 2002). Observational studies have been employed to investigate the role of biofilms in causing infections by studying biofilms extracted from infected tissues or contaminated materials recovered from patients. However, in many cases of biofilm chronic infections, it was very difficult to culture bacteria or to recover bacteria from biofilms by using traditional microbiological methods. Therefore, diagnosis and treatment of these infections may prove difficult (Costerton et al. 2003; Lynch and Robertson 2008).

4 Biofilms in Veterinary Medicine and Zoonotic Infections

Microorganisms of veterinary and medical importance are frequently found in biofilms. In animals, biofilm infections might be caused by environmental and even commensal microorganisms such as *S. aureus*. *S. aureus* has been reported to be a concern in postoperative wound biofilm infections (Galuppo et al. 1999) and mastitis (Melchior et al. 2006a, b). In some cases, the same microorganisms can be responsible for biofilm infections in animals and humans. Such bacteria have included *Acinetobacter baumannii* which have been reported to be responsible for wound infections in humans, dogs, cats and horses (Boerlin et al. 2001; Tomaras et al. 2003). *A. baumannii* has also been isolated and the cause of catheter-related infection in horses (Vaneechoutte et al. 2000). *Actinobacillus equuli* is a Gram-negative bacterium commensal of equine oral cavity and upper respiratory tract found to be responsible for sleepy foal disease (Rycroft and Garside 2000) and postoperative wound infections in horses resistant to antibiotics (Smith and Ross 2002). *Actinobacillus lignieresii*, *A. equuli* and *Actinobacillus suis* can be present in the oropharyngeal flora of cattle, horses and pigs, respectively, and therefore may cause bite wound infections in humans (Weyant et al. 1996). In fact, *A. equuli* was isolated from a 53-year-old butcher affected with septicaemia and presented at hospital suffering acute septic shock (Ashhurst-Smith et al. 1998).

Aeromonas hydrophila is a Gram-negative organism causing septicaemia and pneumonia in humans, pigs, cattle and horses (Lallier and Higgins 1988; Zong et al. 2002). Pasteurellosis is considered one of the most important zoonosis; *Pasteurella haemolytica* has been involved in human infections (Takeda et al. 2003). Pasteurella can cause infections in animals. Aspiration of *P. haemolytica* biofilms by feedlot cattle results in severe respiratory infections that might be lethal in 2% of the animals (Morck et al. 1987; Morck et al. 1990).

Aspergillosis is a term used to define a range of infections caused by the fungus of the *Aspergillus* spp. that mainly affects mainly birds but also other animal species and man. It is considered a zoonosis although transmission to humans usually occurs through contaminated soil or organic material in the environment.

Aspergillosis infections in humans can have serious consequences especially in inmunocompromised patients (Seidler et al. 2008). Aspergillus biofilms has been reported on cardiac devices resulting in endocarditis in numerous immunocompromised patients (Lynch and Robertson 2008).

Candida spp. has been recovered from humans, animals and the environment (Beran and Steele 1994) and zoonotic concerns have been raised (Marshall 2003). *Candida* spp. have been responsible for hospital-acquired infections and associated with catheter-related infections (Hawser and Islam 1999; Chandra and Ghannoum 2004). *C. albicans*, in particular, has also been associated with biofilm formation in medical devices and high mortality (Kojic and Darouiche 2004). Kuhn et al. (2002) suggested that *C. albicans* produces quantitatively more biofilm than other *Candida* species. *Candida* forms complex three-dimensional biofilms offering optimal spatial construction for the establishment of microniches throughout the biofilm. *Candida* biofilms present multiple resistance mechanisms making the organism resistant to a range of antifungal products (Douglas 2003; Ramage et al. 2005). Candida and Aspergillus are emerging as dangerous pathogens (patient survival rate as low as 50%) although their prevalence of implant infections has been shown to be only around 8% (Anderson and Marchant 2000).

M. avium and *M. intracellulare* (*M. avium-intracellulare* complex, MAIC) are slow-growing, atypical mycobacteria, ubiquitous in the environment. *M. avium* produces granulomatous lesions that are indistinguishable from the tubercular lesions produced by *M. tuberculosis* and *M. bovis* (Greene and Gunn-Moore 1990; Zeiss et al. 1994). *M. avium* and *M. intracellulare* are potentially zoonotic and have been found growing in biofilms in drinking water systems (Falkinham et al. 2001). Therefore, samples should be collected from pipe biofilms when testing water systems. *M. avium* adheres to surfaces, grows at low levels of oxygen and is resistant to heavy metals forming biofilms on metallic surfaces (Falkinham et al. 2004). *Mycobacterium ulcerans* causes necrotising skin lesions in humans, a disease known as Buruli ulcer (BU) considered an emerging disease (Walsh et al. 2009). *M. ulcerans* has been found in biofilms attached to aquatic plants (Marsollier et al. 2002). It has been suggested that insect bites may play a role in transmitting *M. ulcerans* (Marsollier et al. 2005). *Mycobacterium marinum* and *M. ulcerans* sequences for the 16S rRNA gene are highly similar (Portaels et al. 1996). *M. marinum* causes disease in fish usually called "fish tank granuloma" (Walsh et al. 2009). *M. marinum* can infect humans through skin lesions and produce superficial and self-limiting lesions in hands, forearms, elbows and knees (Steitz et al. 1997). Demangel et al. (2009) propose that *M. ulcerans* originated from *M. marinum* by transfer of a virulence plasmid carrying genes for mycolactone production. Recently, some closely related mycolactone-producing mycobacteria have been discovered causing public health concern (Chemlal et al. 2002; Gauthier and Rhodes 2009). *Mycobacterium haemophilum* causes bone, joint, skin and pulmonary infections in immunocompromised adult humans and lymphadenitis in children (Kiehn and White 1994; Samra et al. 1999). *M. haemophilum* infections in animals have been reported including a case of pulmonary infection in a royal python (Hernandez-Divers and Shearer 2002) and zebrafish infections (Kent et al.

2004; Whipps et al. 2007). *M. haemophilum* was isolated from infected fish at a research facility (University of Georgia) where biofilm samples obtained were also positive for *M. haemophilum* (Whipps et al. 2007). The possibility of the biofilm acting as a reservoir for infection was considered. *M. haemophilum* has been isolated from biofilms growing in water distribution systems (Falkinham et al. 2001). Overall, aquatic Mycobacteria can affect many species of fish and represent a potential zoonotic risk to humans. Water and associated biofilms have been recognised as natural habitats for *Mycobacterium* spp. (Pedley et al. 2004). Therefore, the implementation of programmes for the prevention, reduction or elimination of these pathogens (living free or as part of biofilms) in aquatic facilities is paramount. *M. avium paratuberculosis* (MAP) causes Johne's disease in animals including primates; however, its role in Chrones disease in humans is still controversial. The identification of MAP in blood extracted from patients suffering from Chrones disease in a case-control study suggested that MAP has a role in the pathogenesis of this disease although more research seems necessary to clarify its role (Naser et al. 2004). Bull et al. (2003) observed that the identification of MAP in the majority of tested individuals with chronic intestinal inflammation suggested causality and the fact that MAP causes chronic inflammation of the intestine of a broad range of animal species made this organism the ideal suspect for Chrones disease. Animals can be infected with MAP for years without showing clinical symptoms and shed MAP in their milk and in faeces contaminating the environment. MAP may survive in the environment for a very long time and possible amplify within environmental protozoa (Grant et al. 2002; Ayele et al. 2005). In this way, MAP may be transmitted to humans through drinking water or aerosols (Hermon-Taylor et al. 2000; Fazakerley et al. 2001; Percival et al. 2000; Bull et al. 2003). Aerosol droplets can concentrate bacteria and in that way mycobacteria can spread via aerosols (Pickup et al. 1999; Beard et al. 2001; Whittington et al. 2004). *M. avium* subsp. *paratuberculosis* infection in livestock is endemic in areas of Wales; furthermore, this microorganism has been isolated from the river Taff in Wales. Epidemiological data suggested that the inhalation of *M. avium* subsp. *paratuberculosis* from the river Taff might be responsible for the clustering pattern of Crohn's disease observed in Cardiff (Pickup et al. 2005). *M. avium* subspecies has been isolated from environments worldwide (Falkinham 1996); they are known to persist in biofilms in drinking water distribution systems (Falkinham et al. 2001; Torvinen et al. 2007). It has also been suggested that potable hot water systems may contain *M. avium* concentrations greater than expected (du Moulin et al. 1988). In fact, *M. avium* may survive traditional water disinfection treatments and can be more resistant to chlorine when growing in biofilms (Taylor et al. 2000; Steed and Falkinham 2006). MAP ability to form biofilms seems to vary between isolates and under different conditions (Carter et al. 2003; Johansen et al. 2009). MAP has also been found to survive in biofilms in food-producing animals watering systems. Therefore, the control of biofilms on farms should be included in any farm management programmes (Cook et al. 2010).

Commensal and biofilm-forming environmental microorganisms such as *Listeria monocytogenes*, *Campylobacter*, *E. coli*, *Salmonella*, *S. aureus*, *S. epidermidi*,

Pasteurella multocida, *P. haemolytica*, *Streptococcus suis*, *S. agalactiae*, *Actinobacillus pleuropneumoniae* and Mycoplasmas cause more than half of the infections in mildly compromised individuals (Costerton et al. 1999; Donlan 2001; Prakash et al. 2003).

Treatment of biofilm infections can prove difficult. Antibiotics can be effective against planktonic cells released from biofilms but cannot eliminate biofilms (Marrie et al. 1982). This fact could explain recurring symptoms, after cycles of antibiotic treatments, until the sessile bacterial population is removed or eliminated (Costerton et al. 1985). Studies of polymicrobic biofilms revealed symbiotic interactions and enhanced transfer of antimicrobial resistance genes (Hansen et al. 2007; Seidler et al. 2008).

Some of the major zoonotic microorganisms that are known to be avid biofilm formers and therefore of public health importance are presented below.

4.1 Listeria monocytogenes

L. monocytogenes is a Gram-positive pathogen that can cause severe infections among pregnant women and immunocompromised patients (Farber and Peterkin 1991). *L. monocytogenes* can form biofilms on a variety of surfaces, such as stainless steel and rubber (Blackman and Frank 1996; Meyer 2003). Borucki et al. (2003) demonstrated that *L. monocytogenes* are able to form biofilms in static conditions. Rieu et al. (2008) using laser-scanning confocal microscopy (LSCM) compared *L. monocytogenes* biofilms grown under two different environmental conditions (static growth media and flow conditions). They reported that *L. monocytogenes* formed biofilms under flow conditions which appeared to be more organised with rounded microcolonies surrounded by a network of knitted chains compared to static conditions. Furthermore, biofilms formed by *L. monocytogenes* under different conditions depended on different patterns of gene expression (Hefford et al. 2005; Rieu et al. 2008).

4.2 Helicobacter

Helicobacter species have been isolated from the stomachs of several animals such as cats, dogs, pigs, birds, mice, chickens, ferrets and monkeys (Mirkin 2009). *Helicobacter* spp. DNA has been detected in the oral cavity of dogs representing a risk factor for *Helicobacter* spp. infections in humans (Recordati et al. 2007). Vertical faecal-oral transmission of *H. pylori* infection of Mongolian gerbil pups from an infected mother has been demonstrated (Oshio et al. 2009).

Helicobacter spp. cause stomach problems in humans and may also cause other conditions such as liver disease, clotting, heart attacks and certain skin conditions (Meining et al. 1998). A large number of different species of *Helicobacter* have been isolated from animals with transmission likely to human. These have included

H. helmannii (Mention et al. 1999), *H. rappini*, *H. felis*, *H. cinaedi*, *Helicobacter* sp. strain Mainz (Vandamme et al. 2000), *H. fennelliae*, *H. pullorum*, *H. hills*, *H. hepaticus*, *H. billis* and *H. canis* (Ferenci 2000). One of the most researched *Helicobacter* spp. in both animals and humans is *Helicobacter pylori* a bacterium well known to form biofilms both within in vivo and within in vitro conditions.

H. pylori is a Gram-negative microaerobic rod of public health importance because it causes gastric ulcers, gastritis and contributes to the development of gastric cancer (Amieva & El-Omar 2008; Kandulski et al. 2008). In fact, *H. pylori* has been considered as carcinogenic since 1994 and can be found all over the world but seems more prevalent (90% prevalence) in developing countries (van Duynhoven and de Jonge 2001). Human infection by *H. pylori* can occur through multiple pathways.

Cole et al. (2004) observed that the presence of mucin increased the number of planktonic *H. pylori* and suggested that *H. pylori* biofilm formation might be more difficult in the mucus-rich stomach of humans and animals. However, Carron et al. (2006) photographically documented the presence of mature dense *H. pylori* biofilms on samples of human gastric mucosa obtained from patients undergoing esophagogastroduodenoscopies. Animals constitute a reservoir for *H. pylori* and the possibility of transmission through the food chain has been considered. In fact, *H. pylori* might survive in foods in a viable but non-culturable form (VBNC) and therefore difficult to isolate from foods leading to underestimation of its prevalence (Dimola and Caruso 1999; van Duynhoven and de Jonge 2001). Quaglia et al. (2008) have detected *H. pylori* DNA in raw goat, sheep and cow milk by using a Nested Polymerase Chain Reaction (Nested-PCR) assay. In addition these flies have been shown to potentially transmit *H. pylori* mechanically through excreta. It has been suggested therefore that flies might contaminate food with *H. pylori* (Grübel et al. 1997).

H. pylori can survive up to 1 year in fresh water as viable coccoid forms that are non-culturable but represent a public health hazard (Shahamat et al. 1989; Adams et al. 2003; Konishi et al. 2007). It has been demonstrated that the consumption of environmental water or dirty water is a risk factor for human infections with *H. pylori* (Goodman et al. 1996; Ahmed et al. 2007). In fact, *H. pylori* infections in humans seem to be correlated with biofilm formation and access to contaminated water (Percival and Thomas 2009). Biofilms can provide a protective environment for *H. pylori* to survive in water and even to reach concentrations that could cause harm to humans (Gião et al. 2008). They have also been documented to survive with amoeba and as such its transmission could be similar to that of *M. avium*.

4.3 Campylobacter

Campylobacter spp. are Gram negative, microaerophilic but reported to be unable to grow in air and outside an animal or human host (Park 2002). However, *Campylobacter* are very robust and can survive environmental stresses. In fact

Campylobacter sp. are considered to be the main pathogen causing human gastrointestinal infections in developed countries (Kalmokoff et al. 2006; Murphy et al. 2006). The majority of human *Campylobacter* infections are caused by *C. jejuni* and *C. coli*, with *C. jejuni* being the more common species isolated from human cases of campylobacteriosis (Skirrow 1994).

Trachoo et al. (2002) suggested that *C. jejuni* might form VBNC forms within biofilms isolated from chicken houses. As with all microorganisms, biofilm formation protects *Campylobacter* organisms from the environment and therefore enhances their survival. Trachoo and Frank (2002) demonstrated that the effectiveness of sanitizers against *C. jejuni* was decreased by the presence of biofilms. However, *C. jejuni* biofilms are inactivated by chlorine (Dykes et al. 2003).

Campylobacter jejuni can form different types of biofilms when surviving in different environmental conditions. This variability in biofilm formation could partly explain the survival of *Campylobacter* in the environment and the high incidence of *Campylobacter*-related infections (Gaynor et al. 2005; Joshua et al. 2006). *C. jejuni* biofilms have been isolated from water and in fact, *C. jejuni* can form biofilms in a variety of materials used in animal production watering systems (Reeser et al. 2007).

Biofilms with low levels of oxygen will promote the growth of microaerophilic bacteria such as *Campylobacter*. Furthermore, the presence of organic material will also assist on the survival and growth of *Campylobacter* in environmental biofilms (Humphrey et al. 1995; Kusamaningrum et al. 2003).

4.4 Salmonella

Salmonella causes abdominal pain, diarrhoea and fever. Most individuals infected with *Salmonella* recover without treatment and require oral fluid replacement in 4–7 days. However, some patients need to be hospitalised and treated with antibiotics when suffering from persistent diarrhoea and sometimes bacteraemia or septicaemia can develop. Salmonellosis can be fatal in immunocompromised patients (CDC 2008).

Salmonella species are known to colonise the intestines of mammals, birds and reptiles and when shed into the environment they have been reported to survive for long periods in water, soil and foods. Most human infections with *Salmonella* in developed countries are related to the consumption of contaminated foods (Angulo et al. 1999). *Salmonella* serotype *typhimurium* and *Salmonella* serotype *enteritidis* seem to be the most common serotypes involved in human infections (Rodrigue et al. 1990; Rubino et al. 1998; Herikstad et al. 2002; Galanis et al. 2006).

Salmonella enteritidis are known to form biofilms under a number of conditions and on different materials (Austin et al. 1998; Bradshaw and Marsh 1999). Interestingly *Salmonella* have been reported to form biofilms on gallstones which is known to enhance their proliferation in these organs. The use of antibiotics has been shown not to eradicate *Salmonella* from infected gallstones (Lai et al. 1992) and surgery is considered an unfeasible solution for individuals with *Salmonella* infected gallstones in developing countries (Crawford et al. 2008).

It has been demonstrated that *S. enterica* serovar typhimurium can compete and outgrow even displace *E. coli* when forming biofilms on HEp-2 epithelial cells (Esteves et al. 2005). Research directed to the study of "biofilm genes" has been conducted with a number of potentially pathogenic Gram-negative bacteria including *Salmonella* (Solano et al. 2002), *E. coli* (Pratt and Kolter 1998) and *Vibrio cholera* (Watnick and Kolter 1999).

4.5 Shigella

Shigella is a Gram-negative, rod-shaped bacteria usually transmitted via faecal contamination through humans, food, water and flies and therefore associated with poor hygienic conditions (Troller 1993; ICMSF 1996). Poor staff hygiene in food processing establishments may lead to food contamination. Gunduz and Tuncel (2006) have isolated *Shigella* from a feeding unit and an aging tank in an ice-cream processing plant which were shown to pose a risk to public health.

4.6 Giardia *and* Cryptosporidium

Giardia and *Cryptosporidium* are parasitic protozoa causing disease in humans and animals frequently transmitted through contaminated water (Brandonisio 2006; Cheng et al. 2009). Wildlife and livestock can contribute to the maintenance of these parasites in the environment (Paziewska et al. 2007) although in a study conducted by Heitman et al. (2002) genetic differences that may indicate host specificity were discovered by genetic characterisation of *Cryptosporidium* isolates collected from humans, dogs, cats, pigs, steer, calfs and beaver hosts.

Cryptosporidium has been responsible for disease outbreaks related to contaminated water in developed countries. The interaction of *Cryptosporidium* oocysts with biofilms present in water distribution systems has been reported but further investigation is required for this (Angles et al. 2007). Biofilms can provide a protective environment for the accumulation of *C. parvum* oocysts assisting on the propagation of this pathogen and the contamination of the environment and water systems (Searcy et al. 2006; Wolyniak et al. 2009). Epidemiological studies revealed that water contamination with animal faeces was the main suspected source of an outbreak of *Cryptosporidium* in Lancashire (UK) in 2000. Furthermore, the persistence of *Cryptosporidium* oocysts after switching to another water source indicated the possibility of oocysts being protected by biofilm formation (Howe et al. 2002). Ecological studies, epidemiological data and risk assessment form the basis for the implementation of effective water treatment to protect public health (Szewzyk et al. 2000). *Giardia* are known to attach to the intestinal epithelial surface forming part of biofilms. Pathogens need to attach to the intestinal lumen and overcome the forces produced by intestinal mucus to remove not properly attached microorganisms (Costerton et al. 2003).

4.7 E. coli

Reisner et al. (2006) conducted an extensive analysis of *E. coli* biofilm formation and reported very different responses to various environmental conditions and great variation between diverse *E. coli* isolates to form biofilms in vitro. In fact, an association between pathogenic *E. coli* strains and increased biofilm formation capabilities was not observed like it has been reported in other microorganisms such as *Enterococcus faecalis* (Mohamed et al. 2004).

4.8 E. coli *0157*

E. coli 0157 infections in humans can be severe in immunocompromised patients and children. Animals are known to act as reservoirs for *E. coli* 0157 contaminating the environment. Humans can become ill through direct contact with carrier animals or contaminated food or water. Investigations into an outbreak of *E. coli 0157* in people visiting a farm in Pennsylvania revealed that 13% of the farmed cattle were carrying *E. coli 0157* and these isolates had the same distinct molecular pattern that was found in isolates from 51 patients tested in the case-control study (Crump et al. 2002). The possibility of *E. coli 0157* requiring another microorganism such as *Pseudomonas aeuroginosa* to assist in its ability to form biofilms has been investigated (Klayman et al. 2009).

4.9 Yersinia

Yersinia can cause illness in humans through the consumption of contaminated raw or undercooked meat, seafood, tofu and contaminated water. *Y. pseudotuberculosis* can cause lymphadenitis and septicaemia and *Y. enterocolitica* causes gastrointestinal syndromes (Naktin and Beavis 1999; Gerald 2009). *Yersinia entercolitica* can form biofilms and its flagella play an important role in biofilm formation. *Y. enterocolitica* and *Y. pseudotuberculosis* are motile but *Y. pestis* is non-motile (Kim et al. 2008).

Yersinia pestis is responsible for plague syndromes in humans. *Y. pestis* infects rodents and is transmitted to humans by fleas. *Y. pestis* produces dense biofilms on the hydrophobic surface of spines inside the proventriculus in the flea's foregut (Jarrett et al. 2004). However, it has been suggested that *Y. pestis* strains that are unable to form biofilms can also cause plague (Eisen et al. 2007).

4.10 Clostridium botulinum

C. botulinum is a Gram-positive, anaerobic, spore-forming rod responsible for botulism. This microorganism produces a potent neurotoxin which causes flaccid paralysis. Different *Clostridium* species are known to produce seven types of toxins (A–G). *Clostridium* produce spores and these are known to be resistant to antimicrobials and aids survival in the environment and in foods (Hirsh and Birbenstein 2004). Foodborne botulism is a severe condition (high mortality rate if not treated properly) caused by ingestion of contaminated foods. The botulinum toxin is only destroyed at high temperatures (80°C for 10 or more minutes) and therefore represents a highly significant public health concern. *C. botulinum* type C forms biofilms and survives in biofilms within grass; this has been suggested to be the cause of equine dysautonomia (Grass Sickness) (Hirsh and Birbenstein 2004).

Clostridium spores have been implicated in food poisoning cases. The hydrophobicity of *Clostridium* spores plays a key role in their adhesion to surfaces, biofilm formation and increased resistance to sterilisation treatments (Wiencek et al. 1990). Hydrophobic interactions have been associated with the adhesion of bacteria to surfaces and biofilm formation (Rosenberg and Kjelleberg 1986).

4.11 Clostridium perfringens

C. perfringens is an ubiquitous bacteria, Gram-positive, anaerobic that can be found in the environment, in animals (Narayan 1982; Songer 1997), as part of the microbiota in human intestine (Carman et al. 2008) and has the ability to form biofilms enabling it to adapt to different environments (Varga et al. 2008). *C. perfringens* biofilms have been shown to provide recalcitrance to antibiotics and may contribute to antibiotic-associated diarrhoea (Asha et al. 2006; Varga et al. 2008). Some strains of *C. perfringens* such as type C, D and E can colonise the guts of mammals and cause enteric infections in livestock (Songer 1997) and can also form biofilms (Varga et al. 2008).

C. perfringens biofilms exhibited a dense extracellular matrix containing carbohydrates and type IV pilin proteins (Varga et al. 2008). These authors observed that biofilm formation increased in absence of glucose or carbohydrates which could represent a survival mechanism for *C. perfringens* in low nutrient or starvation conditions. However, other survival mechanisms have been observed in stressful conditions such as endospore production and enhanced motility (Varga et al. 2004; Mendez et al. 2008).

4.12 Streptococcus *sp.*

Some members of the genus *Streptococcus* are part of the normal microflora in the human body but sometimes they might produce opportunistic infections such as dental caries. However, exogenous pathogens of the genus Streptococcus can cause a wide range of infections from mild conditions to life-threatening illnesses (Cvitkovitch et al. 2003).

Streptococcus-related infections are considered zoonotic with some *Streptococcus* sp. more commonly associated with human infections than others (Acha and Szyfres 2003). The most frequently detected Streptococci in human infections have included *Streptococcus equi* subsp. zooepidemicus and *S. equi* subsp. *equi* (Krauss et al. 2003). Other Streptococci of human significance have included *S. pyogenes* which are known to produce several illnesses including a toxic shock-like syndrome (Demers et al. 1993). *S. pyogenes* are known to form biofilms which assits in their tolerance to antimicrobial treatments (Baldassarri et al. 2006).

Streptococcus pneumoniae and *Enterococcus* (*Streptococcus*) *faecalis* have been shown to have increased resistance to antibiotics and are known to cause serious problems in immunocompromised and hospitalised people (Appelbaum 1992). Raw milk and eggs are considered a source for *Streptococcus* infections in humans (Gerald 2009).

4.13 Streptococcus suis

S. suis type 2 can be isolated from carrier adult pigs' upper respiratory tract and tonsils and may cause disease in young pigs and humans. *S. suis* is usually isolated from infected pig carcases but it can also be found in the faeces of infected herds (Huang et al. 2005). *S. suis* type 2 has been isolated from human patients (associated to the pig industry) which suggests an occupational zoonosis route. However, a few cases have been detected in humans with no known contact with the swine industry (Zanen and Engel 1975; Clifton-Hadley 1983; Sriskandan and Slater 2006). *S. suis* human infections have been reported in New Zealand, Australia, Argentina, several Asian and European countries and Canada (Lun et al. 2007). In fact, *S. suis* is one of the most common causes of bacterial meningitis in Hong Kong (Hui et al. 2005). The possibility of the emergence of a new more virulent *S. suis* strain has been raised based on epidemiological studies of outbreaks caused by this particular strain in China (Sriskandan and Slater 2006).

S. suis can form biofilms as reported by Grenier et al. (2009). Grenier et al. (2009) characterised a biofilm formed by *S. suis* type 2, isolated from a pig with meningitis, and observed that *S. suis* strain 95–8242 produced a dense biofilm when sucrose, glucose or fructose was used as the primary nutrient source. Within this study *S. suis* 95–8242 was shown to be more resistant to penicillin G and ampicillin when grown as a biofilm compared to its planktonic counterpart.

Fittipaldi et al. (2007) have used selective capture of transcribed sequences (SCOTS) and identified 28 genes preferentially expressed by *S. suis* when interacting with cells of the porcine brain microvascular endothelial cells, some of these genes were considered potential new virulence factors.

4.14 Vibrio

V. cholera can form biofilms on crustacean shells, aquatic organisms and aquatic plants to reach an infective dose. Consequently, these biofilms have been reported to act as reservoir for *V. cholera* in a non-culturable coccoid form (Hall-Stoodley and Stoodley 2005; Alam et al. 2007). In addition to *V. cholera*, *Vibrio parahaemolyticus* has been isolated from seafood and associated with foodborne illness (D'Mello 2003).

4.15 Aeromonas

Aeromonas sp. are Gram-negative and rod shaped. They are motile aquatic bacteria considered important pathogens in reptiles, amphibians and fish. In particular, they are known to be a major problem in fish farming. Fish are thought to act as a reservoir of *A. hydrophila* possibly leading to infection in mammals (Lallier and Higgins 1988; Lynch et al. 2002).

In humans, *Aeromonas* sp. are known to cause gastroenteritis (from mild to cholera-like symptoms) and other infections such as endocarditis, septicaemia, haemolytic uraemic syndrome, peritonitis, respiratory infections, myonecrosis, osteomyelitis, ocular infections and meningitis. *A. hydrophila* and *E. coli* have been reported to grow well in biofilms detected in drinking water systems (Walker et al. 2000). As well as water *Aeromonas* sp. have also been isolated from foods (seafood, raw milk, meat and vegetables) as they are well known to form biofilms utilising QS mechanisms. By using an N-acylhomoserine lactone (AHL)-dependent QS system (bacterial communication system) the ability of Aeromonas to form biofilm has been shown to be significantly enhanced.

5 Control of Food-Borne and Water-Borne Biofilms of Zoonotic Importance

Biofilms are known to provide a protective environment for pathogenic bacteria, parasites and viruses aiding their dissemination in water systems leading to disease in animals and humans (Howe et al. 2002; Helmi et al. 2008). Microbial cells in biofilms can easily detach voluntary or involuntary from biofilms to aid their dispersal which represents a very important survival strategy (Sauer et al. 2002). Consequently, bacterial cells which reside in the planktonic phase are thought to be in a phase of moving from one surface to another (Parsek and Fuqua 2004). It is plausible to suggest that these dispersal strategies are therefore the cause of food and water contamination and therefore animal and human infection/disease (Zottola and Sasahara 1994; Piriou et al. 1997).

A dynamic equilibrium between attachment and detachment processes occurs in wastewater biofilms which are composed of bacteria, viruses and parasites. Therefore, the microbial concentrations within a biofilm may be regulated purposely to enhance the survival of the biofilm community (Skraber et al. 2007). In fact biofilms have been compared to tissues of higher evolutioned organisms. If pathogenic bacteria detach from biofilms this is potentially dangerous particularly if the infective dose reaches immunocompromised hosts (Storey et al. 2004). Understanding better the particular phases of biofilm formation, proliferation and detachment could prove very useful for investigating the control of biofilms (Ghannoum and O'Toole 2004; Hall-Stoodley et al. 2004; Sawhney and Berry 2009).

As mentioned previously, biofilm formation causes public health problems in food processing and water systems (Lelieveld 2005; Alakomi et al. 2002). However, there are positive applications for biofilms, in particular, they may be used for water treatment (Bryers 2000; Wuertz et al. 2003). However, water-borne pathogens can form part of biofilms and exchange genetic material with other microorganisms present. The use of markers to detect biofilms in water has been suggested (Committee on Indicators for Waterborne Pathogens 2004). The use of chlorine dioxide for biofilm control and general disinfection in water distribution systems has been proposed (O'Leary et al. 2002).

The formation of biofilms has been observed in various food processing industries such as raw and cooked/fermented meats, raw and smoked fish, seafood, chicken and turkey, milk and dairy products and yeast (Sharma and Anand 2002; Bagge-Ravn et al. 2003; Carpentier and Chassaing 2004; O'Brien et al. 2004).

Pathogenic organisms such as *Listeria* and *Shigella* can form biofilms in food processing establishments (Gunduz and Tuncel 2006). These authors isolated a wide range of microorganisms such as *Proteus*, *Enterobacter*, *Shigella*, *Escherichia*, *Edwardsiella*, *Aeromonas*, *Pseudomonas*, *Staphyloccus*, *Bacillus*, *Listeria* and others from an ice-cream processing plant. These authors also observed that the microbiological burden decreased after the cleaning of food producing areas. However, they neglected to remove biofilms and *Shigella* was still present in one of the tanks after cleaning and disinfection, proving this constitutes a significant public health issue. In this same study, *L. monocytogenes* was found on the conveyor belt of the packaging machine indicating the possibility of food contamination. Again despite cleaning in place because of the existence of recalcitrant biofilm the effectiveness of these cleaning agents was clearly proved.

Important public health pathogens such as *S. typhimurium*, *C. jejuni*, *E. coli O157:H7*, *L. monocytogenes* and *Yersinia enterocolitica* have been found in biofilms causing severe problems in the food industry (Griffiths 2003). In fact, some pathogens of zoonotic importance such as *L. monocytogenes* and *S. aureus* are difficult to remove from biofilms (Lelieveld 2005). *L. monocytogenes* has been isolated from dairy and meat processing plants (Wong 1998). *Listeria* has been found to be particularly resistant to disinfectants and temperatures (Wirtanen and Salo 2004). In fact, *Listeria* can survive for up to 10 years in food processing establishments due to the presence of persistent strains with enhanced attachment properties (Lundén et al. 2002). A dramatic and lethal foodborne epidemic caused by *Listeria* forming biofilms in a food plant and contaminating food products occurred in the USA in 1998. In 2000, USDA required *Listeria* testing in food processing establishments (Drexler 2002). Bjerklie (2003) tested 41 meat processing establishments and found *L. monocytogenes* on or in 39% of the floors tested, 29% of floor drains, 34% of cleaning equipment, 24% of wash areas and 20% of food-contact surfaces. *Listeria* and other pathogens of public health importance forming part of biofilms can survive sanitation procedures in food premises (Arnold 1998). In fact, resistance to chlorine and heat treatments increases when microorganisms such as *Listeria* and *Salmonella* attach to surfaces (Wirtanen and Mattila-Sandholm 1992; Dhir and Dodd 1995). *C. jejuni*, *E. coli O157:H7*, *L. monocytogenes* and *S. typhimurium* forming biofilms were more resistant to trisodium phosphate treatment (Somers et al. 1994). Some sanitizers have been shown to be ineffective against bacteria in biofilms formed during milk processing (Mosteller and Bishop 1993). Biofilms may also form in pasteurisation equipment contaminating pasteurised product (Langeveld et al. 1995).

The presence of biofilms on materials and equipment used in the food industry poses a risk to human health. In fact, biofilms may recontaminate previously treated and/or cooked food products (Lelieveld 2005). Food contamination may result in human foodborne illness and lowers product shelf-life producing economic losses (Criado et al. 1994).

The presence of biofilms in crates used to transport poultry has been documented. The biofilms in crates are difficult to remove by pressure washing, even from well-designed smooth crates (Mead 2005). It seems very important to understand biofilm formation conditions and properties in order to successfully implement sanitation strategies in the poultry industry and protect public health. Poultry production can be an intensive and heavily contaminated operation due to the high number of animals being processed in commercial processing establishments. Furthermore, the use of water at several stages of the process will facilitate the formation of biofilms. Therefore, biofilms can be very prevalent and efforts should be directed to prevent and/or eliminate biofilms in poultry processing. Biofilm control and elimination strategies should be considered as part of HACCP implementation and food safety systems in the food industry. Several factors such as plant and equipment design, maintenance, cleaning and disinfection products and techniques will influence the formation and/or elimination of biofilms in food establishments. Control measures to prevent and eliminate biofilms in the food industry involve adequate design, good manufacturing practices including effective cleanliness of surfaces and proper staff training (Husu et al. 1990; Eklund et al. 1995; Autio et al. 1999; Miettinen et al. 1999, 2001; Lyytikäinen et al. 2000; Wirtanen 2002; Aarnisalo et al. 2003; Lundén 2004; Miettinen and Wirtanen 2006; Wirtanen and Salo 2004).

Some surfaces are preferred by microorganisms to form biofilms. The identification of factors that make different surfaces resistant or susceptible to biofilm formation seems crucial to control biofilms in the food industry. Some types of rubber materials seem to inhibit microbial attachment. Although the presence of some elements such as zinc and sulphur can partially explain this fact, there are other factors associated with rubber that contribute to the inhibitory effect (Arnold and Silvers 2000). However, proper maintenance of rubber material will also influence the effect on biofilm formation (Arnold and Bailey 2000). The effects of corrosion on metallic surfaces and biofilm formation have also been investigated. Attachment of bacteria to metallic surfaces is facilitated by corrosive treatments (Arnold and Suzuki 2003). Therefore, proper maintenance of surfaces and equipment used in food processing and the use of materials resistant to corrosion seem important measures to reduce and control biofilm formation. Effective sanitation strategies are crucial for the control of biofilms in the food industry. Some materials such as Teflon seem easier to effectively clear of biofilms than others like stainless steel (Marriott 1999). Thorough cleaning before disinfection of surfaces and equipment is crucial to removed attached microorganisms and for the effectiveness of sanitation procedures in the food industry. The use of mechanical energy has been shown to be most effective against biofilms (Gibson et al. 1999). Food producers should use mechanical equipments which do not create aerosols. The use of non-aerosol generating detergents and sanitizers will be more effective when used together with mechanical methods (Meyer 2003). The use of sanitizers such as acidic quaternary ammonia, chlorine dioxide and peracetic acid has been found more effective to remove attached microorganisms (Krysinski et al. 1992). The use of antimicrobial coatings on equipment surfaces, such as bacteriocins and silver ions, will help in controlling the formation of biofilms (Kumar and Anand 1998; Meyer 2003). Special treatments such as the use of specific enzymes might be necessary to eliminate pathogens from biofilms in the food industry (Oulahal-Lagsir et al. 2003). However, some decontamination technologies used in or on foods might potentially provide stressful conditions promoting biofilm formation and increasing pathogens' virulence (Samelis and Sofos 2003). It has also been suggested that the use of chemical sanitizers may exert a selective pressure on attached pathogens selecting for resistant strains (Langsrud et al. 2003).

Furthermore, a steady increase in the prevalence of antibiotic-resistant strains isolated from food processing environments has been observed (Teuber et al. 2003; Nayak et al. 2004) and antibiotic-resistant strains also exhibited cross-resistance to other antimicrobial agents, such as sanitizers (Langsrud et al. 2003; Lundén et al. 2003). This phenomenon could be linked to the transfer of genetic material such as plasmids coding for antibiotic resistance between microorganisms present in biofilms. Biofilms might then represent a reservoir for antimicrobial-resistance genes (Watnick and Kolter 2000; Jefferson 2004; Parsek and Fuqua 2004; Weigel et al. 2007). Human infections with antibiotic-resistant pathogens represent a serious public health problem. *Salmonella enterica* serovars are increasingly resistant to commonly administered antibiotics (Boyle et al. 2007). Recommendations have been made to limit the use of antibiotics in farm animals and to adopt non-antimicrobial farm management strategies (Angulo et al. 1999).

More research seems necessary in the control and prevention of biofilm formation including a balance between benefits and risks of using particular decontamination technologies in order to select optimum treatments and improve food safety. In general, decontamination technologies currently used reduce pathogen prevalence on fresh meat (Sofos 2005).

6 Further Research and Biofilm Control Strategies

The use of new technologies for the study of biofilms can provide interesting and useful information for the control of biofilm infections. Pathogens such as *L. pneumophila*, *C. jejuni* and *C. parvum* have been identified in biofilms by using episcopic differential interference contrast (EDIC) microscopy together with epi-fluorescence using gold nanoparticles or fluorophores (EDIC/EF) revealing 3D biofilm structure (Keevil 2003). LSCM is also a powerful technology to study biofilm architecture and 3D structure and used to study *P. aeruginosa* and *S. aureus* biofilms (Klausen et al. 2003; Jefferson et al. 2005; Pamp and Tolker-Nielsen 2007). However, biofilm structure can be diverse in response to environmental conditions (Wimpenny et al. 2000).

New molecular technologies such as the SCOTS can offer advantages for the study of genetic regulation of biofilm formation (Osaki et al. 2002).

Bioengineering has also been successfully used to produce bacteriophages that can induce the production of biofilm-degrading enzymes during infection to act against the biofilm matrix and cells (Lu and Collins 2007). The use of artificially engineered biofilms has been proposed to capture pathogens present in water to monitor the hygienic status of drinking water (Bauman et al. 2009). Biofilm engineering can contribute to the elimination of biofilms and to the control of biofilm infections. Nanotechnologies such as the use of fluoride nanomaterials might present a solution to inhibit biofilm formation (Lellouche et al. 2009). The use of current electric fields and ultrasonic radiation can render biofilms more susceptible to conventional antibiotic treatments (Costerton et al. 1994; Rediske et al. 1998).

The use of microemulsions and nanoemulsions has also been tested to control biofilms. Teixeira et al. (2007) used emulsions to inactivate biofilms formed by *P. aeruginosa*, *Salmonella* spp., *S. aureus*, *E. coli* 0157:H7 (VT-) and *L. monocytogenes*. All biofilms were inhibited by the used emulsions except the one formed by *L. monocytogenes*. Essential oil compounds have been proved successful against biofilms of *L. monocytogenes* and *E. coli* 0157:H7 representing a potential solution for the treatment of biofilms in the food industry (Pérez-Conesa et al. 2006).

The study of biofilms' structure, architecture, cellular spatial arrangement (Davey and O'Toole 2000) and synergistic and antagonistic or inhibitory interactions between biofilm cells will provide new insights into biofilm infections and the development of effective treatments (Reisner et al. 2006). A synergistic effect was observed in mixed biofilms of *P. aeruginosa* and MRSA involved in catheter-associated urinary tract infections (CAUTI). In fact, *P. aeruginosa* was producing more exotoxin A when forming biofilms with MRSA (Goldsworthy 2008).

Adhesion to surfaces is the first necessary step for the production of biofilms (Klemm and Schembri 2004) and therefore actions taken to prevent adhesion of microorganisms could provide interesting solutions for the prevention of biofilms. Several factors play a role in biofilm formation and bacterial adhesion such as environmental conditions, the capacity of microorganisms to adhere to surfaces and also the nature of the surface (Katsikogianni and Missirlis 2004). Bacterial attachment to inert surfaces can be reduced by using aqueous extract of fish muscle proteins (FMPs) to precondition the surface (Bernbom et al. 2006). This type of proteinaceous coating from fish muscle was used against biofilm formation by UTI *E. coli* and *Klebsiella* strains on different materials. A 100-fold reduction in biofilm formation was observed although the effect depended also on other variables such as the growth medium, the particular bacterial strain and the extract (Vejborg and Klemm 2008).

Antibiotic resistance is one of the main global public health concerns (Croft et al. 2007). Antibiotic resistance associated with biofilm infections can be explained by several factors associated with biofilm formation such as the different growth phases of cells forming biofilms and the presence of the extracellular matrix (Stewart and Franklin 2008).

Further research into genetic control of MRSA biofilm formation will increase our understanding of the pathogenesis of *MRSA* infections involving biofilms and will be useful for the development of novel therapies to treat persistent infections (O'Neill et al. 2007).

Oritavancin is a semisynthetic lipoglycopeptide with bactericidal properties against Gram-positive microorganisms such as methicillin-resistant *S. aureus* (MRSA), vancomycin-resistant *S. aureus* (VRSA) and others (Belley et al. 2009). Oritavacin can produce loss of membrane integrity in susceptible biofilm cells (McKay et al. 2006) being this its principal antibacterial mechanism against stationary phase *S. aureus* cells (Belley et al. 2009).

The selection of the right antibiotics to treat human and animal infections caused by biofilms is paramount (Clutterbuck et al. 2007). Furthermore, more research into biofilm infections in humans and animals, biofilm resistance mechanisms and new strategies for effective treatment should be developed. New therapies need to overcome significant difficulties encountered when treating biofilm infections (Belley et al. 2009). It has been suggested that prophylactic use of microbicides in medical devices can be effective against biofilm formation and associated infections.

The use of antibiotics as part of any prophylactic strategy is controversial but they are routinely used in immunocompromised patients. It has been suggested that the most effective treatment of biofilm infections associated with medical devices combines removal of an infected material and systemic antibiotic and/or antifungal therapy. However, fungal infections can be very difficult to treat. New experimental therapies to prevent and/or treat biofilm-related infections of medical devices are being developed such as new materials and intelligent implants. Future research into biofilm-related infections and the use of new imaging technologies and biomarkers will assist on the fight against biofilm infections (Lynch and Robertson 2008). Compounds that can interfere with QS systems or cellular communication systems represent an important alternative for the treatment of biofilm infections. One of the main advantages is that these products are very unlikely to create selective pressure for more virulent or resistant microbial strains (Morten et al. 2003).

7 The Future

Future research should include genetic investigations into regulatory genes involved in biofilm formation and survival mechanisms, bioengineering and new technologies for the study of biofilms. Biofilm control strategies should be based on risk assessment and assessment of the effectiveness of control methods. Furthermore, new treatments and therapies should be developed and their effectiveness on the fight against biofilms adequately assessed.

References

Aarnisalo K, Autio T, Sjöberg A-M, Lundén J, Korkeala H, Suihko M-L (2003) Typing of *Listeria monocytogenes* isolates originating from the food processing industry with automated ribotyping and pulsed-field gel electrophoresis. J Food Prot 66:249–255

Acha PN, Szyfres B (2003) Zoonoses and communicable diseases common to man and animals. In: PAHO (ed) Chlamydioses, rickettsioses, and viruses, vol 2, 3rd edn. PAHO, Washington, DC, pp 246–276

Adams BL, Bates TC, Oliver JD (2003) Survival of *Helicobacter pylori* in a natural freshwater environment. Appl Environ Microbiol 69:7462–7466

Ahmed KS, Khan AA, Ahmed I, Tiwari SK, Habeed A, Ahi JD, Abid Z, Ahmed N, Habibullah CM (2007) Impact of household hygiene and water source on the prevalence and transmission of *Helicobacter pylori:* a South Indian perspective. Singapore Med J 48:543–549

Alakomi H-L, Kujanpa ÈAÈK, Partanen L, Suihko M-L, Salo S, Siika-Aho M, Saarela M, Mattila-Sandholm T, Raaska L (2002) Microbiological problems in paper machine environments. VTT Research Notes 2152. Otamedia Oy, Espoo

Alam M, Sultana M, Nair GB, Siddique AK, Hasan NA, Sack RB, Sack DA, Ahmed KU, Sadique A, Watanabe H, Grim CJ, Huq A, Colwell RR (2007) Viable but nonculturable *Vibrio cholerae* O1 in biofilms in the aquatic environment and their role in cholera transmission. Proc Natl Acad Sci USA 104:17801–17806

Al-Ani FK, Roberson J (2007) Glanders in horses: a review of the literature. Veterinarski Arhiv 77:203–218

Allison DG (2003) The biofilm matrix. Biofouling 19:139–150

Amieva MR, El-Omar EM (2008) Host–bacterial interactions in *Helicobacter pylori* infection. Gastroenterology 134:306–323

Amorena B, Gracia E, Monzon M, Leiva J, Oteiza C, Perez M, Alabart JL, Hernandez-Yago J (1999) Antibiotic susceptibility assay for *Staphylococcus aureus* in biofilms developed in vitro. J Antimicrob Chemother 44:43–55

Anderson JM, Marchant RE (2000) Biomaterials: factors favoring colonization and infection. In: Waldvogel FA, Bisno AL (eds) Infections associated with indwelling medical devices, 3rd edn. American Society of Microbiology, Washington, DC, pp 89–109

Angles ML, Chandy JP, Cox PT, Fisher IH, Warnecke MR (2007) Implications of biofilm-associated waterborne *Cryptosporidium* oocysts for the water industry. Trends Parasitol 23:352–356

Angulo FJ, Tauxe RV, Cohen ML (1999) Significance and sources of antimicrobial-resistant nontyphoidal *Salmonella* infections in humans in the United States: the need for prudent use of antimicrobial agents, including restricted use of fluoroquinalones, in food animals. http://www.omafra.gov.on.ca/english/livestock/animalcare/amr/facts/angulo.htm. Accessed 14 Sept 2009

Appelbaum PC (1992) Antimicrobial resistance in *Streptococcus pneumoniae*: an overview. Clin Infect Dis 15:77–83

Arnold JW (1998) Development of bacterial biofilms during poultry processing. Poult Avian Biol Rev 9:1–9

Arnold JW, Bailey GW (2000) Surface finishes on stainless steel reduce bacterial attachment and early biofilm formation: scanning electron and atomic force microscopy study. Poult Sci 79:1839–1845

Arnold JW, Silvers S (2000) Comparison of poultry processing equipment surfaces for susceptibility to bacterial attachment and biofilm formation. Poult Sci 79:1215–1221

Arnold JW, Suzuki O (2003) Effects of corrosive treatment on stainless steel surface finishes and bacterial attachment. Trans ASABE 46:1595–1602

Asha NJ, Tompkins D, Wilcox MH (2006) Comparative analysis of prevalence, risk factors, and molecular epidemiology of antibiotic-associated diarrhea due to *Clostridium difficile*, *Clostridium perfringens*, and *Staphylococcus aureus*. J Clin Microbiol 44:2785–2791

Ashhurst-Smith C, Norton R, Thoreau W, Peel MM (1998) *Actinobacillus equuli* septicemia: an unusual zoonotic infection. J Clin Microbiol 36:2789–2790

Austin JW, Sanders G, Kay WW, Collinson SK (1998) Thin aggregative fimbriae enhance *Salmonella enteritidis* biofilm formation. FEMS Microbiol Lett 162:295–301

Autio T, Hielm S, Miettinen M, Sjöberg A-M, Aarnisalo K, Björkroth J, Mattila-Sandholm T, Korkeala H (1999) Sources of *Listeria monocytogenes* contamination in a cold-smoked rainbow trout processing plant detected by pulsed-field gel electrophoresis typing. Appl Environ Microbiol 65:150–155

Ayele WY, Svastovar P, Roubal P, Bartos M, Pavlik I (2005) *Mycobacterium avium* subsp. *paratuberculosis* cultured from locally and commercially pasteurized cow's milk in the Czech Republic. Appl Environ Microbiol 71:1210–1214

Baba T, Takeuchi F, Kuroda M, Yuzawa H, Aoki K, Oguchi A, Nagai Y, Iwama N, Asano K, Naimi T, Kuroda H, Cui L, Yamamoto K, Hiramatsu K (2002) Genome and virulence determinants of high virulence community-acquired MRSA. Lancet 359:1819–1827

Bagge-Ravn D, Ng Y, Hjelm M, Christiansen JN, Johansen C, Gram L (2003) The microbial ecology of processing equipment in different industries – analysis of the microflora during processing and following cleaning and disinfection. Int J Food Microbiol 87:239–250

Baldassarri L, Creti R, Recchia S, Imperi M, Facinelli B, Giovanetti E, Pataracchia M, Alfarone G, Orefici G (2006) Therapeutic failures of antibiotics used to treat macrolide-susceptible *Streptococcus pyogenes* infections may be due to biofilm formation. J Clin Microbiol 44:2721–2727

Bannoehr J, Ben Zakour NL, Waller AS, Guardabassi L, Thoday KL, van den Broek AH, Fitzgerald JR (2007) Population genetic structure of the *Staphylococcus* intermedius group: insights into agr diversification and the emergence of methicillin-resistant strains. J Bacteriol 189:8685–8692

Baptiste KE, Williams K, Willams NJ, Wattret A, Clegg PD, Dawson S, Corkill JE, O'Neill T, Hart CA (2005) Methicillin-resistant staphylococci in companion animals. Emerg Infect Dis 11:1942–1944

Bartram J (2007) Legionella and the prevention of legionellosis. World Health Organization, Geneva

Bauman WJ, Nocker A, Jones WL, Camper AK (2009) Retention of a model pathogen in a porous media biofilm. Biofouling 25:229–240

Beard PM, Rhind SM, Buxton D, Daniels MJ, Henderson D, Pirie A, Rudge K, Greig A, Hutchings MR, Stevenson K, Sharp JM (2001) Natural paratuberculosis infection in rabbits in Scotland. J Comp Pathol 124:290–299

Beenken KE, Blevins JS, Smeltzer MS (2003) Mutation of *sarA* in *Staphylococcus aureus* limits biofilm formation. Infect Immun 71:4206–4211

Beenken KE, Dunman PM, McAleese F, Macapagal D, Murphy E, Projan SJ, Blevins JS, Smeltzer MS (2004) Global gene expression in *Staphylococcus aureus* biofilms. J Bacteriol 186:4665–4684

Belley A, Neesham-Grenon E, McKay G, Arhin FF, Harris R, Beveridge T, Parr TR Jr, Moeck G (2009) Oritavancin kills stationary-phase and biofilm *Staphylococcus aureus* cells in vitro. Antimicrob Agents Chemother 53:918–925

Beran GW, Steele JH (1994) Handbook of zoonoses: section A. Bacterial, rickettsial, chlamydial, and mycotic. CRC, Boca Raton, FL

Berk SG, Ting RS, Turner GW, Ashburn RJ (1998) Production of respirable vesicles containing live *Legionella pneumophila* cells by two *Acanthamoeba* spp. Appl Environ Microbiol 64:279–286

Bernbom N, Jørgensen RL, Ng YY, Meyer RL, Kingshott P, Vejborg RM, Klemm P, Besenbacher F, Gram L (2006) Bacterial adhesion to stainless steel is reduced by aqueous fish extract coatings. Biofilms 3:25–36

Bjarnsholt T, Jensen PØ, Fiandaca MJ, Pedersen J, Hansen CR, Andersen CB, Pressler T, Givskov M, Høiby N (2009) *Pseudomonas aeruginosa* biofilms in the respiratory tract of cystic fibrosis patients. Pediatr Pulmonol 44:547–558

Bjerklie S (2003) Controlling biofilms and controlling Listeria. Meat Process 26–33

Blackman IC, Frank JF (1996) Growth of *Listeria monocytogenes* as a biofilm an various food-processing surfaces. J Food Prot 59:827–831

Boerlin P, Eugster S, Gaschen F, Straub R, Schawalder P (2001) Transmission of opportunistic pathogens in a veterinary teaching hospital. Vet Microbiol 82:347–359

Booij-Vrieling HE, van der Reijden WA, Houwers DJ, de Wit WE, Bosch-Tijhof CJ, Penning LC, van Winkelhoff AJ, Hazewinkel HA (2010) Comparison of periodontal pathogens between cats and their owners. Vet Microbiol 144:147–152

Boost MV, O'Donoghue MM, Siu KH (2007) Characterisation of methicillin-resistant *Staphylococcus aureus* isolates from dogs and their owners. Clin Microbiol Infect 13:731–733

Borucki MK, Peppin JD, White D, Loge F, Call DR (2003) Variation in biofilm formation among strains of *Listeria monocytogenes*. Appl Environ Microbiol 69:7336–7342

Boyle N, Bishop JL, Grass GA, Finlay BB (2007) Meeting review *Salmonella*: from pathogenesis to therapeutics. J Bacteriol 189:1489–1495

Boyle-Vavra S, Ereshefsky B, Wang C-C, Daum RS (2005) Successful multiresistant community-associated methicillin-resistant *Staphylococcus aureus* lineage from Taipei, Taiwan, that carries either the novel staphylococcal chromosome cassette *mec* (SCC*mec*) type V_T or SCC*mec* type IV. J Clin Microbiol 43:4719–4730

Bradshaw DJ, Marsh PD (1999) Use of continuous flow techniques in modeling dental plaque biofilms. Methods Enzymol 310:279–296

Bradshaw DJ, Marsh PD, Allison C, Schilling KM (1996) Effect of oxygen, inoculum composition and flow rate on development of mixed-culture oral biofilms. Microbiology 142:623–629

Brandonisio O (2006) Waterborne transmission of *Giardia* and *Cryptosporidium*. Parassitologia 48:91–94

Bryers J (ed) (2000) Process analysis and applications. Wiley, New York

Bull TJ, McMinn EJ, Sidi-Boumedine K, Skull A, Durkin D, Neild P, Rhodes G, Pickup R, Hermon-Taylor J (2003) Detection and verification of *Mycobacterium avium* subsp. *paratuberculosis* in fresh ileocolonic mucosal biopsy specimens from individuals with and without Crohn's disease. J Clin Microbiol 41:2915–2923

Carman RJ, Sayeed S, Li J, Genheimer CW, Hiltonsmith MF, Wilkins TD, McClane BA (2008) *Clostridium perfringens* toxin genotypes in the feces of healthy North Americans. Anaerobe 14:102–108

Carpentier B, Chassaing D (2004) Interactions in biofilms between *Listeria monocytogenes* and resident microorganisms from food industry premises. Int J Food Microbiol 97:111–122

Carron MA, Tran VR, Sugawa C, Coticchia JM (2006) Identification of *Helicobacter pylori* biofilms in human gastric mucosa. J Gastrointest Surg 10:712–717

Carter G, Wu M, Drummond DC, Bermudez LE (2003) Characterization of biofilm formation by clinical isolates of *Mycobacterium avium*. J Med Microbiol 52:747–752

CDC (2008) Salmonellosis. CDC, Atlanta, GA. http://www.cdc.gov/nczved/dfbmd/disease_listing/salmonellosis_gi.html. Accessed 14 Sept 2009

Chaieb K, Mahdouani K, Bakhrouf A (2005) Detection of *icaA* and *icaD* loci by polymerase chain reaction and biofilm formation by *Staphylococcus epidermidis* isolated from dialysate and needles in a dialysis unit. J Hosp Infect 61:225–230

Chandra J, Ghannoum A (2004) Fungal biofilms. In: Ghannoum M, O'Toole GA (eds) Microbial biofilms. ASM, Washington, DC, pp 30–42

Chemlal K, Huys G, Laval F, Vincent V, Savage C, Gutierrez C, Laneelle M-A, Swings J, Meyers WM, Daffe M, Portaels F (2002) Characterization of an unusual mycobacterium: a possible missing link between *Mycobacterium marinum* and *Mycobacterium ulcerans*. J Clin Microbiol 40:2370–2380

Cheng HW, Lucy FE, Graczyk TK, Broaders MA, Tamang L, Connolly M (2009) Fate of *Cryptosporidium parvum* and *Cryptosporidium hominis* oocysts and *Giardia duodenalis* cysts during secondary wastewater treatments. Parasitol Res 105:689–696

Clifton-Hadley FA (1983) Zoonoses in practice–Streptococcus suis type 2 infection. Br Vet J 139:1–5

Cloete TE, Smith F, Steyn PL (1989) The use of planktonic bacterial populations in open and closed recirculating water cooling systems for the evaluation of biocides. Int Biodeterior 25:115–122

Clutterbuck AL, Woods EJ, Knottenbelt DC, Clegg PD, Cochrane CA, Percival SL (2007) Biofilms and their relevance to veterinary medicine. Vet Microbiol 121:1–17

Cochrane CA, Freeman K, Woods E, Welsby S, Percival SL (2009) Biofilm evidence and the microbial diversity of horse wounds. Can J Microbiol 55:197–202

Cole SP, Harwood J, Lee R, She R, Guiney DG (2004) Characterization of monospecies biofilm formation by *Helicobacter pylori*. J Bacteriol 186:3124–3132

Committee on Indicators for Waterborne Pathogens (CB) (2004) Indicators for waterborne pathogens. National Academies, Washington, DC

Cook KL, Britt JS, Bolster CH (2010) Survival of *Mycobacterium avium* subsp. *paratuberculosis* in biofilms on livestock watering trough materials. Vet Microbiol 141:103–109

Corpe WA (1980) Microbial surface components involved in adsorption of microorganisms onto surfaces. In: Bitton G, Marshall KC (eds) Adsorption of microorganisms to surfaces. Wiley, New York, pp 105–144

Costerton JW, Marrie TJ, Cheng K-J (1985) Phenomena of bacterial adhesion. In: Savage DC, Fletcher M (eds) Bacterial adhesion: mechanisms and physiological significance. Plenum, New York, pp 650–654

Costerton JW, Ellis B, Lam K, Johnson F, Khoury AE (1994) Mechanism of electrical enhancement of efficacy of antibiotics in killing biofilm bacteria. Antimicrob Agents Chemother 38:2803–2809

Costerton JW, Lewandowski Z, Caldwell DE, Korber DR, Lappin-Scott HM (1995) Microbial biofilms. Annu Rev Microbiol 49:711–745

Costerton JW, Stewart PS, Greenberg EP (1999) Bacterial biofilms: a common cause of persistent infections. Science 284:1318–1322

Costerton W, Veeh R, Shirtliff M, Pasmore M, Post C, Ehrlich G (2003) The application of biofilm science to the study and control of chronic bacterial infections. J Clin Invest 112:1466–1477

Cramton SE, Gerke C, Schnell NF, Nichols WW, Cotz F (1999) The intercellular adhesion (*ica*) locas is present in *Staphylococcus aureus* and is required for biofilm formation. Infect Immun 67:5427–5433

Crawford RW, Gibson DL, Kay WW, Gunn JS (2008) Identification of a bile-induced exopolysaccharide required for *Salmonella* biofilm formation on gallstone surfaces. Infect Immun 76:5341–5349

Criado M-T, Suarez B, Ferreio CM (1994) The importance of bacterial adhesion in the dairy industry. Food Technol 48:123–126

Croft AC, D'Antoni AV, Terzulli SL (2007) Update on the antibacterial resistance crisis. Med Sci Monit 13:103–118

Crump JA, Sulka AC, Langer AJ, Schaben C, Crielly AS, Gage R, Baysinger M, Moll M, Withers G, Toney DM, Hunter SB, Hoekstra RM, Wong SK, Griffin PM, Van Gilder TJ (2002) An outbreak of *Escherichia coli* O157:H7 infections among visitors to a dairy farm. N Engl J Med 347:555–560

Cucarella C, Tormo MA, Ubeda C, Trotonda MP, Monzon M, Peris C, Amorena B, Lasa I, Penades JR (2004) Role of biofilm-associated protein bap in the pathogenesis of bovine *Staphylococcus aureus*. Infect Immun 72:2177–2185

Cvitkovitch DG, Yung-Hua L, Richard PE (2003) Quorum sensing and biofilm formation in streptococcal infections. J Clin Invest 112:1626–1632

D'Mello JPF (2003) Food safety. CABI, Wallingford

Darouiche RO (2001) Device-associated infections: a macroproblem that starts with microadherence. Clin Infect Dis 33:1567–1572

Davey ME, O'Toole GA (2000) Microbial biofilms: from ecology to molecular genetics. Microbiol Mol Biol Rev 64:847–867

Declerck P (2010) Biofilms: the environmental playground of *Legionella pneumophila*. Environ Microbiol 12:557–566

Declerck P, Behets J, Delaedt Y, Margineanu A, Lammertyn E, Ollevier F (2005) Impact of non-*Legionella* bacteria on the uptake and intracellular replication of *Legionella pneumophila* in *Acanthamoeba castellanii* and *Naegleria lovaniensis*. Microb Ecol 50:536–549

Declerck P, Behets J, van Hoef V, Ollevier F (2007) Detection of *Legionella pneumophila* and some of its amoeba hosts in floating biofilms from anthropogenic and natural aquatic environments. Water Res 41:3159–3167

Demangel C, Stinear TP, Cole ST (2009) Buruli ulcer: reductive evolution enhances pathogenicity of *Mycobacterium ulcerans*. Nat Rev Microbiol 7:50–60

Demers B, Simor AE, Vellend H, Schlievert PM, Byrne S, Jamieson F, Walmsley S, Low DE (1993) Severe invasive group A streptococcal infections in Ontario, Canada: 1987–1991; editorial response by DL Stevens. Clin Infect Dis 16:792–802

Dhir VK, Dodd CER (1995) Susceptibility of suspended and surface-attached *Salmonella enteritidis* to biocides and elevated temperatures. Appl Environ Microbiol 61:1731–1738

Dimola S, Caruso ML (1999) *Helicobacter pylori* in animals affecting the human habitat through the food chain. Anticancer Res 19:3889–3894

Donlan RM (2001) Biofilms and device-associated infections. Emerg Infect Dis 7:277–281

Donlan RM, Costerton JW (2002) Biofilms: survival mechanisms of clinically relevant microorganisms. Clin Microbiol Rev 15:167–193

Douglas LJ (2003) Candida biofilms and their role in infection. Trends Microbiol 11:30–36

Drexler M (2002) Secret agents: the menace of emerging infections. Joseph Henry, Washington, DC

du Moulin GC, Stottmeier KD, Pelletier PA, Tsang AY, Hedley-Whyte J (1988) Concentration of *Mycobacterium avium* by hospital hot water systems. JAMA 260:1599–1601

Dunman PM, Murphy E, Haney S, Palacios D, Tucker-Kellogg G, Wu S, Brown EL, Zagursky RJ, Shlaes D, Projan SJ (2001) Transcription profiling-based identification of *Staphylococcus aureus* genes regulated by the *agr* and/or *sarA* loci. J Bacteriol 183:7341–7353

Dykes GA, Sampathkumar B, Korber DR (2003) Planktonic or biofilm growth effects survival, hydrophobicity and protein expression patterns of a pathogenic *Campylobacter jejuni* strain. Int J Food Microbiol 89:1–10

Eisen RJ, Wilder AP, Bearden SW, Montenieri JA, Gage KL (2007) Early-phase transmission of *Yersinia pestis* by unblocked *Xenopsylla cheopis* (Siphonaptera: Pulicidae) is as efficient as transmission by blocked fleas. J Med Entomol 44:678–682

Eklund MW, Poysky FT, Paranjpye RN, Lashbrook LC, Peterson ME, Pelroy GA (1995) Incidence and sources of *Listeria monocytogenes* in cold-smoked fishery products and processing plants. J Food Prot 58:502–508

Epstein CR, Yama WC, Peiris JSM, Epstein RJ (2009) Methicillin-resistant commensal staphylococci in healthy dogs as a potential zoonotic reservoir for community-acquired antibiotic resistance. Infect Genet Evol 9:283–285

Esteves CL, Jones BD, Clegg S (2005) Biofilm formation by *Salmonella enterica* serovar typhimurium and *Escherichia coli* on epithelial cells following mixed inoculations. Infect Immun 73:5198–5203

Falkinham JO III (1996) Epidemiology of infection by nontuberculous mycobacteria. Clin Microbiol Rev 9:177–215

Falkinham JO III, Norton CD, LeChevallier MW (2001) Factors influencing numbers of *Mycobacterium avium*, *Mycobacterium intracellulare*, and other mycobacteria in drinking water distribution systems. Appl Environ Microbiol 67:1225–1231

Falkinham JO, Nichols G, Bartram J, Dufour A, Portaels F (2004) Natural ecology and survival in water of mycobacteria of potential public health significance. In: Bartram J (ed) Pathogenic mycobacteria in water: a guide to public health consequences, monitoring and management. World Health Organization, Albany, NY

Farber JM, Peterkin PI (1991) *Listeria monocytogenes*, a food-borne pathogen. Microbiol Rev 55:476–511

Fardini Y, Chung P, Dumm R, Joshi N, Han YW (2010) Transmission of diverse oral bacteria to murine placenta: evidence for the oral microbiome as a potential source of intrauterine infection. Infect Immun 78:1789–1796

Fazakerley C, Pryor M, Percival SL (2001) The isolation of *Mycobacterium* spp. from groundwater, chlorinated distribution systems and biofilms. In: Gilbert PG, Allison D, Walker JT, Brading M (eds) Biofilm community interactions: chance or necessity? Bioline, Cardiff, pp 53–57

Ferenci P (2000) The importance of *Helicobacter* – also beyond the stomach. Acta Med Aust 27:109–111

Fittipaldi N, Gottschalk M, Vanier G, Daigle F, Harel J (2007) Use of selective capture of transcribed sequences to identify genes preferentially expressed by *Streptococcus suis* upon interaction with porcine brain microvascular endothelial cells. Appl Environ Microbiol 73:4359–4364

Fitzpatrick F, Humphreys H, O'Gara JP (2005) Evidence for *icaADBC*-independent biofilm development mechanism in methicillin-resistant *Staphylococcus aureus* clinical isolates. J Clin Microbiol 43:1973–1976

Fox LK, Zadoks RN, Gaskins CT (2005) Biofilm production by *Staphylococcus aureus* associated with intramammary infection. Vet Microbiol 107:295–299

Futagawa-Saito K, Ba-Thein W, Sakurai N, Fukuyasu T (2006) Prevalence of virulence factors in *Staphylococcus intermedius* isolates from dogs and pigeons. BMC Vet Res 2:4

Galanis E, Lo Fo Wong DM, Patrick ME, Binsztein N, Cieslik A, Chalermchikit T, Aidara-Kane A, Ellis A, Angulo FJ, Wegener HC, World Health Organization Global Salm-Surv (2006) Web-based surveillance and global *Salmonella* distribution, 2000–2002. Emerg Infect Dis 12:381–388

Galuppo LD, Pascoe JR, Jang SS, Willits NH, Greenman SL (1999) Evaluation of iodophor skin preparation techniques and factors influencing drainage from ventral midline incisions in horses. J Am Vet Med Assoc 215:963–969

Gauthier DT, Rhodes MW (2009) Mycobacteriosis in fishes: a review. Vet J 180:33–47

Gaynor EC, Wells DH, MacKichan JK, Falkow S (2005) The *Campylobacter jejuni* stringent response controls specific stress survival and virulence-associated phenotypes. Mol Microbiol 56:8–27

Gerald BL (2009) Foodborne-illness causing pathogens. In: Edelstein S (ed) Food and nutrition at risk in America, Chap. 2. Jones and Barlett, Sudbury, MA

Ghannoum MA, O'Toole GA (2004) Microbial biofilms. American Society for Microbiology, Washington, DC

Gião MS, Azevedo NF, Wilks SA, Vieira MJ, Keevil CW (2008) Persistence of *Helicobacter pylori* in heterotrophic drinking-water biofilms. Appl Environ Microbiol 74:5898–5904

Gibson H, Taylor JH, Hall KE, Holah JT (1999) Effectiveness of cleaning techniques used in the food industry in terms of the removal of bacterial biofilms. J Appl Microbiol 87:41–48

Giron JA, Torres AG, Freer E, Kaper JB (2002) The flagella of enteropathogenic *Escherichia coli* mediate adherence to epithelial cells. Mol Microbiol 44:361–379

Goldsworthy MJH (2008) Gene expression of *Pseudomonas aeruginosa* and MRSA within a catheter-associated urinary tract infection biofilm model. Biosci Horiz 1:28–37

Goodman KJ, Correa P, Tengana AJ, Ramírez H, DeLany JP, Pepinosa OG, Quiñones ML, Parra TC (1996) *Helicobacter pylori* infection in the Colombian Andes: a population-based study of transmission pathways. Am J Epidemiol 144:290–299
Grant IR, Ball HJ, Rowe MT (2002) Incidence of *Mycobacterium paratuberculosis* in bulk raw and commercially pasteurized cows' milk from approved dairy processing establishments in the United Kingdom. Appl Environ Microbiol 68:2428–2435
Greene CE, Gunn-Moore DA (1990) Mycobacterial infections. In: Green CE (ed) Infectious diseases of the dog and cat, 2nd edn. WB Saunders, Philadelphia, pp 313–315
Grenier D, Grignon L, Gottschalk M (2009) Characterisation of biofilm formation by a *Streptococcus suis* meningitis isolate. Vet J 179:292–295
Gresham HD, Lowrance JH, Caver TE, Wilson BS, Cheung AL, Lindberg FP (2000) Survival of *Staphylococcus aureus* inside neutrophils contributes to infection. J Immunol 164:3713–3722
Greub G, Raoult D (2004) Microorganisms resistant to free-living amoebae. Clin Microbiol Rev 17:413–433
Griffiths MW (2003) Listeria. In: Caballero B, Trugo LC, Finglas PM (eds) Encyclopedia of food sciences and nutrition, vol 6. Academic Press, London, pp 3562–3573
Grübel P, Hoffman JS, Chong FK, Burstein NA, Mepani C, Cave DR (1997) Vector potential of houseflies (*Musca domestica*) for *Helicobacter pylori*. J Clin Microbiol 35:1300–1303
Guardabassi L, Schwarz S, Lloyd DH (2004) Pet animals as reservoirs of antimicrobial resistant bacteria. J Antimicrob Chemother 54:321–332
Guerrieri E, Bondi M, Sabia C, de Niederhäusern S, Borella P, Messi P (2008) Effect of bacterial interference on biofilm development by *Legionella pneumophila*. Curr Microbiol 57:532–536
Gunduz GT, Tuncel G (2006) Biofilm formation in an ice cream plant. Antonie Leeuwenhoek 89:329–336
Hall-Stoodley L, Stoodley P (2005) Biofilm formation and dispersal and the transmission of human pathogens. Trends Microbiol 13:7–10
Hall-Stoodley L, Costerton JW, Stoodley P (2004) Bacterial biofilms: from the natural environment to infectious diseases. Nat Rev Microbiol 2:95–108
Hansen SK, Rainey PB, Haagensen JA, Molin S (2007) Evolution of species interactions in a biofilm community. Nature 445:533–536
Hawser S, Islam K (1999) Comparisons of the effects of fungicidal and fungistatic antifungal agents on the morphogenetic transformation of *Candida albicans*. J Antimicrob Chemother 43:411–413
Hefford MA, D'Aoust S, Cyr TD, Austin JW, Sanders G, Kheradpir E, Kalmokoff ML (2005) Proteomic and microscopic analysis of biofilms formed by *Listeria monocytogenes*. Can J Microbiol 51:197–208
Heitman TL, Frederick LM, Viste JR, Guselle NJ, Morgan UM, Thompson RC, Olson M (2002) Prevalence of *Giardia* and *Cryptosporidium* and characterization of *Cryptosporidium* spp. isolated from wildlife, human, and agricultural sources in the North Saskatchewan River Basin in Alberta, Canada. Can J Microbiol 48:530–541
Helmi K, Skraber S, Gantzer C, Willame R, Hoffmann L, Cauchie HM (2008) Interactions of *Cryptosporidium parvum*, *Giardia lamblia*, vaccinal poliovirus type 1, and bacteriophages phiX174 and MS2 with a drinking water biofilm and a wastewater biofilm. Appl Environ Microbiol 74:2079–2088
Herikstad H, Motarjemi Y, Tauxe RV (2002) *Salmonella* surveillance: a global survey of public health serotyping. Epidemiol Infect 129:1–8
Hermon-Taylor J, Bull TJ, Sheridan JM, Cheng J, Stellakis ML, Sumar N (2000) Causation of Crohn's disease by *Mycobacterium avium* subspecies *paratuberculosis*. Can J Gastroenterol 14:521–539
Hernandez-Divers SJ, Shearer D (2002) Pulmonary mycobacteriosis caused by *Mycobacterium haemophilum* and *M. marinum* in a royal python. J Am Vet Med Assoc 220:1661–1663
Hirsh DC, Birbenstein EL (2004) Clostridium. In: Hirsh DC, Maclachlan NJ, Walker RL (eds) Veterinary microbiology, Chap. 36. Blackwell, Oxford

Howe AD, Forster S, Morton S, Marshall R, Osborn KS, Wright P, Hunter PR (2002) *Cryptosporidium* oocysts in a water supply associated with a cryptosporidiosis outbreak. Emerg Infect Dis 8:619–624

Huang YT, Teng LJ, Ho SW, Hsueh PR (2005) *Streptococcus suis* infection. J Microbiol Immunol Infect 38:306–313

Hui AC, Ng KC, Tong PY, Mok V, Chow KM, Wu A, Wong LK (2005) Bacterial meningitis in Hong Kong: 10-years' experience. Clin Neurol Neurosurg 107:366–370

Hume RD, Hann WD (1984) Growth relationships of *Legionella pneumophila* with green algae (*Chlorophyta*). In: Thornsberry C, Balows A, Feely JC, Jakubowski W (eds) Legionellae. Proceedings of the 2nd international symposium. American Society for Microbiology, Washington, DC, pp 323–324

Humphrey T, Mason M, Martin K (1995) The isolation of *Campylobacter jejuni* from contaminated surfaces and its survival in diluents. Int J Food Microbiol 26:295–303

Husu JR, Seppänen JT, Sivelä SK, Rauramaa AL (1990) Contamination of raw milk by *Listeria monocytogenes* on dairy farms. J Vet Med 37:268–275

ICMSF (1996) Microorganisms in foods, vol. 5. Characteristics of microbial pathogens. Blackie Academic and Professional, London

Jarrett CO, Deak E, Isherwood KE, Oyston PC, Fischer ER, Whitney AR, Kobayashi SD, DeLeo FR, Hinnebusch BJ (2004) Transmission of *Yersinia pestis* from an infectious biofilm in the flea vector. J Infect Dis 190:783–792

Jefferson KK (2004) What drives bacteria to produce a biofilm? FEMS Microbiol Lett 236: 163–173

Jefferson KK, Goldmann DA, Pier GB (2005) Use of confocal microscopy to analyse the rate of vancomycin penetration through *Staphylococcus aureus* biofilms. Antimicrob Agents Chemother 49:2467–2473

Jemili-Ben Jomaa M, Boutiba-Ben Boubaker I, Ben Redjeb S (2006) Identification of staphylococcal cassette chromosome mec encoding methicillin resistance in *Staphylococcus aureus* isolates at Charles Nicolle Hospital of Tunis. Pathol Biol 54:453–455

Jennings SS, Moran AP, Carroll CV (2003) Bioaerosols and biofilms. In: Lens P, Moran AP, Mahony T, Stoodley P, O'Flaherty V (eds) Biofilms in medicine, industry and environmental biotechnology. IWA, London, pp 160–178

Johansen TB, Agdestein A, Olsen I, Nilsen SF, Holstad G, Djønne B (2009) Biofilm formation by *Mycobacterium avium* isolates originating from humans, swine and birds. BMC Microbiol 9:159

Joshua GWP, Guthrie-Irons C, Karlyshev AV, Wren BW (2006) Biofilm formation in *Campylobacter jejuni*. Microbiology 152:387–396

Kalmokoff M, Lanthier P, Tremblay TL, Foss M, Lau PC, Sanders G, Austin J, Kelly J, Szymanski CM (2006) Proteomic analysis of *Campylobacter jejuni* 11168 biofilms reveals a role for the motility complex in biofilm formation. J Bacteriol 188:4312–4320

Kandulski A, Selgrad M, Malfertheiner P (2008) *Helicobacter pylori* infection: a clinical overview. Digest Liver Dis 40:619–626

Karlsson A, Arvidson S (2002) Variation in extracellular protease production among clinical isolates of *Staphylococcus aureus* due to different levels of expression of the protease repressor *sar*A. Infect Immun 70:4239–4246

Katsikogianni M, Missirlis YF (2004) Concise review of mechanisms of bacterial adhesion to biomaterials and of techniques used in estimating bacteria-material interactions. Eur Cells Mater 8:37–57

Keevil CW (2003) Rapid detection of biofilms and adherent pathogens using scanning confocal laser microscopy and episcopic differential interference contrast microscopy. Water Sci Technol 47:105–116

Kent ML, Whipps CM, Mathews JL, Florio D, Watral V, Bishop-Stewart JK, Poort M, Bermudez L (2004) Mycobacteriosis in zebrafish (*Danio rerio*) research facilities. Comp Biochem Physiol C Pharmacol Toxicol Endocrinol 138:383–390

Khoury AE, Lam K, Ellis BD, Costerton JW (1992) Prevention and control of bacterial infections associated with medical devices. ASAIO J 38:174–178

Kiehn TE, White M (1994) *Mycobacterium haemophilum*: an emerging pathogen. Eur J Clin Microbiol Infect Dis 13:925–931

Kim TJ, Young BM, Young GM (2008) Effect of flagellar mutations on *Yersinia enterocolitica* biofilm formation. Appl Environ Microbiol 74:5466–5474

Kite P, Eastwood K, Sugden S, Percival SL (2004) Use of in vivo-generated biofilms from hemodialysis catheters to test the efficacy of a novel antimicrobial catheter lock for biofilm eradication in vitro. J Clin Microbiol 42:3073–3076

Klausen M, Aaes-Jorgensen A, Molin S, Tolker-Nielsen T (2003) Involvement of bacterial migration in the development of complex multicellular structures in *Pseudomonas aeruginosa* biofilms. Mol Microbiol 50:61–68

Klayman BJ, Volden PA, Stewart PS, Camper AK (2009) *Escherichia coli O157:H7* requires colonizing partner to adhere and persist in a capillary flow cell. Environ Sci Technol 43: 2105–2111

Klemm P, Schembri M (2004) Type 1 fimbriae, curli, and antigen 43: adhesion, colonization, and biofilm formation. In: Curtiss R III et al (eds) EcoSal – *Escherichia coli* and *Salmonella*: cellular and molecular biology. ASM, Washington, DC

Klingenberg C, Glad GT, Olsvik R, Flaegstad T (2001) Rapid PCR detection of the methicillin resistance gene, mecA, on the hands of medical and non-medical personnel and healthy children and on surfaces in a neonatal intensive care unit. Scand J Infect Dis 33:494–497

Knulst JC, Rosenberger D, Thompson B, Paatero J (2003) Intensive sea surface microlayer investigations of open leads in the pack ice during Arctic Ocean 2001 expedition. Langmuir 19:10194–10199

Kojic EM, Darouiche RO (2004) *Candida* infections of medical devices. Clin Microbiol Rev 17:255–267

Konishi K, Saito N, Shoji E, Takeda H, Kato M, Asaka M, Ooi HK (2007) *Helicobacter pylori:* longer survival in deep ground water and sea water than in a nutrient-rich environment. APMIS 115:1285–1291

Korber DR, Lawrence JR, Sutton B, Caldwell DE (1989) Effect of laminar flow velocity on the kinetics of surface recolonization by Mot^+ and Mot^- *Pseudomonas fluorescens*. Microb Ecol 18:1–19

Krauss H, Weber A, Appel M, Enders B, Schiefer HG, Slenczka W, Zahner H (2003) Zoonoses: infectious diseases transmissible from animals to humans, 3rd edn. ASM, Washington, DC

Krepsky N, Rocha Ferreira RB, Ferreira Nunes AP, Casado Lins UG, Costa e Silva Filho F, de Mattos-Guaraldi AL, Netto-dosSantos KR (2003) Cell surface hydrophobicity and slime production of *Staphylococcus epidermidis* Brazilian isolates. Curr Microbiol 46:280–286

Krysinski EP, Brown LJ, Marchisello TJ (1992) Effect of cleaners and sanitizers on *Listeria monocytogenes* attached to product contact surfaces. J Food Prot 55:246–251

Kuhn DM, Chandra J, Mukherjee PK, Ghannoum MA (2002) Comparison of biofilm formation *Candida albicans* and *Candida parapsilosis* on bioprosthetic surfaces. Infect Immun 70: 878–888

Kumar CG, Anand SK (1998) Significance of microbial biofilms in food industry: a review. Int J Food Microbiol 42:9–27

Kumon H (2000) Management of biofilm infections in the urinary tract. World J Surg 24: 1193–1196

Kusamaningrum HD, Riboldi G, Hazeleger WC, Beumer RR (2003) Survival of foodborne pathogens on stainless steel surfaces and cross-contamination to foods. Int J Food Microbiol 85:227–236

Lai CW, Chan RC, Cheng AF, Sung JY, Leung JW (1992) Common bile duct stones: a cause of chronic salmonellosis. Am J Gastroenterol 87:1198–1199

Lallier R, Higgins R (1988) Biochemical and toxigenic characteristics of *Aeromonas* spp. isolated from diseased mammals, moribund and healthy fish. Vet Microbiol 18:63–71

Lambe DW Jr, Fergeson KP, Mayberry-Carson KJ, Tober-Meyer B, Costerton JW (1991) Foreign-body-associated experimental osteomyelitis induced with *Bacteroides fragilis* and *Staphylococcus epidermidis* in rabbits. Clin Orthop Relat Res 266:285–294

Langeveld LPM, Van Montfort-Quasig RMGE, Weerkamp AH, Waalewijn R, Wever JS (1995) Adherence, growth and release of bacteria in a tube heat exchanger for milk. Neth Milk Dairy J 49:207–220

Langsrud S, Sidhu MS, Heir E, Holck AL (2003) Bacterial disinfectant resistance – a challenge for the food industry. Int Biodeterior Biodegrad 51:283–290

Lebeer S, Verhoeven TL, Perea Vélez M, Vanderleyden J, De Keersmaecker SC (2007) Impact of environmental and genetic factors on biofilm formation by the probiotic strain *Lactobacillus rhamnosus* GG. Appl Environ Microbiol 73:6768–6775

Lee YK, Lim CY, Teng WL, Ouwehand AC, Tuomola EM, Salminen S (2000) Quantitative approach in the study of adhesion of lactic acid bacteria to intestinal cells and their competition with *Enterobacteria*. Appl Environ Microbiol 66:3692–3697

Lelieveld H (ed) (2005) Improving hygiene in the food industry. Woodhead, Cambridge

Lelieveld H, Mostert T, White B (2001) Hygiene in food processing: principles and practice. Woodhead, Cambridge

Lellouche J, Kahana E, Elias S, Gedanken A, Banin E (2009) Antibiofilm activity of nanosized magnesium fluoride. Biomaterials 30:5969–5978

Leriche V, Carpentier B (1995) Viable but nonculturable *Salmonella typhimurium* in singleand binary-species biofilms in response to chlorine treatment. J Food Prot 58:1186–1191

Leuckfeld I, Paster BJ, Kristoffersen AK, Olsen I (2010) Diversity of *Veillonella spp.* from subgingival plaque by polyphasic approach. APMIS 118:230–242

Lu TK, Collins JJ (2007) Dispersing biofilms with engineered enzymatic bacteriophage. Proc Natl Acad Sci USA 104:11197–11202

Lun ZR, Wang QP, Chen XG, Li AX, Zhu XQ (2007) *Streptococcus suis*: an emerging zoonotic pathogen. Lancet Infect Dis 7:201–209

Lundén J (2004) Persistent *Listeria monocytogenes* contamination in food processing plants. Academic dissertation, Faculty of Veternary, University of Helsinki. Yliopistopaino, Helsinki

Lundén JM, Autio TJ, Korkeala HJ (2002) Transfer of persistent *Listeria monocytogenes* contamination between food-processing plants associated with a dicing machine. J Food Prot 65: 1129–1133

Lundén J, Autio T, Markkula A, Hellström S, Korkeala H (2003) Adaptive and cross-adaptive responses of persistent and non-persistent *Listeria monocytogenes* strains to disinfectants. Int J Food Microbiol 82:265–272

Lynch AS, Robertson GT (2008) Bacterial and fungal biofilm infections. Annu Rev Med 59: 415–428

Lynch MJ, Swift S, Kirke DF, Keevil CW, Dodd CE, Williams P (2002) The regulation of biofilm development by quorum sensing in *Aeromonas hydrophila*. Environ Microbiol 4:18–28

Lyytikäinen O, Autio T, Maijala R, Ruutu P, Honkanen-Buzalski T, Miettinen M, Hatakka M, Mikkola J, Anttila VJ, Johansson T, Rantala L, Aalto T, Korkeala H, Siitonen A (2000) An outbreak of *Listeria monocytogenes* serotype 3a from butter in Finland. J Infect Dis 181: 1838–1841

Malik S, Coombs GW, O'Brien FG, Peng H, Barton MD (2006) Molecular typing of methicillin-resistant staphylococci isolated from cats and dogs. J Antimicrob Chemother 58:428–431

Marrie TJ, Nelligan J, Costerton JW (1982) A scanning and transmission electron microscopic study of an infected endocardial pacemaker lead. Circulation 66:1339–1342

Marriott N (1999) Principles of food sanitation, 4th edn. Aspen, Gaithersburg, MD

Marshall K (2003) Fungal diseases in small mammals: therapeutic trends and zoonotic considerations. Vet Clin North Am Exot Anim Pract 6:415–427

Marsollier L, Robert R, Aubry J, Saint Andre JP, Kouakou H, Legras P, Manceau AL, Mahaza C, Carbonnelle B (2002) Aquatic insects as a vector for *Mycobacterium ulcerans*. Appl Environ Microbiol 68:4623–4628

Marsollier L, Aubry J, Coutanceau E, André JP, Small PL, Milon G, Legras P, Guadagnini S, Carbonnelle B, Cole ST (2005) Colonization of the salivary glands of *Naucoris cimicoides* by *Mycobacterium ulcerans* requires host plasmatocytes and a macrolide toxin, mycolactone. Cell Microbiol 7:935–943

Mashiba K, Hamamoto T, Torikai K (1993) A case of Legionnaires' disease due to aspiration of hot spring water and isolation of *Legionella pneumophila* from hot spring water. Kansenshogaku Zasshi 67:163–166

McBain AJ (2009) Chapter 4: in vitro biofilm models: an overview. Adv Appl Microbiol 69:99–132

McKay GA, Fadhil I, Beaulieu S, Ciblat S, Far AR, Moeck G, Parr TR Jr (2006) Oritavancin disrupts transmembrane potential and membrane integrity concomitantly with cell killing in *Staphylococcus aureus* and vancomycin-resistant enterococci, abstract C1–682. Abstract of the 46th Interscience conference on antimicrobial agents and chemother, San Francisco, CA, 27–30 Sept 2006

McKenney D, Hubner J, Muller E, Wang Y, Goldmann DA, Pier GB (1998) The *ica* locus of *Staphylococcus epidermidis* encodes production of the capsular polysaccharide/adhesin. Infect Immun 66:4711–4720

Mead G (ed) (2005) Food safety control in the poultry industry. Woodhead, Cambridge

Meining A, Kroher G, Stolte M (1998) Animal reservoirs in the transmission of *Helicobacter heilmannii* – results of a questionnaire-based study. Scand J Gastroenterol 33:795–798

Melchior MB, Fink-Gremmels J, Gaastra W (2006a) Comparative assessment of the antimicrobial susceptibility of *Staphylococcus aureus* isolates from bovine mastitis in biofilm versus planktonic culture. J Vet Med B Infect Dis Vet Pub Health 53:326–332

Melchior MB, Vaarkamp H, Fink-Gremmels J (2006b) Biofilms: a role in recurrent mastitis infections? Vet J 171:398–407

Mendez M, Huang IH, Ohtani K, Grau R, Shimizu T, Sarker MR (2008) Carbon catabolite repression of type IV pilus-dependent gliding motility in the anaerobic pathogen *Clostridium perfringens*. J Bacteriol 190:48–60

Mention K, Michaud L, Guimber D, DeLasalle EM, Vincent P, Turck D, Gottrand F (1999) Characteristics and prevalence of *Helicobacter heilmannii* infection in children undergoing upper gastrointestinal endoscopy. J Pediatr Gastroenterol Nutr 29:533–539

Meyer B (2003) Approaches to prevention, removal and killing of biofilms. Int Biodeterior Biodegrad 51:249–253

Miettinen H, Wirtanen G (2006) Ecology of *Listeria* spp. in a fish farm and molecular typing of *L. monocytogenes* from fish farming and processing companies. Int J Food Microbiol 112: 138–146

Miettinen MK, Björkroth KJ, Korkeala HJ (1999) Characterization of *Listeria monocytogenes* from an ice-cream plant by serotyping and pulsed-field gel electrophoresis. Int J Food Microbiol 46:187–192

Miettinen MK, Palmu L, Björkroth KJ, Korkeala H (2001) Prevalence of *Listeria monocytogenes* in broilers at the abattoir, processing plant, and the retail level. J Food Prot 64:994–999

Mirkin G (2009) Helicobacter and stomach ulcers. http://www.drmirkin.com/morehealth/G123.htm, Accessed on 22nd of August 2009

Mohamed JA, Huang W, Nallapareddy SR, Teng F, Murray BE (2004) Influence of origin of isolates, especially endocarditis isolates, and various genes on biofilm formation by *Enterococcus faecalis*. Infect Immun 72:3658–3663

Morck DW, Raybould TJ, Acres SD, Babiuk LA, Nelligan J, Costerton JW (1987) Electron microscopic description of glycocalyx and fimbriae on the surface of *Pasturella haemolytica*. Can J Vet Res 51:83–88

Morck DW, Costerton JW, Bolingbroke DO, Ceri H, Boyd ND, Olson ME (1990) A guinea pig model of bovine pneumonic *Pasteurellosis*. Can J Vet Res 54:139–145

Moretro T, Hermansen L, Holck AL, Sidhu MS, Rudi K, Langsrud S (2003) Biofilm formation and the presence of the intercellular adhesion locus *ica* among staphylococci from food and food processing environments. Appl Environ Microbiol 69:5648–5655

Morris DO, Rook KA, Shofer FS, Rankin SC (2006) Screening of *Staphylococcus aureus*, *Staphylococcus intermedius*, and *Staphylococcus schleiferi* isolates obtained from small companion animals for antimicrobial resistance: a retrospective review of 749 isolates (2003–04). Vet Dermatol 17:332–337

Morten H, Eberl L, Nielsen J, Givskov M (2003) Quorum sensing: a novel target for the treatment of biofilm infections. BioDrugs 4:241–250

Mosteller TM, Bishop JR (1993) Sanitizer efficacy against attached bacteria in a milk biofilm. J Food Prot 56:34–41

Murphy C, Carroll C, Jordan KN (2006) Environmental survival mechanisms of the foodborne pathogen *Campylobacter jejuni*. J Appl Microbiol 100:623–632

Naktin J, Beavis KG (1999) *Yersinia enterocolitica* and *Yersinia pseudotuberculosis*. Clin Lab Med 19:523–536

Narayan KG (1982) Food borne infection with *Clostridium perfringens* type A. Int J Zoonoses 9:12–32

Naser S, Ghobrial G, Romero C, Valentine J (2004) Culture of subspecies from the blood of patients with Crohn's disease. Lancet 364:1039–1044

Nayak R, Stewart T, Wang RF, Lin J, Cerniglia CE, Kenney PB (2004) Genetic diversity and virulence gene determinants of antibiotic-resistant *Salmonella* isolated from preharvest turkey production sources. Int J Food Microbiol 91:51–62

O'Brien SS, Lindsay D, von Holy A (2004) The presence of *Enterococcus*, coliforms and *E. coli* in a commercial yeast manufacturing process. Int J Food Microbiol 94:23–31

O'Leary KC, Gagnon GA, Chauret CP, Andrews RC (2002) Evaluation of chlorine dioxide for biofilm control in a model distribution system. Proceedings of the Water Environment Federation Disinfection, vol 9. Water Environment Federation, Orlando, FL, pp 297–305

Olson ME, Ceri H, Morck DW, Buret AG, Read RR (2002) Biofilm bacteria: formation and comparative susceptibility to antibiotics. Can J Vet Res 66:86–92

O'Neill E, Pozzi C, Houston P, Smyth D, Humphreys H, Robinson DA, O'Gara JP (2007) Association between methicillin susceptibility and biofilm regulation in *Staphylococcus aureus* isolates from device-related infections. J Clin Microbiol 45:1379–1388

Osaki M, Takamatsu D, Shimoji Y, Sekizaki T (2002) Characterization of *Streptococcus suis* genes encoding proteins homologous to sortase of gram-positive bacteria. J Bacteriol 184:971–982

Oshio I, Osaki T, Hanawa T, Yonezawa H, Zaman C, Kurata S, Kamiya S (2009) Vertical *Helicobacter pylori* transmission from Mongolian gerbil mothers to pups. J Med Microbiol 58:656–662

O'Toole GA, Kolter R (1998) Flagellar and twitching motility are necessary for *Pseudomonas aeruginosa* biofilm development. Mol Microbiol 30:295–304

Oulahal-Lagsir N, Martial-Gros A, Bonneau M, Blom LJ (2003) "*Escherichia coli*-milk" biofilm removal from stainless steel surfaces: synergism between ultrasonic waves and enzymes. Biofouling 19:159–168

Pamp SJ, Tolker-Nielsen T (2007) Multiple roles of biosurfactants in structural biofilm development by *Pseudomonas aeruginosa*. J Bacteriol 189:2531–2539

Panlilio AL, Culver DH, Gaynes R, Tolson JS, Martone WJ (1992) Methicillin-resistant *Staphylococcus aureus* in U.S. hospitals, 1975–1991. Infect Control Hosp Epidemiol 13:582

Park SF (2002) The physiology of *Campylobacter* species and its relevance to their role as foodborne pathogens. Int J Food Microbiol 74:177–188

Parsek MR, Fuqua C (2004) Biofilms 2003: emerging themes and challenges in studies of surface-associated microbial life. J Bacteriol 186:4427–4440

Paziewska A, Bednarska M, Niewęgłowski H, Karbowiak G, Bajer A (2007) Distribution of *Cryptosporidium* and *Giardia* spp. in selected species of protected and game mammals from North-Eastern Poland. Ann Agric Environ Med 14(2):265–270

Pedley S, Bartram J, Rees G, Dufour A, Cotruvo J (eds) (2004) Pathogenic mycobacteria in water: a guide to public health consequences, monitoring, and management. IWA, London

Percival SL, Thomas JG (2009) Transmission of *Helicobacter pylori* and the role of water and biofilms. Water Health 7:469–477
Percival SL, Walker JT, Hunter P (2000) Microbiological aspects of biofilms and drinking water. CRC, Boca Raton, FL, ISBN 084930590
Percival SL, Kite P, Eastwood K, Murga R, Carr J, Arduino MJ, Donlan RM (2005) Tetrasodium EDTA as a novel central venous catheter lock solution against biofilm. Infect Control Hosp Epidemiol 26:515–519
Percival SL, Sabbuba NA, Kite P, Stickler DJ (2009) The effect of EDTA instillations on the rate of development of encrustation and biofilms in Foley catheters. Urol Res 37:205–209
Pérez-Conesa D, McLandsborough L, Weiss J (2006) Inhibition and inactivation of *Listeria monocytogenes* and *Escherichia coli O157:H7* colony biofilms by micellar-encapsulated eugenol and carvacrol. J Food Prot 69:2947–2954
Pickup RW, Mallinson HEH, Rhodes G (1999) Sampling the water bodies: tangential flow filtration. Environ Monitor Bacteria 12:29–34
Pickup RW, Rhodes G, Arnott S, Sidi-Boumedine K, Bull TJ, Weightman A, Hurley M, Hermon-Taylor J (2005) *Mycobacterium avium* subsp. *paratuberculosis* in the catchment area and water of the River Taff in South Wales, United Kingdom, and its potential relationship to clustering of Crohn's disease cases in the city of Cardiff. Appl Environ Microbiol 71: 2130–2139
Piriou P, Dukan S, Levi Y, Jarrige PA (1997) Prevention of bacterial growth in drinking water distribution systems. Water Sci Technol 35:283–287
Portaels F, Fonteyene PA, de Beenhouwer H, de Rijk P, Guédénon A, Hayman J, Meyers MW (1996) Variability in 3′ end of 16S rRNA sequence of *Mycobacterium ulcerans* is related to geographic origin of isolates. J Clin Microbiol 34:962–965
Prakash B, Veeregowda BM, Krishnappa G (2003) Biofilms: a survival strategy of bacteria. Curr Sci 85:1299–1307
Pratt LA, Kolter R (1998) Genetic analysis of *Escherichia coli* biofilm formation: roles of flagella, motility, chemotaxis and type I pili. Mol Microbiol 30:285–293
Quaglia NC, Dambrosio A, Normanno G, Parisi A, Patrono R, Ranieri G, Rella A, Celano GV (2008) High occurrence of *Helicobacter pylori* in raw goat, sheep and cow milk inferred by *glm*M gene: a risk of food-borne infection? Int J Food Microbiol 124:43–47
Raad I, Hanna H, Jiang Y, Dvorak T, Reitzel R, Chaiban G, Sherertz R, Hachem R (2007) Comparative activity of daptomycin. Linezolid and tigecycline against catheter-related methicillin-resistant *Staphylococcus* bacteremic isolates embedded in biofilm. Antimicrob Agents Chemother 51:1656–1660
Ramage G, Saville SP, Thomas DP, Lopez-Ribot JL (2005) Candida biofilms: an update. Eukaryot Cell 4:633–638
Raus J, Love DN (1983) Characterization of coagulase-positive *Staphylococcus intermedius* and *Staphylococcus aureus* isolated from veterinary clinical specimens. J Clin Microbiol 18:789–792
Recordati C, Gualdi V, Tosi S, Facchini RV, Pengo G, Luini M, Simpson KW, Scanziani E (2007) Detection of *Helicobacter* spp. DNA in the oral cavity of dogs. Vet Microbiol 119:346–351
Rediske AM, Hymas WC, Wilkinson R, Pitt WG (1998) Ultrasonic enhancement of antibiotic action on several species of bacteria. J Gen Appl Microbiol 44:283–288
Reeser RJ, Medler RT, Billington SJ, Jost BH, Joens LA (2007) Characterization of *Campylobacter jejuni* biofilms under defined growth conditions. Appl Environ Microbiol 73:1908–1913
Reisner A, Krogfelt KA, Klein BM, Zechner EL, Molin S (2006) In vitro biofilm formation of commensal and pathogenic *Escherichia coli* strains: impact of environmental and genetic factors. J Bacteriol 188:3572–3581
Rieu A, Briandet R, Habimana O, Garmyn D, Guzzo J, Piveteau P (2008) Listeria monocytogenes EGD-e biofilms: no mushrooms but a network of knitted chains. Appl Environ Microbiol 74:4491–4497

Rodrigue DC, Tauxe RV, Rowe B (1990) International increase in *Salmonella enteritidis*: a new pandemic? Epidemiol Infect 105:21–27
Rosenberg M, Kjelleberg S (1986) Hydrophobic interactions: role in bacterial adhesion. Adv Microbiol Ecol 9:353–393
Rubino S, Muresu E, Solinas M, Santona M, Paglietti B, Azara A, Schiaffino A, Santona A, Maida A, Cappuccinelli P (1998) IS*200* fingerprint of *Salmonella enterica* serotype Typhimurium human strains isolated in Sardinia. Epidemiol Infect 120:215–222
Rycroft AN, Garside LH (2000) *Actinobacillus* species and their role in animal disease. Vet J 159:18–36
Samelis J, Sofos JN (2003) Strategies to control stress-adapted pathogens. In: Yousef AE, Juneja VK (eds) Microbial stress adaptation and food safety. CRC, Boca Raton, FL, pp 303–351
Samra Z, Kaufmann L, Zeharia A, Ashkenazi S, Amir J, Bahar J, Reischl U, Naumann L (1999) Optimal detection and identification of *Mycobacterium haemophilum* in specimens from pediatric patients with cervical lymphadenopathy. J Clin Microbiol 37:832–834
Sauer K, Camper AK, Ehrlich GD, Costerton JW, Davies DG (2002) *Pseudomonas aeruginosa* displays multiple phenotypes during development as a biofilm. J Bacteriol 184:1140–1154
Sawhney R, Berry V (2009) Bacterial biofilm formation, pathogenicity, diagnostics and control: an overview. Indian J Med Sci 63:313–321
Schelonka RL, Ascher DP, McMahon DP, Drehner DM, Kuskie M (1994) Catheter related sepsis caused by *Mycobacterium avium* complex. Pediatr Infect Dis J 13:236–238
Searcy KE, Packman AI, Atwill ER, Harter T (2006) Capture and retention of *Cryptosporidium parvum* oocysts by *Pseudomonas aeruginosa* biofilms. Appl Environ Microbiol 72:6242–6247
Seidler MJ, Salvenmoser S, Müller FM (2008) *Aspergillus fumigatus* forms biofilms with reduced antifungal drug susceptibility on bronchial epithelia cells. Antimicrob Agents Chemother 52:4130–4136
Shahamat M, Paszko-Kolva C, Yamamoto H, Colwell R (1989) Ecological studies of *Campylobacter pylori*. Klin Wochenschr 67:62–63
Sharma S, Anand SK (2002) Characterization of constitutive microflora of biofilms in dairy processing lines. Food Microbiol 19:627–636
Silva FR, Mattos EM, Coimbra MV, Ferreira-Carvalho BT, Figueiredo AM (2001) Isolation and molecular characterization of methicillin-resistant coagulase-negative staphylococci from nasal flora of healthy humans at three community institutions in Rio de Janeiro City. Epidemiol Infect 127:57–62
Skirrow MB (1994) Diseases due to *Campylobacter*. Helicobacter and related bacteria. J Compar Pathol 111:113–149
Skraber S, Helmi K, Willame R, Ferréol M, Gantzer C, Hoffmann L, Cauchie HM (2007) Occurrence and persistence of bacterial and viral faecal indicators in wastewater biofilms. Water Sci Technol 55:377–385
Smith MA, Ross MW (2002) Postoperative infection with *Actinobacillus* spp. in horses: 10 cases (1995–2000). J Am Vet Med Assoc 221:1306–1310
Sofos JN (2005) Improving the safety of fresh meat. Woodhead, Cambridge
Solano C, García B, Valle J, Berasain C, Ghigo JM, Gamazo C, Lasa I (2002) Genetic analysis of *Salmonella enteritidis* biofilm formation: critical role of cellulose. Mol Microbiol 43:793–808
Somers EB, Schoeni JL, Wong AC (1994) Effect of trisodium phosphate on biofilm and planktonic cells of *Campylobacter jejuni*, *Escherichia coli O157:H7*. *Listeria monocytogenes* and *Salmonella typhimurium*. Int J Food Microbiol 22:269–276
Songer JG (1997) Clostridial diseases of animals. In: Rood JI, McClane BA, Songer JG, Titball RW (eds) The clostridia: molecular biology and pathogenesis. Academic Press, San Diego, CA, pp 153–182
Srinivasan A, Kraus CN, DeShazer D, Becker PM, Dick JD, Spacek L, Bartlett JG, Byrne WR, Thomas DL (2001) Glanders in a military research microbiologist. N Engl J Med 345:256–258
Sriskandan S, Slater JD (2006) Invasive disease and toxic shock due to zoonotic *Streptococcus suis*: an emerging infection in the east? Public Libr Sci Med 3:e187

Steed KA, Falkinham JOIII (2006) Effect of growth in biofilms on chlorine susceptibility of *Mycobacterium avium* and *Mycobacterium intracellulare*. Appl Environ Microbiol 72:4007–4011

Steitz A, Feddersen A, Freytag C, Daniello S, Schopf RE, Böcher WO, Bhakdi S, Husmann M (1997) Rapid identification of *Mycobacterium marinum* by comparative 16S-rRNA-gene analysis in five cases of progredient cutaneous infections. Eur J Dermatol 7:295–299

Stewart PS, Franklin MJ (2008) Physiological heterogeneity in biofilms. Nat Rev Microbiol 6:199–210

Stickler D, Ganderton L, King J, Nettleton J, Winters C (1993) *Proteus mirabilis* biofilms and the encrustation of urethral catheters. Urol Res 21:407–411

Stone R (2007) Racing to defuse a bacterial time bomb. Science 317:1022–1024

Stoodley P, Wilson S, Hall-Stoodley L, Boyle JD, Lappin-Scott HM, Costerton JW (2001) Growth and detachment of cell clusters from mature mixed species biofilms. Appl Environ Microbiol 67:5608–5613

Stoodley P, Sauer K, Davies DG, Costerton JW (2002) Biofilms as complex differentiated communities. Annu Rev Microbiol 56:187–209

Storey MV, Ashbolt NJ, Stenström TA (2004) Biofilms, thermophilic amoebae and *Legionella pneumophila*, a quantitative risk assessment for distributed water. Water Sci Technol 50: 77–82

Sung JY, Leung JW, Shaffer EA, Lam K, Olson ME, Costerton JW (1992) Ascending infection of the biliary tract after surgical sphincterotomy and biliary stenting. J Gastroenterol Hepatol 7:240–245

Szewzyk U, Szewzyk R, Manz W, Schleifer KH (2000) Microbiological safety of drinking water. Annu Rev Microbiol 54:81–127

Takeda S, Arashima Y, Kato K, Ogawa M, Kono K, Watanabe K, Saito T (2003) A case of *Pasteurella haemolytica* sepsis in a patient with mitral valve disease who developed a splenic abscess. Scand J Infect Dis 35:764–765

Talan DA, Staatz D, Staatz A, Goldstein EJC, Singer K, Overturf GD (1989) *Staphylococcus intermedius* in canine gingiva and canine-inflicted human wound infections: laboratory characterization of a newly recognized zoonotic pathogen. J Clin Microbiol 27:78–81

Tanner MA, Everett CL, Youvan DC (2000) Molecular phylogenetic evidence for noninvasive zoonotic transmission of *Staphylococcus intermedius* from a canine pet to a human. J Clin Microbiol 38:1628–1631

Taponen S, Jantunen A, Pyorala E, Pyorala S (2003) Efficacy of targeted 5-day combined parenteral and intramammary treatment of clinical mastitis caused by penicillin-susceptible or penicillin-resistant *Staphylococcus aureus*. Acta Vet Scand 44:53–62

Taylor RH, Falkinham JO III, Norton CD, LeChevallier MW (2000) Chlorine, chloramine, chlorine dioxide, and ozone susceptibility of *Mycobacterium avium*. Appl Environ Microbiol 66:1702–1705

Teixeira PC, Leite GM, Domingues RJ, Silva J, Gibbs PA, Ferreira JP (2007) Antimicrobial effects of a microemulsion and a nanoemulsion on enteric and other pathogens and biofilms. Int J Food Microbiol 118:15–19

Temmerman R, Vervaeren H, Noseda B, Boon N, Verstraete W (2006) Necrotrophic growth of *Legionella pneumophila*. Appl Environ Microbiol 72:4323–4328

Teuber M, Schwarz F, Perreten V (2003) Molecular structure and evolution of the conjugative multiresistance plasmid pRE25 of *Enterococcus faecalis* isolated from a raw-fermented sausage. Int J Food Microbiol 88:325–329

Tomaras AP, Dorsey CW, Edelmann RE, Actis LA (2003) Attachment to and biofilm formation on abiotic surfaces by *Acinetobacter baumannii*: involvement of a novel chaperoneusher pili assembly system. Microbiology 149:3473–3484

Tormo MA, Martí M, Valle J, Manna AC, Cheung AL, Lasa I, Penadés JR (2005) SarA is an essential positive regulator of *Staphylococcus epidermidis* biofilm development. J Bacteriol 187:2348–2356

Torvinen E, Lehtola MJ, Martikainen PJ, Miettinen IT (2007) Survival of *Mycobacterium avium* in drinking water biofilms as affected by water flow velocity, availability of phosphorus and temperature. Appl Environ Microbiol 73:6201–6207

Trachoo N, Frank JF (2002) Effectiveness of chemical sanitizers against *Campylobacter jejuni*-containing biofilms. J Food Prot 65:1117–1121

Trachoo N, Frank JF, Stern NJ (2002) Survival of *Campylobacter jejuni* in biofilms isolated from chicken houses. J Food Prot 65:1110–1116

Tremoulet F, Duche O, Namane A, Martinie B, Labadie JC (2002) A proteomic study of *Escherichia coli* O157:H7 NCTC 12900 cultivated in biofilm or in planktonic growth mode. FEMS Microbiol Lett 215:7–14

Troller JA (1993) Sanitation in food processing, 2nd edn. Academic Press, San Diego, pp 124–126

Uehling DT (1991) Current concepts of the urinary bladder defenses against infection. Int Urogynecol J 2:32–35

Valle J, Toledo-Arana A, Berasain C, Ghigo JM, Amorena B, Penadés JR, Lasa I (2003) SarA and not sigmaB is essential for biofilm development by *Staphylococcus aureus*. Mol Microbiol 48:1075–1087

Van Der Wende E, Characklis WG (1990) Biofilms in potable water distribution systems. In: McFeters GA (ed) Drinking water microbiology: progress and recent developments, Brock Springer Series in Contemporary Bioscience. Springer, New York, pp 249–268

van Duijkeren E, Houwers DJ, Schoormans A, Broekhuizen-Stins MJ, Ikawaty R, Fluit AC, Wagenaar JA (2008) Transmission of methicillin-resistant Staphylococcus intermedius between humans and animals. Vet Microbiol 128:213–215

van Duynhoven YTHP, de Jonge R (2001) Transmission of *Helicobacter pylori*: a role for food? Bull World Health Org 79(5). http://www.who.int/bulletin/archives/79(5)455.pdf. Accessed 13 Sept 2009

Van Loosdrecht MC, Heijnen JJ, Eberl H, Kreft J, Picioreanu C (2002) Mathematical modelling of biofilm structures. Antonie Leeuwenhoek 81:245–256

Vandamme P, Harrington CS, Jalava K, On SLW (2000) Misidentifying helicobacters: the *Helicobacter cinaedi* example. J Clin Microbiol 38:2261–2266

Vandenesch F, Célard M, Arpin D, Bes M, Greenland T, Etienne J (1995) Catheter-related bacteremia associated with coagulase-positive *Staphylococcus intermedius*. J Clin Microbiol 33:2508–2510

Vaneechoutte M, Devriese LA, Dijkshoorn L, Lamote B, Deprez P, Verschraegen G, Haesebrouck F (2000) *Acinetobacter baumannii*-infected vascular catheters collected from horses in an equine clinic. J Clin Microbiol 38:4280–4281

Varga J, Stirewalt VL, Melville SB (2004) The CcpA protein is necessary for efficient sporulation and enterotoxin gene (*cpe*) regulation in *Clostridium perfringens*. J Bacteriol 186:5221–5229

Varga JJ, Therit B, Melville SB (2008) Type IV Pili and the CcpA protein are needed for maximal biofilm formation by the Gram-positive anaerobic pathogen *Clostridium perfringens*. Infect Immun 76:4944–4951

Vasudevan P, Nair MK, Annamalai T, Venkitanarayanan KS (2003) Phenotypic and genotypic characterization of bovine mastitis isolates of *Staphylococcus aureus* for biofilm formation. Vet Microbiol 92:179–185

Vejborg RM, Klemm P (2008) Blocking of bacterial biofilm formation by a fish protein coating. Appl Environ Microbiol 74:3551–3558

Vengust M, Anderson ME, Rousseau J, Weese JS (2006) Methicillin-resistant staphylococcal colonization in clinically normal dogs and horses in the community. Lett Appl Microbiol 43:602–606

Vorachit M, Lam K, Jayanetra P, Costerton JW (1995) Electron microscopy study of the mode of growth of *Pseudomonas pseudomallei* in vitro and in vivo. J Trop Med Hyg 98: 379–391

Vuong C, Otto M (2002) *Staphylococcus epidermidis* infections. Microbes Infect 4:481–489

Wagner VE, Bushnell D, Passador L, Brooks AI, Iglewski BH (2003) Microarray analysis of *Pseudomonas aeruginosa* quorum-sensing reulons: effects of growth phase and environment. J Bacteriol 185:2080–2095

Walker JT, Bradshaw DJ, Bennett AM, Fulford MR, Martin MV, Marsh PD (2000) Microbial biofilm formation and contamination of dental-unit water systems in general dental practice. Appl Environ Microbiol 66:3363–3367

Walker TS, Tomlin KL, Worthen GS, Poch KR, Lieber JG, Saavedra MT, Fessler MB, Malcolm KC, Vasil ML, Nick JA (2005) Enhanced *Pseudomonas aeruginosa* biofilm development mediated by human neutrophils. Infect Immun 73:3693–3701

Walsh D, Portaels F, Meyers W (2009) Buruli ulcer (*Mycobacterium ulcerans* infection): a re-emerging disease. Clin Microbiol Newsl 31:119–127

Watnick PI, Kolter R (1999) Steps in the development of a *Vibrio cholerae* El Tor biofilm. Mol Microbiol 34:586–595

Watnick P, Kolter R (2000) Biofilm, city of microbes. J Bacteriol 182:2675–2679

Webb JS, Givskov M, Kjelleberg S (2003) Bacterial biofilms: prokaryotic adventures in multicellularity. Curr Opin Microbiol 6:578–585

Weigel LM, Donlan RM, Shin DH, Jensen B, Clark NC, McDougal LK, Zhu W, Musser KA, Thompson J, Kohlerschmidt D, Dumas N, Limberger RJ, Patel JB (2007) High-level vancomycin-resistant *Staphylococcus aureus* isolates associated with a polymicrobial biofilm. Antimicrob Agents Chemother 51:231–238

Weissenberger CA, Cazalet C, Buchrieser C (2007) *Legionella pneumophila* – a human pathogen that co-evolved with fresh water protozoa. Cell Mol Life Sci 64:432–448

Weyant RS, Moss CW, Weaver RE, Hollis DG, Jordan JJ, Cook EC, Daneshvar MI (1996) Identification of unusual pathogenic gram-negative aerobic and facultatively anaerobic bacteria, 2nd edn. Williams & Wilkins, Baltimore, MD

Wheelis M (1998) First shots fired in biological warfare. Nature 395:213

Whipps CM, Dougan ST, Kent ML (2007) *Mycobacterium haemophilum* infections of zebrafish (*Danio rerio*) in research facilities. FEMS Microbiol Lett 270:21–26

Whitchurch CB, Tolker-Nielsen T, Ragas PC, Mattick JS (2002) Extracellular DNA required for bacterial biofilm formation. Science 295:1487

Whittington RJ, Marshall DJ, Nicolls PJ, Marsh IB, Raddacliff LA (2004) Survival and dormancy of *Mycobacterium avium* subsp. *paratuberculosis* in the environment. Appl Environ Microbiol 70:2989–3004

WHO (2005) Legionellosis. WHO, Geneva. http://www.who.int/mediacentre/factsheets/fs285/en/index.html. Accessed 24 Oct 2009

Wiencek KM, Klapes NA, Foegeding PM (1990) Hydrophobicity of *Bacillus* and *Clostridium* spores. Appl Environ Microbiol 56:2600–2605

Wimpenny J (2000) An overview of biofilms as functional communities. In: Allison D (ed) Community structure and co-operation in biofilms. Cambridge University Press, West Nyack, NY, p 1

Wimpenny J, Manz W, Szewzyk U (2000) Heterogeneity in biofilms. FEMS Microbiol Rev 24:661–671

Wirtanen G (2002) Equipment hygiene in the food processing industry ± hygiene problems and methods of controlling *Listeria monocytogenes*, VTT Publications 480. Otamedia Oy, Espoo. In Finnish

Wirtanen G, Mattila-Sandholm T (1992) Effect of the growth phase of food-borne biofilms on their resistance to a chlorine sanitizer Part II. Food Sci Technol 25:50–54

Wirtanen G, Salo S (2004) DairyNET ± hygiene control in Nordic dairies, VTT Publication 545. Otamedia Oy, Espoo

Witte W, Strommenger B, Stanek C, Cuny C (2007) Methicillin-resistant *Staphylococcus aureus* ST398 in humans and animals, Central Europe. Emerg Infect Dis 13:255–258

Wolyniak EA, Hargreaves BR, Jellison KL (2009) Retention and release of *Cryptosporidium parvum* oocysts by experimental biofilms composed of a natural stream microbial community. Appl Environ Microbiol 75:4624–4626

Wong ACL (1998) Biofilms in food processing environments. J Dairy Sci 81:2765–2770
Wright JB, Athar MA, van Olm TM, Wootliff JS, Costerton JW (1989) Atypical legionellosis: isolation of *Legionella pneumophila* serogroup 1 from a patient with aspiration pneumonia. J Hosp Infect 13:187–190
Wuertz S, Bishop PL, Wilderer PA (2003) Biofilms in wastewater treatment: an interdisciplinary approach. IWA, London
Yao Y, Sturdevant DE, Otto M (2005) Genomewide analysis of gene expression in *Staphylococcus epidermidis* biofilms: insights into the pathophysiology of *S. epidermidis* biofilms and the role of phenol-soluble modulins in formation of biofilms. J Infect Dis 191:289–298
Yarwood JM, Bartels DJ, Volper EM, Greenberg EP (2004) Quorum sensing in *Staphylococcus aureus* biofilms. J Bacteriol 186:1838–1850
Yu VL, Plouffe JF, Pastoris MC, Stout JE, Schousboe M, Widmer A, Summersgill J, File T, Heath CM, Paterson DL, Chereshsky A (2002) Distribution of *Legionella* species and serogroups isolated by culture in patients with sporadic community-acquired legionellosis: an international collaborative survey. J Infect Dis 186:127–128
Zanen HC, Engel HWB (1975) Porcine streptococci causing meningitis and septicemia in man. Lancet 1:1286–1288
Zeiss CJ, Jardine J, Huchzermeyer H (1994) A case of disseminated tuberculosis in a dog caused by *Mycobacterium avium-intracellulare*. J Am Anim Hosp Assoc 30:419–424
Zhang GH, Sun YX, Gu P, Ding SY (2006) Inactivation of pathogenic microorganisms by ultraviolet. Technol Water Treat 32:5–8
Zong ZY, Lu XJ, Gao YY (2002) *Aeromonas hydrophilia* infection: clinical aspects and therapeutic options. Rev Med Microbiol 13:151
Zottola EA, Sasahara KC (1994) Microbial biofilms in the food industry – should they be a concern? Int J Food Microbiol 23:125–148

Biofilms and Role to Infection and Disease in Veterinary Medicine

Alice J. Gardner, Steven L. Percival, and Christine A. Cochrane

Abstract Biofilms play an increasing role within the medical and veterinary community. Due to the increased resistance of a biofilm, they can have direct and indirect effects upon a range of infections and diseases including chronic non-healing wounds, implant/prosthesis infection and mastitis. These problems can have significant effects on other industries, for example mastitis can have a detrimental effect on milk yield in the dairy industry. The degree of severity biofilms can cause increases the pressure on the veterinary industry to diagnose and treat infections and diseases quicker and with more effective results. With maturity, biofilms may become more resistant to the effects of antimicrobials which make the infection harder to treat. As elaborated on in previous chapters, many antibiotherapy treatments currently used to treat bacterial infections are aimed at planktonic bacterial cells as opposed to cells encased in a biofilm; this makes their treatment increasingly problematic. Without adequate diagnostic and treatment protocols to treat veterinary biofilms, their impact will remain a significant challenge.

A.J. Gardner
Institute of Biological, Environmental and Rural Sciences, Aberystwyth University, Ceredigion SY23 3AL, UK

S.L. Percival
Department of Pathology, Medical School, West Virginia University, Morgantown, WV, USA
e-mail: biofilms@hotmail.co.uk

C.A. Cochrane (✉)
Faculty of Health and Life Sciences, Institute of Ageing and Chronic Disease, University of Liverpool, Leahurst, Neston. CH64 7TE, United Kingdom
e-mail: c.a.cochrane@liverpool.ac.uk

S.L. Percival et al. (eds.), *Biofilms and Veterinary Medicine*,
Springer Series on Biofilms 6, DOI 10.1007/978-3-642-21289-5_4,

1 Introduction

As mentioned previously in this book, bacteria evident in the natural environment (animal host, rivers, water systems) exist in two phenotypic states: planktonic (free floating) and sessile (attached – biofilm) (Clutterbuck et al. 2007). The bacterial state of existence specifically determines susceptibility, in particular to eradication and destruction by the host immune system and responses to antibiotherapeutic measures. Compared with their planktonic counterparts, sessile bacteria within biofilms, are known to have heightened resistance against lethal destruction from immune complexes and antimicrobial drugs. This increased resistance can be attributed to a number of reasons including a reduced bacterial metabolism, a self-produced extracellular matrix, which can sequester many antimicrobial agents, a specific biofilm phenotype, a biofilm structure that can be altered depending on outside perturbation and the specific growth response to on a particular surface (Costerton et al. 1999).

Biofilms can cause significant health problems in both humans and animals, including chronic non-healing wounds, mastitis and medical implant/prosthesis infection. Polymeric substances, exo-enzymes and toxins found within a biofilm may have a direct effect on body repair, for example, wound healing. They also have an indirect effect on the length of the inflammatory state. Typically, biofilms prolong inflammation in the host culminating in the formation of a chronic infection (Costerton et al. 1999; Donlan and Costerton 2002; Parsek and Singh 2003). A prolonged inflammatory stage is particularly pertinent in relation to failure of wound healing.

In animal species, the risk of infection is probably considerably greater than the risk in humans. This is due to the difference in animal housing and living environments – animals naturally frequent environments with a large and much more diverse microbial community. Where contamination is defined as "the presence of non-replicating microorganisms", Krahwinkel and Boothe (2006) identified three potential sources of bacterial contamination. These were environment, cutaneous and endogenous (Krahwinkel and Boothe 2006). The suggestion behind the "hygiene hypothesis" is that a certain level of environmental, cutaneous and endogenous bacteria is safe and indeed, important to general health; these are commonly referred to as commensal or "friendly bacteria". However, when the populations of bacteria increase or when the bacteria gain access to an area where they are not usually present or where they are given the opportunity to form a pathogenic biofilm, bacterial colonisation and proliferation then becomes a health threat.

In addition to their ability to colonise body surfaces, biofilms are able to colonise artificial surfaces including tubing and implants, such as intravenous catheters, the teeth and gingivae, the lungs, the ears, the urogenital tract and course wounds (Potera 1999). Perhaps, the two most prominent issues related to biofilm infection are chronic wound infections and sub-clinical and clinical mastitis. To a lesser extent in veterinary species medical device infection is an additional potential danger. The ability to colonise a variety of surfaces within a host has important clinical implications in both human and veterinary medicine.

1.1 Biofilm Development

A biofilm is described as a cluster of bacterial cells that are surrounded by a self-produced exo-polymeric matrix that is irreversibly attached to an abiotic or biotic surface (Costerton 2004). The development of a biofilm can vary between different bacterial species; some organisms such as *Pseudomonas* spp. are able to form a biofilm rapidly in a variety of different environmental conditions, whereas other species need more specific environmental condition, such as a specific pH, the availability of defined nutrients and specific environmental temperatures (Fera et al. 1989; Ferris et al. 1989; Cowan et al. 1991; O'Toole et al. 2000). If the invading bacterial species is capable of rapidly and easily forming a biofilm then, where a biofilm is present, the treatment of an infection will become more complicated. This added difficulty can be attributed at least partially to the inherent ability of a biofilm to evade the host's defence mechanisms and to resist the effects of antimicrobial drugs. Biofilm-related infections are typically slow to develop and are usually persistent. They rarely resolve under the host immune defences alone and are only transiently susceptible to antibiotherapy (Parsek and Singh 2003).

Biofilms are frequently associated with chronic infections and disease. A good example is cystic fibrosis in humans (Lee et al. 2005). Chronic infections tend to be less aggressive than acute infections but by definition persist for months and even years if the underlying cause is not eliminated. Due to the nature of their development, the biofilms can cause periods of quiescence and acute exacerbation during the course of an infection/disease (Costerton et al. 2003).

Biofilm development can be divided into five stages;

1. Initial microbe attachment to a surface
2. Formation of the exopolymeric matrix
3. Early development of biofilm architecture
4. Maturation of biofilm architecture
5. Dispersion of bacterial cells from the biofilm (Lappin-Scott and Bass 2001; Lasa 2006)

Once the biofilm has reached "maturity" and has an optimal cell mass, dispersion of the outer layer of cells is initiated. This may provide an explanation for the acute exacerbation of clinical signs encountered in biofilm infections and disease (Hall-Stoodley and Stoodley 2005). The dispersal stage of biofilm development is exacerbated after any assault on the biofilm, for example, exacerbation may follow antibiotherapy. The ability of the biofilm to cause a persistent infection makes it important to diagnose the presence of a biofilm and to prevent its maturation.

1.2 Diagnosis

Biofilm-related infections can cause significant difficulties in the initial diagnosis of a complaint and may explain some circumstances of a wrong diagnosis or a missed diagnoses. In many circumstances swab or blood culture yields negative results even in the face of a strong clinical impression of a significant infection. These negative results might suggest that the underlying causes are not bacteriological in origin when in fact they are. However, many negative culture results are due to the presence of a biofilm; cells within a biofilm are protected by their extracellular polymeric matrix. The percentage of matrix material to each bacterial cell is substantial. A biofilm is made up of around 15% bacterial cells and 85% matrix material (Donlan and Costerton 2002). The genes involved in the formation of biofilm matrices are rapidly up-regulated by the adherent cells on a surface; this up-regulation may be significant within 12 min of the initial adherence of the cells to the surface (Davies et al. 1993; Davies and Geesey 1995).

The rapid formation and strength of the extracellular matrix provide protection to the bacterial cells from the mechanical action of swabbing. The extracellular matrix can also lose water from its highly hydrated form to create a harder, less permeable outer "shell" thereby affording further protection for the cells within it (Sutherland 2001). Although these are those that will cause the most morbidity and long-term health issues, the organisms within the biofilm are not represented in cultures taken from the infected site. The more time the biofilm has to mature the harder the biofilm becomes to diagnose and eradicate and therefore rapid, early detection of a developing biofilm is essential.

When presented with bacterial infection or disease, the presence of a developing or mature biofilm should be identified as quickly as possible. Many of the most accurate diagnostic tests rely on the immune response of the host. Enzyme-linked immunosorbent assays (ELISA) have proven useful in the early detection of biofilm infection. These tests are used to detect the host antibodies against biofilm-specific epitopes that are common to all staphylococcal species, a major perpetrator of biofilm growth. ELISA tests are amongst the most sensitive methods of detection, since they directly detect the antibody or antigen of interest as opposed to agglutination tests, which are secondary binding tests.

1.3 Immune Response to Infection

The successful control and eradication of an infectious disease requires an "understanding of the characteristics and behaviour of the agent causing the disease and its interaction with the host and the environment" (Nicoll et al. 2005). When a host is initially challenged by a bacterial infection, the immune system responds through a combination of innate immune responses (non-specific response provided by complement proteins and phagocytic cells including neutrophils and macrophages) and the adaptive (humoral) immune response (specific response to an antigen provided by various lymphocyte branches). These two branches of the immune system are linked through the macrophages role as antigen-presenting cells (Carrick and Begg 2008).

In the event of a bacterial invasion, phagocytic cells of the innate immune response migrate to the site of infection and begin to engulf and phagocytose the foreign cells. Neutrophils are usually the first host cells to reach the infected area and to begin phagocytosis. Once neutrophils begin to phagocytose the invading microorganism, the respiratory burst is initiated and the toxic products of the neutrophil are released and these are usually effective at killing the pathogen. Toxic products released by neutrophils upon interaction with a bacterial cell (or other foreign antigen) contain antimicrobial and cytotoxic substances that are able to kill phagocytosed bacterial cells and therefore neutrophil degranulation is an essential part of the inflammatory response. However, in a biofilm-forming infection, this stage of the immune response is ineffective; the bacterial cells remain unharmed and viable. The respiratory burst attracts more neutrophils and macrophages to the area. Macrophages are phagocytic cells alongside neutrophils; however they also act as antigen-presenting cells which have specific receptors that recognise pathogen-associated molecular patterns (PAMPs) on the surface of bacterial cells. Once bound to the PAMP, macrophages synthesise and secrete a mixture of products, including interleukin-1 (Il-1) and tumour necrosis factor-α (TNF-α), which trigger inflammation and activate the adaptive immune response.

Once activated, the adaptive immune responds through a T-lymphocytes cell-mediated immune (CMI) response and a B-cell-mediated humoral response. CMI initiation is triggered once the bacterial cells have been engulfed by the macrophages and the PAMP is presented alongside major histocompatability complexes (MHC) class I and II molecules. $CD4^+$ molecules on T cells recognise MHC class II molecules whilst CD8 molecules recognise MHC class I molecules. Once the combination of PAMP and MHC class II on a macrophage is recognised by a T cell, the T cell differentiates into T-helper (Th) cells of two subtypes; Th1 and Th2 for cell-mediated immunity and antibody production, respectively. MHC class I molecules process cytotoxic T cells which have antimicrobial killing actions. The humoral response is responsible for initiating production of mature B cells which are able to produce antibodies directed against the invading microorganism. The humoral response provides the secondary wave of immune defence to support the initial phagocytosing cells.

The humoral response is particularly important in biofilm-related infection due to the ability of the biofilm to evade the initial immune response of neutrophils and macrophages. It has been demonstrated that in *Pseudomonas aeruginosa* biofilms quorum sensing molecules synthesise a specific lipid called rhamnolipid. This effectively eliminates polymorphonuclear neutrophils which are essential to the innate immune response to infection (Jensen et al. 2007; Bjarnsholt et al. 2008). Due to the persistence of a biofilm, the host immune response continues its assault with neutrophils and macrophages in an attempt to destroy the biofilm. However, continued assault results in degradation of the surrounding host tissue as a result of the prolonged inflammatory response, prolonged exposure to high levels of cytokines and proteases and toxic substances released by effete neutrophils (Carrick and Begg 2008).

The ability of immune cells to phagocytose the invading bacterial cells differs between planktonic cells and cells that have originated from a biofilm. Jesaitis et al.

(2003) studied the hosts' neutrophil response to a *Pseudomonas aeruginosa* biofilm and concluded that the neutrophils were, in fact, able to penetrate the biofilm. However, upon contact with the biofilm neutrophils had limited degranulation capacity and reduced motility. Furthermore, the neutrophil was only partially activated at the neutrophil–biofilm interface (Jesaitis et al. 2003). When neutrophils are unable to efficiently interact with a biofilm and release their antimicrobial products the biofilm remains intact and can continue to cause infection. Cerca et al. (2006) also found that the host immune cells were able to successfully penetrate *Staphylococcus epidermidis* biofilms but were unable to engulf and phagocytose the bacterial cells within the biofilm. They also reported that excessive extracellular bacterial antigens were released from the biofilm (Cerca et al. 2006) and suggested that this would contribute to the escalation of the response of immune cells.

Ordinarily, mucosal surfaces that are at risk from microbial adhesion are protected by the antibody fractions of the host innate immune system that destroys bacterial cells deposited on these surfaces (Ganz 2001). However, cells with a biofilm phenotype have the ability to bypass this fundamental first line host protection barrier and are able to adhere to these surfaces anyway. They may also take advantage of hosts with an impaired immune function and take the opportunity to colonise mucosal surfaces; opportunistic infection is common in immunocompromised animals. With a deficient immune defence against biofilm-associated infections and disease, additional defence mechanisms such as antimicrobials are usually required.

1.4 The Role of Antimicrobials in the Treatment of a Biofilm

The majority of available antimicrobials are aimed at planktonic bacterial cells as opposed to cells encased within a biofilm that have a different phenotype in respect of growth and resistance compared with their planktonic counterparts (Lewis 2001). Continued use of ineffective antimicrobials will have no impact on the biofilm and may only result in increased selection for resistant bacteria (Morley et al. 2005). Collection of swabs from the wound and subsequent culture for aerobes and anaerobes can confirm the bacterial species within the wound and their biofilm-forming capacity. Some bacterial isolates have strong biofilm-forming abilities whilst others have more limited ability or none at all. Part of the resistance mechanism of a biofilm is the presence of persister cells. These comprise around 1% of the total biofilm bacterial population (Brooun et al. 2000; Lewis 2008) and they are able to rapidly reproduce the original bacterial population of the biofilm after an assault from an antimicrobial agent (Lewis 2005, 2008). Therefore, where a biofilm is present, for example, in the treatment of chronic, non-healing wounds, it is paramount that all necrotic tissue is removed; biofilms are commonly present in necrotic tissue and so debridement will minimise the presence of resistant bacteria and persister cells. This approach supports both the host immune system and the effects of antimicrobial drugs in overcoming any remaining infection.

Antimicrobial drugs have minimal effects on eradicating biofilms. Instead they often only suppress possible symptoms of infection by eliminating the planktonic cells that are released from the biofilm (Stewart and Costerton 2001). As a result of this failure, the biofilm may act as a source of continued infection for months or even years if not correctly treated (Costerton et al. 1999). Biofilms release planktonic bacteria systematically during the detachment phase of the biofilm life cycle (Costerton et al. 1995); dissolution of the EPM by enzymes allows the release of bacterial cells from the biofilm into the surrounding area as planktonic cells (Allison et al. 1998). These planktonic cells can rapidly multiply and disperse (Costerton et al. 1999); if they cannot be destroyed by the hosts' antibodies they are free to colonise other surfaces and cause infection. Therefore, the biofilm acts as a reservoir for repeated acute infections after antibiotic or antimicrobial treatments. Due to the co-existence of biofilms and planktonic bacteria within the host, any antimicrobial agent that is used to treat the infection or disease must be capable of eliminating both the biofilm and the free-floating planktonic bacterial cells (Chaw et al. 2005).

When it comes to choosing antibiotherapy, it is important to take into account the state of growth of the cells within the biofilm. The majority of cells encased within the biofilm are in a state of limited growth known as the stationary growth phase where the rate of bacterial cell growth is at maintenance level and cell division has ceased. The effect of an antimicrobial agent is negligible during this state of limited bacterial growth even if the drug has managed to diffuse through the extracellular matrix. Penicillins and cephalosporins are ineffective against non-dividing bacteria and will therefore have no effect against bacteria that are encased within the polymeric matrix (Bishop 2005). Other drugs such as aminoglycosides and fluroquinolones are effective against non-dividing bacterial cells and are able to equilibrate across the surface of the biofilm (Brooun et al. 2000). However, whether the antimicrobial compound has maintained full effectiveness after diffusing through the matrix is questionable.

The correct dosage of an antimicrobial drug is also an important factor in increasing the chances of a positive effect against the biofilm. Brooun et al. (2000) studied *Pseudomonas aeruginosa* biofilms and the dose-dependent killing of its biofilm using the fluroquinolone antibiotics, ofloxacin and ciprofloxacin. They demonstrated that the majority of cells within the biofilm were destroyed by concentrations similar to those that destroyed planktonic cells of the same species. However results revealed that a proportion of the biofilm was left intact. These indestructible cells which are present in a biofilm are the persister cells. Persister cells cannot be destroyed due to their inactivated cell death function. This is a major reason why antimicrobial drug treatments have limited effect at biofilm eradication. To be effective against the entire population of the biofilm increased concentrations of the drugs are necessary. However, the concentrations that would be effective against the biofilm may have adverse effects on the host.

The use of some antimicrobials (tetracyclines, quinopristine/dalfopristins and erythromycin) may have a detrimental effect on the host as they may actually stimulate the formation of a biofilm (Anderson and O'Toole 2010). Instead of lysing the bacterial cells these drugs stimulate the expression of certain genes within the bacterial cell, namely the intercellular adhesion (*ica*) locus in *Staphylococcus epidermidis*. The *ica* locus (including *ica*A, *ica*B, *ica*C, *ica*D genes) is an essential gene for biofilm formation and increased expression of the *ica* locus has been associated with an increased adherence ability of the bacterial cell to a biological surface and as a result increased biofilm formation (Melchior et al. 2006a, b). The choice of antimicrobial is even more essential to ensure no deleterious effects.

1.5 Biofilms and Their Relevance to Infection and Disease in Veterinary Medicine

The ability to treat many bacterial infections such as chronic non-healing wounds and mastitis is related to the ability of bacteria to form a biofilm. In human medicine, the Centre for Disease Control and Prevention (CDC) suggests that as much as 65% of bacterial infections are biofilm related (Potera 1999). Published figures are not available for the prevalence of biofilms within animal bacterial infections but it is likely that the percentage occurrence is similar to that seen in humans, if not more. Unlike planktonic bacterial infections which are typically rapid and acute in onset, in biofilm-related infections and disease there is a temporal delay in the clinical appearance of the evidence for an infection (Ward et al. 1992). In extreme cases of biofilm-related infection and disease, radical surgical removal may be the only option; there have been cases of amputation in human cases of biofilm infection (Wieman 2005).

The following section provides a description of three of the most common issues associated with biofilm infection occurring in both human and veterinary patients. By no means are these the only issues faced by clinicians with regards to biofilm infections and disease; however, these are the issues that are predominate in current research.

1.6 Wounds

Wounds, whether of surgical or traumatic origin, are a frequent occurrence in veterinary clinics and are common sites for biofilm formation. Bacterial infection of wounds is an important aspect of patient care in both human (Burke 2003) and veterinary practice (Orsini et al. 2004). Most wounds support a stable polymicrobial community without any adverse effects ($<10^5$ colony forming units (CFU)/gram of tissue) and without showing any clinical signs of infection (Hansson et al. 1995; Waldron and Zimmerman-Pope 2003). It is important to treat wounds efficiently and effectively so as to avoid escalation of the bacterial burden within the wound and consequent compromise in both the repair process and the overall health status of the patient. Open wounds can provide an environment conducive to the formation of a biofilm (Xu et al. 2000; Krahwinkel and Boothe 2006). Once a biofilm has colonised the wound bed, the repair process is interrupted and fails to progress through the sequential stages of wound healing. Wound healing is not a linear process; the recognised phases run concurrently and are influenced by both intrinsic and extrinsic factors, such as bacteria, bleeding and health of the patient (Waldorf and Fewkes 1995). Wound healing can typically be divided into three phases: (1) haemostasis and acute inflammation, (2) proliferation and (3) remodelling (Cochrane 1997; Theoret 2006). Interruption of the wound repair process due to the presence of a biofilm is primarily identified by a prolonged inflammatory response that results in a chronic non-healing wound. Biofilms also have a more direct effect on the wound repair process due to the presence of polymeric substances, exo-enzymes and toxins present in the biofilm.

At particular risk from chronic non-healing wounds is the equid. Horses are particularly at risk from chronic non-healing wounds of the lower limb similar to venous leg ulcers seen in humans (Theoret 2004). The horse's environment is naturally associated with a plethora of bacteria that have the potential to colonise wounds and form an impenetrable biofilm, particularly wounds of the lower limbs. Once the biofilm has become established within the wound bed, the wound becomes difficult to treat with traditional antibiotic treatments. Successful treatment of chronic non-healing wounds should ideally involve a combined treatment effort using surgical debridement of the biofilm and necrotic tissue, thorough lavage with physiologic saline and topical and/or systemic administration of antibiotics. Debridement is a vital part of the treatment protocol for chronic non-healing wounds as it physically removes the biofilm and reduces the chances of the resistant population of bacteria, the persister cells, replicating to re-colonising and re-infecting the wound.

Ordinarily the host immune system is capable of coping with a bacterial infection when the bacteria are in planktonic form. However, when bacteria adhere to the surface and form biofilms the host immune cells have a limited capacity to destroy the bacterial cells encased within its protective matrix. There are also issues with those cells that have been released from the biofilm by dispersal mechanisms but have not reverted to a true planktonic phenotype in which they would be truly susceptible (Kharazmi 1991; Parsek and Singh 2003; Fedtke et al. 2004).

The issue for the clinician is whether the antimicrobials to be used to combat the wound infection are effective against biofilms. Most commonly used antimicrobials are ineffective in destroying a biofilm as the drugs were developed with planktonic bacterial infections in mind (Lewis 2001; Percival and Bowler 2004; Clutterbuck et al. 2007; Percival et al. 2008). Swabs taken from the wound for culture for aerobis and anaerobic bacteria can confirm the bacterial species within the wound and their biofilm-forming capacity can usually then be predicted.

Different bacteria may predominate in wounds on different host species, for example, *Sarcocystis felis* is commonly associated with cat wounds (Songer and Post 2005). Although horses have a particularly high risk of biofilm-infected lower limb wounds, biofilms are also present in other wounds in different species. However, similar treatment protocols are used with any biofilm-related wound infection irrespective of the host. The antibiotherapy choice and dosage would be the only differing aspect to the treatment of the infection.

1.7 Medical Implant and Prostheses

Biofilm infection in humans with medical implant devices have been frequently documented and pose a particular risk to individuals with an impaired immune system (Khoury et al. 1992; Pawlowski et al. 2005). Occlusion of the lumen of catheters has also been documented, in particular urinary catheters (Donlan 2001; Suller et al. 2005). Typically the microorganism associated with human urinary catheter occlusion is *Proteus mirabilis* either as a homogenous colony or as a heterogeneous colony (Stickler et al. 1993). Although there is less evidence to support the same event in animals, it is reasonable to assume that similar occurrences occur. The capacity of many bacteria to adhere and colonise to the surface of medical and prosthesis implants such as intravenous catheters is a major contributing factor to post-surgical bacterial infections (Costerton et al. 1999). Microscopic damage and imperfections to the surface of tubing during manufacture and damage during use can also increase the risk of biofilm formation; bacteria come to reside in the indentations on the surface (Khoury et al. 1992).

Exposure to potential pathogenic microorganisms increases the risk of medical implant infection. A recent study reported that the presence of an infected wound significantly increased the risk for prosthetic vascular graft infection ($p = 0.001$) (Antonios et al. 2006). Compromised immune function due to the initial bacterial infection may also provide an opportunity for bacteria to freely colonise the surface of the medical implant without strong resistance from the host's immune system. Three of the most common microorganisms associated with human catheter-related infections are *Staphylococcus aureus, Staphylococcus epidermidis* and *Escherichia coli* (Choong et al. 2001; Warren 2001; Pascual 2002; Tenke et al. 2006). These species are also the ones most frequently isolated in veterinary patients (Mathews et al. 1996; Marsh-Ng et al. 2007; Edwards et al. 2008) and significantly they are the species commonly isolated as normal skin commensals in both humans and animals (Nagase et al. 2002). However, they are opportunistic pathogens. Where they gain entry into an unusual site, for example, entering the body through catheter implantation and entering systemic circulation they can become clinically significant.

Research to identify a material that is able to resist bacterial adhesion on its surface is ongoing; recently, antimicrobial urological catheters, which are able to inhibit the growth of a variety of microorganisms, have been introduced. In particular, there are reports of benefits on microorganism inhibition when using heparin and sparfloxacin in the manufacture of catheters (Kowalczuk et al. 2010) and serum (Hammond et al. 2010). These developments should have a positive influence on the reduction of bacterial infection occurrence seen in hospitalised human and animal patients.

There is a growing bank of information about the role that common biofilm-forming bacteria such as *Staphylococcus aureus* have in indwelling medical device infections. However, little is known about the biofilm-forming ability of other less well-documented species such as *Chryseobacterium meningosepticum* which has been shown recently to have a strong biofilm-forming capacity. Lin et al. (2010)

also described patients who had an intravascular indwelling catheter and patients who had had an initial inappropriate antibiotic treatment program for *Chryseobacterium meningosepticum* infection had an increased risk of fatality. This study focused on blood stream infection in humans; however, it remains possible that similar problems may occur in veterinary patients with catheter implantation.

There is also a potential risk for bacteria to withstand the chemical and mechanical cleaning processes used for medical equipment. The potential risk of infection from biofilm-contaminated equipment such as endoscope tubing is a highly relevant problem in the clinical environment. Biofilms have been shown to survive and remain viable after cleaning methods that include disinfectants and mechanical brushing. The biofilm retains the potential to infect subsequent patients when the equipment is re-used. Newer techniques and biofilm-removal systems have proved more effective than previous protocols using detergents (Nicoll et al. 2005; Marion et al. 2006). It is important that techniques that are effective at removing biofilm from re-usable devices are utilised to ensure that bacteria encased within a biofilm do not remain on the re-usable devices.

1.8 Mastitis

Mastitis is another significant health issue involving biofilm infection. It affects many species including sheep, pigs, dogs, cats, goats and horses but is most commonly isolated in the cow where the importance placed on the commercial value of dairy products is the highest. Adhesion of bacterial cells (e.g. *Staphylococcus aureus*) to the mammary gland epithelium has been considered the primary step in the pathogenesis of mastitis (Cifrian et al. 1994).

Staphylococcus aureus and *Staphylococcus epidermidis* are commonly associated with bovine mastitis (Oliveira et al. 2007; Melchoir et al. 2009). More recently *Streptococcus uberis* has become a significant isolated species (Wyder et al. 2011). Oliveira et al. (2007) demonstrated that the ability of *Staphylococcus aureus* and *Staphylococcus epidermidis* to form a biofilm increases over time (24, 48 and 72 h). In order to tailor the type and dosage of the antimicrobial used to treat the infection, the accurate identification of the bacteria responsible and its potential ability to form a biofilms should be made.

Initial biofilm formation in mastitis cases occurs approximately 24 h after exposure to the infecting microorganism. Bacterial clusters appear in the mammary alveoli (Fig. 1) and lactiferous ducts and also within the interstitial tissue (Hensen et al. 2000). If the bacterial species is capable of forming a biofilm the cells will begin to proliferate whilst producing their protective shield of polymeric substances. Dispersal of planktonic cells from the biofilm prolongs the state of infection in the animal and presents as chronic mastitis that is difficult to eradicate using typical treatment protocols. Research has confirmed that bacterial strains growing as a biofilm in mastitis are less susceptible to current mastitis therapies (Melchior et al. 2006a, b, 2007).

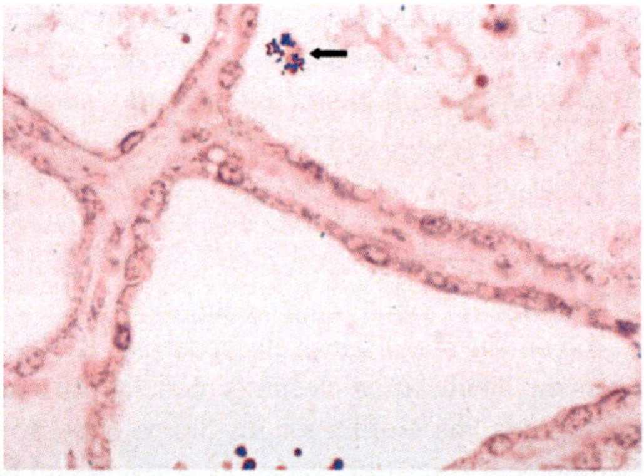

Fig. 1 A cluster of *Staphylococci* (*arrow*) within the mammary alveolus 24 h after intramammary exposure to *Staphylococcus aureus*. From Hensen et al. (2000) with permission

A component of the inflammatory response to mammary infection is an influx of immune cells called somatic cells to the infected gland in order to eliminate the microorganisms (Harmon 1994). Detection of a high somatic cell count (SCC) in milk is suggestive of a mastitis infection and the infection results in a reduction in the quality and amount of milk produced. Somatic cell counts are used to assess the quality of milk produced; a SCC in excess of 400,000 per ml averaged over a 3-month period is indicative of poor-quality milk and unacceptably high rates of infection in the animal (Browing et al. 1995). The milk also dilutes the concentration of immune cells which limits their ability to effectively clear the infected gland of the invading pathogen. Additionally some components of milk (casein and fat) reduce the bactericidal abilities of the immune cells (Targowski 1983). Milk provides a favourable environment for the microorganism to flourish (Chin 1982; Gillespie et al. 2003). It provides sufficient nutrients and the ideal opportunity to form a biofilm due to the high shear forces caused by the milking process. Once the biofilm matures and reaches a critical cell mass the outer layer of cells are dispersed into the surrounding space and are free to colonise other surfaces and may even enter the circulation and cause systemic infection (Melchior et al. 2006a, b). The dispersion of cells from the biofilm is partly responsible for the repeated "flair ups" seen in the clinical mastitis cases; many of the recently released bacterial cells retain at least part of their resistance to host immune cells and to antimicrobial therapy. However many will eventually revert back to their original phenotype and once again become susceptible to any antimicrobials present in the circulation. It is the cells that are released from the biofilm that give rise to clinical signs such as inflammation; therefore the clinical signs will dissipate once they are destroyed by antimicrobial therapy. However there may be a residual sub-clinical infection as a result of a biofilm.

Although there has been a reduction in the number of clinical mastitis cases diagnosed in cattle, the incidence of chronic sub-clinical mastitis cases has risen. This can reasonably be blamed on biofilm formation within the mammary epithelium. Many of the mastitis-causing bacteria are endemic in the environment of cattle, particularly those housed in indoor barns; unhygienic conditions increases the risk of infection. Therefore mastitis cases are encountered more frequently when cattle are typically housed indoors in winter months. Many of the bacteria present are non-pathogenic, however some environmental bacterial species do have the potential to be pathogenic. Group A Streptococci (*Streptococcus* pyogenes) are not natural pathogens of cattle, they are often found in the throat and on the skin of humans. However cows can develop mastitis if they are infected with these particular isolates by transference from a human carrier. This emphasises the need for stringent hygiene.

There is a considerable cost element for the dairy farmer when mastitis is confirmed in the herd. Mastitis causes a reduction in milk production and quality; an increase in the use of antimicrobial drug therapies; and an increase in the use of veterinary services (Melchior et al. 2006a, b; Clutterbuck et al. 2007). When the basis of the infection is a biofilm the costs are further increased due to the persistent resistance of such bacteria and the recurrences of clinical disease that inevitably occur.

Although mastitis-causing isolates remain susceptible (in vitro) to current antibiotherapies the overall cure rates of mastitis remain disappointing (Erskine et al. 2002; Guérin-Faublée et al. 2002). Epidemiological studies have reported a wide range of efficacy from 0 to 80%, by antimicrobial treatments (Sol et al. 1997; Deluyker et al. 1999; Wilson et al. 1999). Where mastitis has developed into a chronic infection following antimicrobial treatment, udder health remains unsatisfactory in many cases and somatic cell counts either remain elevated or after an initial decline a second rise is detected (Melchior et al. 2006a, b). Hillerton and Kliem (2002) suggest that around 40% of cases are a result of reoccurrence from a previous infection. This is associated with the presence of persister cells after antibiotherapy treatment which can repopulate the original biofilm causing a "flair-up" of clinical signs. Elevated somatic cell counts can also be encountered after treatment and this may be attributed to the post-antimicrobial damage to the biofilm causing an increased release of planktonic cells. The increase in planktonic cells causes an influx of phagocytic cells; a rise in the somatic cell count may therefore be identified following treatment.

The ideal time to attempt to eliminate chronic mastitis infections is during the "dry period" – the period at the end of lactation prior to parturition. During the dry period the medium for successful bacterial growth has been removed and therefore the number of bacterial cells should be reduced. Furthermore bacterial cells capable of forming a biofilm have optimum growth in medium with sufficient nutrients; when growing at a solid/liquid interface; and in areas of high shear pressure. During the dry period these ideal environmental conditions are removed and therefore the ability of bacteria to form a biofilm may be reduced. Microorganisms depend on water for their structure and functions and the removal of water critically damages membranes, proteins and nucleic acids and is therefore lethal to most micro-organisms (Pawlowski et al. 2005; Suller et al. 2005). However, the extracellular matrix encasing the bacterial cells within the biofilm may also be able to protect against dehydration and so to allow the biofilm to survive within the "dry" udder. There have been reports that biofilm-associated bacteria can survive desiccation for prolonged periods if they are encased within the polymeric matrix. They can then be successfully revived when rehydrated (Costerton et al. 2003; Suller et al. 2005). Türetgen et al. (2007) grew biofilms for 30 and 60 days, air dried them at 27°C for up to 200 h and found that biofilms grown for 60 days had a thicker matrix and a higher number of viable cells after air drying. At 24 h both heterotrophic bacterial cells from the biofilm–water interface and anaerobic bacteria cells from the deeper layers of the biofilm had the greatest survival of 30- and 60-day-old biofilms. However viable cells were identifiable after 168 h of air drying (Browing et al. 1995). The ability of biofilms to survive what would ordinarily be lethal desiccation to a planktonic bacterial cell is yet another way for biofilms to resist destruction and to persist in a host, causing chronic infections and disease.

The incidence of mastitis in the horse is much lower than in the cow (Jackson 1986). Some have attributed this difference to the high activity of lysozyme in the mares milk (Zou et al. 1998). The lysozymes present in mare's milk afford some protection for the foal from infection by pathogenic bacteria that can cause gastroenteritis. They may also form a protective mechanism for the mare against bacteria that are capable of causing mastitis. The lysozyme present in the milk lyses the bacterial cells by breaking down the cell wall. The concentrations of lysozyme in bovine milk are significantly less and therefore the level of protection afforded by this specific mechanism is proportionately reduced. Once the bacteria begin to adhere to a surface and begin the formation of a biofilm the lysozymes present are unable to damage individual bacterial cells due to the surrounding extracellular matrix. The mare has a second protection barrier to bacteria through high concentrations of immunoglobulin, in particular IgG, which is present in colostrum (LeBlanc et al. 1986). This milk immunoglobulin has a limiting effect on the amount of bacteria present in the milk and this in turn limits the liability to biofilms formation.

2 Conclusion

Biofilms have the potential to cause serious medical complications in both humans and animals alike. There are a variety of ways in which biofilms can cause health issues in humans, as well as animals with wounds and mastitis being two common clinical challenges in animals. Although research is still compiling evidence on the role of biofilms in veterinary medicine, it is safe to say that they do play a significant role particularly in chronic infections. Much of the information regarding biofilm infection and disease has been extrapolated from human research and applied to the veterinary environment as there are still many gaps in veterinary biofilm research.

Biofilms are currently featured highly in research programs with researchers studying the specific architecture in an effort to help determine control strategies (Bridier et al. 2010). However, the search for a successful treatment of biofilm infections that can prevent and eradicate biofilms in the clinical environment is still ongoing.

Timely and accurate identification of infections where a biofilm is present is essential in preventing persisting infections. Once identified, treatment protocols need to be tailored specifically for biofilm infection; biofilm-related infections do not respond to antibiotherapy treatment protocols in the same way as planktonic bacterial populations. Hygiene remains one of the most critical pilates in the control of dissemination of infection and the reduction in the opportunities for biofilm formation.

References

Allison DG, Ruiz B et al (1998) Extracellular products as mediators of the formation and detachment of *Pseudomonas fluorescens* biofilms. FEMS Microbiol Lett 167(2):179–184

Anderson GG, O'Toole GA (2010) Innate and induced resistance mechanisms of bacterial biofilms. *Current Topics in Microbiology and Immunology* 322: 85–105

Antonios VS, Noel AA et al (2006) Prosthetic vascular graft infection: a risk factor analysis using a case-control study. J Infect 53(1):49–55

Bishop Y (ed) (2005) The veterinary formulary. Pharmaceutical Press, London

Bjarnsholt T, Kirketerp-Møller K et al (2008) Why chronic wounds will not heal: a novel hypothesis. Wound Repair Regen 16(1):2–10

Bridier A, Dubois-Brissonnet F et al (2010) The biofilm architecture of sixty opportunistic pathogens deciphered using a high throughput CLSM method. J Microbiol Methods 82:64–70

Brooun A, Liu S et al (2000) a dose-response study of antibiotic resistance in *Pseudomonas aeruginosa* biofilms. Antimicrob Agents Chemother 44(3):640–646

Browing A, Cumberlege et al (1995) The Dairy products (Hygiene) Regulation 1995. 1086. f. a. f. Ministry of Agriculture

Burke JP (2003) Infection control – a problem for patient safety. N Engl J Med 348(7):651–656

Carrick JB, Begg AP (2008) Peripheral blood leukocytes. Vet Clin North Am Equine Pract 24(2): 239–259

Cerca N, Jefferson K et al (2006) Comparative antibody-mediated phagocytosis of *Staphylococcus epidermidis* cells grown in a biofilm or in the planktonic state. Infect Immun 74(8):4849–4855

Chaw KC, Manimaran M et al (2005) Role of silver ions in destabilisation of intermolecular adhesion forces measured by atomic force microscopy in *Staphylococcus epidermis* biofilms. Antimicrob Agents Chemother 49(12):4853–4859

Chin J (1982) Raw milk; a continuing vehicle for the transmission of infectious disease agents in the United States. J Infect Dis 146(3):440–441

Choong S, Wood S et al (2001) Catheter-associated urinary tract infection and encrustation. Int J Antimicrob Agents 17(4):305–310

Cifrian E, Guidry AJ et al (1994) Adherence of *Staphylococcus aureus* to cultured bovine mammary epithelial cells. J Dairy Sci 77(4):970–983

Clutterbuck AL, Woods EJ et al (2007) Biofilms and their relevance to veterinary medicine. Vet Microbiol 121(1–2):1–17

Cochrane CA (1997) Models in vivo of wound healing in the horse and the role of growth factors. Vet Dermatol 8:259–272

Costerton JW (2004) A short history of the development of the biofilm concept. In: Ghannoum MA, O'Toole GA (eds) Microbial biofilms. ASM Press, Washington, pp 4–19

Costerton JW, Lewandowski Z et al (1995) Microbial biofilms. Annu Rev Microbiol 49(1):711–745

Costerton JW, Stewart PS et al (1999) Bacterial biofilms: a common cause of persistent infections. Science 284(5418):1318–1322

Costerton JW, Veeh R et al (2003) The application of biofilm science to the study and control of chronic bacterial infections. J Clin Invest 112(10):1466–1477

Cowan MM, Warren TM et al (1991) Mixed-species colonization of solid surfaces in laboratory biofilms. Biofouling 3(1):23–34

Davies DG, Geesey GG (1995) Regulation of the alginate biosynthesis gene *alg*C in *Pseudomonas aeruginosa* during biofilm development in continuous culture. Appl Environ Microbiol 61(3): 860–867

Davies DG, Chakrabarty AM et al (1993) Exopolysaccharide production in biofilms: substratum activation of alginate gene expression by *Pseudomonas aeruginosa*.. Appl Environ Microbiol 59(4):1181–1186

Deluyker HA, Chester ST et al (1999) A multilocation clinical trial in lactating dairy cows affected with clinical mastitis to compare the efficacy of treatment with intramammary infusions of a lincomycin/neomycin combination with an ampicillin/cloxacillin combination. J Vet Pharmacol Ther 22(4):274–282

Donlan RM (2001) Biofilms and device-associated infections. Emerg Infect Dis 7(2):277–281

Donlan RM, Costerton JW (2002) Biofilms: survival mechanisms of clinically relevant microorganisms. Clin Microbiol Rev 15:167–193

Edwards JF, Lassala AL et al (2008) Staphylococcus-associated abortions in Ewes with long-term central venous catheterization. Vet Pathol 45(6):881–888

Erskine RJ, Walker RD et al (2002) Trends in antibacterial susceptibility of mastitis pathogens during a seven-year period. J Dairy Sci 85(5):1111–1118

Fedtke I, Gotz F et al (2004) Bacterial evasion of innate host defenses – the *Staphylococcus aureus* lesson. Int J Med Microbiol 294:189–194

Fera P, Siebel MA et al (1989) Seasonal variations in bacterial colonisation of stainless steel, aluminium and polycarbonate surfaces in a sea water flow system. Biofouling 1(3):251–261

Ferris FG, Schultze S et al (1989) Metal interactions with microbial biofilms in acidic and neutral pH environments. Appl Environ Microbiol 55(5):1249–1257

Ganz T (2001) Antimicrobial proteins and peptides in host defense. Semin Respir Infect 16(1):4–10

Gillespie IA, Adak GK et al (2003) Milkborne general outbreaks of infectious intestinal disease, England and Wales, 1992–2000. Epidemiol Infect 130(3):461–468

Guérin-Faublée V, Tardy F et al (2002) Antimicrobial susceptibility of Streptococcus species isolated from clinical mastitis in dairy cows. Int J Antimicrob Agents 19(3):219–226

Hall-Stoodley L, Stoodley P (2005) Biofilm formation and dispersal and the transmission of human pathogens. Trends Microbiol 13(1):7–10

Hammond A, Dertien J et al (2010) Serum inhibits *P. aeruginosa* biofilm formation on plastic surfaces and intravenous catheters. J Surg Res 159(2):735–746

Hansson CJ, Hoborn J et al (1995) The microbial flora in venous leg ulcers without signs of clinical infection. Acta Derm Venereol 75:24–30

Harmon RJ (1994) Physiology of mastitis and factors affecting somatic cell counts. J Dairy Sci 77(7):2103–2112

Hensen SM, Pavičić MJ et al (2000) Location of *Staphylococcus aureus* within the experimentally infected Bovine Udder and the expression of capsular polysaccharide type 5 in situ. J Dairy Sci 83(9):1966–1975

Hillerton JE, Kleim KE (2002) Effective treatment of *Streptococcus uberis* clinical mastitis to minimise the use of antibiotics. Journal of Dairy Science 85(4):1009–1014

Jackson PGG (1986) Equine mastitis: comparative lessons. Equine Vet J 18(2):88–89

Jensen PO, Bjarnsholt T et al (2007) Rapid necrotic killing of polymorphonuclear leukocytes is caused by quorum-sensing-controlled production of rhamnolipid by *Pseudomonas aeruginosa*. Microbiology 153(5):1329–1338

Jesaitis AJ, Franklin MJ et al (2003) Compromised host defence on *Pseudomonas aeruginosa* biofilms: characterisation of neutrophil and biofilm interactions. J Immunol 171:4329–4339

Kharazmi A (1991) Mechanisms involved in the evasion of the host defence by *Pseudomonas aeruginosa*. Immunol Lett 30:201–205

Khoury AE, Lam K et al (1992) Prevention and control of bacterial infections associated with medical devices. ASAIO J 38(3):M174–M178

Kowalczuk D, Ginalska G et al (2010) Characterization of the developed antimicrobial urological catheters. Int J Pharm 402(1–2):175–183

Krahwinkel DJ, Boothe HW (2006) Topical and systemic medications for wounds. Vet Clin North Am Small Anim Pract 36:739–757
Lappin-Scott HM, Bass C (2001) Biofilm formation: attachment, growth and detachment of microbes from surfaces. Am J Infect Control 29:250–251
Lasa I (2006) Towards the identification of the common features of bacterial biofilm development. Int Microbiol 9(1):21–28
LeBlanc M, McLaurin BI et al (1986) Relationships among serum immunoglobulin concentration in foals, colostral specific gravity, and colostral immunoglobulin concentration. J Am Vet Med Assoc 189(1):57–60
Lee B, Haagensen AJ et al (2005) Heterogeneity of biofilms formed by nonmucoid *Pseudomonas aeruginosa* isolates from patients with cystic fibrosis. J Clin Microbiol 43(10):5247–5255
Lewis K (2001) Riddle of biofilm resistance. Antimicrob Agents Chemother 45:999–1007
Lewis K (2005) Persister cells and the riddle of biofilm survival. Biochemistry 70(2):267–274
Lewis K (2008) Multidrug tolerance of biofilms and persister cells. Curr Top Microbiol Immunol 322:107–131
Lin P, Chen H-L, Huang C-T, Su L-H, Chiu C-H (2010) Biofilm production, use of intravascular indwelling catheters and inappropriate antimicrobial therapy as predictors of fatality in *Chryseobacterium meningosepticum* bacteraemia. International Journal of Antimicrobial Agents 36(5):436–440
Marion K, Freney J et al (2006) Using an efficient biofilm detaching agent: an essential step for the improvement of endoscope reprocessing protocols. J Hosp Infect 64(2):136–142
Marsh-Ng ML, Burney DP et al (2007) Surveillance of infections associated with intravenous catheters in dogs and cats in an intensive care unit. J Am Anim Hosp Assoc 43(1):13–20
Mathews KA, Brooks MJ et al (1996) A prospective study of intravenous catheter contamination. J Vet Emerg Crit Care 6(1):33–43
Melchior MB, Fink-Gremmels J et al (2006a) Comparative assessment of the antimicrobial susceptibility of *Staphylococcus aureus* isolates from bovine mastitis in biofilm versus planktonic culture. J Vet Med B 53(7):326–332
Melchior MB, Vaarkamp H et al (2006b) Biofilms: a role in recurrent mastitis infections? Vet J 171(3):398–407
Melchior MB, Fink-Gremmels J et al (2007) Extended antimicrobial susceptibility assay for *Staphylococcus aureus* isolates from bovine mastitis growing in biofilms. Vet Microbiol 125(1–2):141–149
Melchoir MB, van Osch MHJ et al (2009) Biofilm formation and genotyping of *Staphylococcus aureus* bovine mastitis isolates: evidence for lack of penicillin-resistance in *Agr*-type II strains. Vet Microbiol 137:83–89
Morley PS, Apley MD et al (2005) Antimicrobial drug use in veterinary medicine. J Vet Intern Med 19(4):617–629
Nagase N, Sasaki M et al (2002) Isolation and species distribution of staphylococci from animal and human skin. J Vet Med Sci 64(3):245–250
Nicoll A, Gay NJ et al (2005) Theory of infectious disease transmission and control. In: Borriello SP, Murray PR, Funke G (eds) Topley and Wilson's microbiology and microbial infections: bacteriology, vol 1. Hodder Arnold, London, pp 335–358
Oliveira R, Nunes SF et al (2007) Time course of biofilm formation by *Staphylococcus aureus* and *Staphylococcus epidermidis* mastitis isolates. Vet Microbiol 124:187–191
Orsini JA, Elce Y et al (2004) Management of severely infected wounds in the equine patient. Clin Tech Equine Prac 3:225–236
O'Toole G, Kaplan HB et al (2000) Biofilm formation as microbial development. Annu Rev Microbiol 54:49–79
Parsek MR, Singh PK (2003) Bacterial biofilms: an emerging link to disease pathogenesis. Annu Rev Microbiol 57:677–701
Pascual A (2002) Pathogenesis of catheter-related infections: lessons for new designs. Clin Microbiol Infect 8(5):256–264

Pawlowski KS, Wawro D et al (2005) Bacterial biofilm formation on a human cochlear implant. Otol Neurotol 26(5):972–975

Percival SL, Bowler PG (2004) Biofilms and their potential role in wound healing. Wounds 16:234–240

Percival SL, Bowler P et al (2008) Assessing the effect of an antimicrobial wound dressing on biofilms. Wound Repair Regen 16:52–57

Potera C (1999) Forging a link between biofilms and disease. Science 283(5409):1837–1839

Sol J, Sampimon OC et al (1997) Factors associated with bacteriological cure during lactation after therapy for subclinical mastitis caused by *Staphylococcus aureus*. J Dairy Sci 80(11): 2803–2808

Songer JG, Post KW (2005) Veterinary microbiology: bacterial and fungal agents of animal disease. Elsevier Saunders, Missouri

Stewart PS, Costerton JW (2001) Antibiotic resistance of bacteria in biofilms. Lancet 358:135–138

Stickler D, Ganderton L et al (1993) *Proteus mirabilis*; biofilms and the encrustation of urethral catheters. Urol Res 21(6):407–411

Suller MTE, Anthony VJ et al (2005) Factors modulating the pH at which calcium and magnesium phosphates precipitate from human urine. Urol Res 33(4):254–260

Sutherland I (2001) Biofilm exopolysaccharides: a strong and sticky framework. Microbiology 147:3–9

Targowski SP (1983) Role of immune factors in protection of mammary gland. J Dairy Sci 66(8): 1781–1789

Tenke P, Kovacs B et al (2006) The role of biofilm infection in urology. World J Urol 24(1):13–20

Theoret CL (2004) Wound repair in the horse: problems and proposed innovative solutions. Clin Tech Equine Prac 3:134–140

Theoret CL (2006) Wound repair: equine surgery. WB Saunders, Saint Louis, pp 44–62

Turetgen I, Ilhan-Sungur E, Cotuk A (2007) Effects of short-term drying on biofilm-associated bacteria. *Annals of Microbiology* 57(2):277–280

Waldorf H, Fewkes J (1995) Wound healing. Adv Dermatol 10:77–96

Waldron DR, Zimmerman-Pope N (2003) Superficial skin wounds. In: Slater D (ed) Textbook of small animal surgery. WB Saurnders, Philadelphia, pp 259–273

Ward KH, Olson ME et al (1992) Mechanism of persistent infection associated with peritoneal implants. J Med Microbiol 36(6):406–413

Warren JW (2001) Catheter-associated urinary tract infections. Int J Antimicrob Agents 17(4): 299–303

Wieman TJ (2005) Principles of management: the diabetic foot. Am J Surg 190(2):295–299

Wilson DJ, Gonzalez RN et al (1999) Comparison of seven antibiotic treatments with no treatment for bacteriological efficacy against bovine mastitis pathogens. J Dairy Sci 82(8):1664–1670

Wyder AB, Boss R et al (2011) *Streptococcus* spp. and related bacteria: their identification and their pathogenic potential for chronic mastitis: a molecular approach. Res Vet Sci (in press)

Xu KD, McFeters GA et al (2000) Biofilm resistance to antimicrobial agents. Microbiology 146(3): 547–549

Zou S, Brady HA et al (1998) Protective factors in mammary gland secretions during the periparturient period in the mare. J Equine Vet Sci 18(3):184–188

Role of Biofilms in the Oral Health of Animals

David W. Williams, Michael A.O. Lewis, Steven L. Percival, Tomoari Kuriyama, Sónia da Silva, and Marcello P. Riggio

Abstract In humans the importance of biofilms in disease processes is now widely recognised together with the difficulties in treating such infections once established. One of the earliest and certainly most studied biofilm in humans is that of dental plaque which is responsible for two of the most prevalent human infections, namely dental caries and periodontal disease. However, comparable studies of dental plaque in animals are relatively limited, despite the fact that similar infections also occur, and in the case of farm animals there is an associated economic impact. In addition, biofilms in the mouths of animals can also be detrimental to human health when transferred by animal bites. As a result, an understanding of both the microbial composition of animal plaque biofilms together with their role in animal diseases is important. Through the use of modern molecular studies, an insight into the oral microflora of animals is now being obtained and, to date, reveals that

D.W. Williams (✉)
Oral Microbiology Group, Tissue Engineering & Reparative Dentistry, School of Dentistry, Cardiff University, Heath Park, Cardiff CF14 4XY, UK
e-mail: williamsdd@cardiff.ac.uk

M.A.O. Lewis
School of Dentistry, Cardiff University, Heath Park, Cardiff CF14 4XY, UK

S.L. Percival
Department of Pathology, Medical School, West Virginia University, Morgantown, WV, USA

T. Kuriyama
Department of Oral and Maxillofacial Surgery, Graduate School of Medical Science, Kanazawa University, Kanazawa 920-8640, Japan

S. da Silva
Centre for Biological Engineering, IBB-Institute for Biotechnology and Bioengineering, Universidade do Minho, Campus de Gualtar, Braga 4710-057, Portugal

M.P. Riggio
Infection and Immunity Research Group, Dental School, College of Medical Veterinary and Life Sciences, University of Glasgow, 378 Sauchiehall Street, Glasgow G2 3JZ, UK

S.L. Percival et al. (eds.), *Biofilms and Veterinary Medicine*,
Springer Series on Biofilms 6, DOI 10.1007/978-3-642-21289-5_5,

despite differences in terms of microbial species and relative proportions occurring between humans and animals, similarities do indeed exist. This information can be exploited in our efforts to both manage and treat infections in animals arising from the presence of an oral biofilm. This Chapter describes our current understanding of the microbial composition of animal plaque, its role in disease and how oral hygiene measures can be implemented to reduce subsequent infection.

1 Introduction

Dental plaque of humans provided the first biofilm microorganisms studied and were described in the seventeenth century by Antonie van Leeuwenhoek (1632–1723) as 'animalcules' living within his own dental plaque (Costerton et al. 1978; Gest 2004; Donlan and Costerton 2002). Our knowledge of biofilms has advanced greatly since plaque microorganisms were first visualised by van Leeuwenhoek, and in recent times the term biofilm has been defined as communities of sessile microorganisms that are encased within an extracellular polymeric substance (EPS) that has been generated by the microorganisms themselves.

As well as being the earliest studied biofilm, dental the plaque is also the most extensively investigated. The microorganisms within human dental plaque have attracted researchers primarily due to their relationship with plaque-mediated infections, dental caries and periodontal disease. In comparison, however, there is relatively little information available about the oral microflora of dental plaque in animals, nor its aetiology with equivalent animal infections. This is significant given that plaque-mediated diseases are considered to be the primary causes of tooth loss (Colmery and Frost 1986) in animals and also a potential cause of systemic disease.

Whilst the dentition of animals varies greatly with different species, what is consistent is that all animals are highly dependent on the correct function of their dentition in order to survive. Reduced function of the dentition following oral disease can lead to a number of associated problems for the animal, including malnutrition, systemic infection as well as physical dysfunction (DeBowes et al. 1996).

Further interest in the oral microflora of animals stems from concerns over possible wound infection following animal bites in humans (Goldstein 1992; Haddad et al. 2008). Animal bites are considered a significant public health problem. In fact it has been estimated that some 2% of the population are bitten in some way each year (Dendle and Looke 2008).

The risk of infection following animal bites is known to vary and is dependent on the animal species involved and its oral microbiota. Numerous studies have shown that a broad range of pathogens can be isolated from infected bite wounds. These have included *Corynebacterium auriscanis* from a dog bite (Bygott et al. 2008), *Aeromonas hydrophila* from a tiger bite (Easow and Tuladhar 2007), *Pasteurella multocida* from a cat bite (Westling et al. 2006), *Staphylococcus hyicus* from a donkey bite (Osterlund and Nordlund 1997) and *Flavobacterium* from a pig bite (Goldstein et al. 1990).

The most predominant genera found in infected animal bites, and particularly those from horses (Escande et al. 1997), dogs (Sin Fai Lam et al. 1993) and cats, is *Pasteurella* (order chronologically Weber et al. 1984; Arons et al. 1982; Tindall and Harrison 1972; Albert and Stevens 2010). Interestingly, the first documented case of *Pasteurella canis* bacteraemia in humans has only recently been reported and thought to originate from a dog lick to an open leg wound (Albert and Stevens 2010). There have also been cases of bat bites, with two fatal cases of Australian bat lyssavirus infection recently reported (Dendle and Looke 2008).

A recent review of the literature on the diagnosis and treatment of bite wound infections found that the most common pathogens associated with bite wounds were *Streptococcus* sp., *Staphylococcus* sp., *P. multocida*, *Capnocytophaga canimorsus* as well as anaerobic bacteria (Yaqub et al. 2004). It was found in this review that human bites differed from animal bites by the higher prevalence of *Staphylococcus aureus* and *Eikenella corrodens*. In this respect, there is particular interest in the oral microbiology of companion animals that are collectively responsible for over 90% of animal bites in humans needing medical intervention (Król et al. 2006).

Animals are frequently used as models of human oral diseases (Fine 2009) and an assessment of disease outcome in such studies is dependent upon knowledge of the existing oral microflora and the ability of any introduced microorganism to successfully colonise the oral cavity.

1.1 Composition of the Oral Microflora of Animals

Determining the composition of the oral microflora of animals and establishing its relationship with oral health is problematic given the complexity and the variation that occurs in oral microbial communities at different locations (Rober et al. 2008) and indeed at different times (Marsh 2006). The reasons for these differences are extensive and include the wide range of habitats within the oral cavity, the host's immune status and the influence of external environmental factors such as changes in diet and receipt of antimicrobial therapies or other agents that can affect microbial composition.

Within the oral cavity, soft tissues such as the buccal and palatal mucosa and the tongue will support the presence and growth of microorganisms. In the case of the buccal mucosa, the actual number of residing microorganisms is generally low, which can in part be explained by the high turnover of the buccal epithelium. This means that microorganisms adhered to epithelial cells will be removed as the cells are shed from the mucosal surface. Higher numbers of microorganisms are often present on the surface of the tongue, where the tongue papillae afford protection to attached microbes from the normally protective flushing activity of saliva.

In addition to soft tissues, hard and non-shedding surfaces for attachment of microorganisms are provided by teeth, and it is on these surfaces where microorganisms reside in their highest numbers in the form of dental plaque.

Dental plaque can be divided into supragingival and subgingival forms and, as these names imply, distinction is based on the plaque location with respect to the gingival margin. In addition, significant variation in terms of microbial composition can occur between these two types of plaque. Generally, Gram-positive, facultatively anaerobic streptococci are the predominant constituents of supragingival plaque, whereas Gram-negative anaerobic bacteria provide the largest component of the microbial community of subgingival plaque.

Differences in the oral microflora between animal species and humans can be pronounced, with most animal studies focussing on the mouths of companion, working and food animals. To exemplify this, bacteria of the genus *Streptococcus* tend to provide the greatest proportion of bacteria in the oral cavity of humans, however much lower incidences of these bacteria occur in the mouths of dogs (Wunder et al. 1976; Takada et al. 2006; Rober et al. 2008). Suggested reasons for this relate to the respective dietary habits of humans and dogs, the generally higher salivary pH encountered in dogs (Harvey and Emily 1993) and the existence of possible bias in commercial diagnostic systems towards species associated with humans.

Microorganisms in both supragingival and subgingival plaque of animals have previously been characterised using culture methods (Harvey et al. 1995; Love et al. 1990). Culture of plaque microorganisms from the gingival margins of cats has revealed mixed populations of facultative and obligate anaerobic bacterial species comprising members of the genera *Actinomyces*, *Bacteroides*, *Fusobacterium*, *Peptostreptococcus* and *Propionibacterium* (Love et al. 1990). The most prevalent facultative anaerobe was *P. multocida*, whilst *Clostridium villosum* was the most commonly isolated obligately anaerobic species. The microflora of dental plaque in dogs has been demonstrated to be similar to that of cats (Harvey et al. 1995). It should be noted that *P. multocida* is the causative agent of human pasteurellosis. This particular bacterium is not a member of human commensal microflora but can enter into the human body through bite wounds by cats or dogs.

Recent developments in molecular techniques have enabled the detection of unculturable microorganisms. A comparative assessment of the microbiology of the supragingival and subgingival plaque, cheek, tongue and tonsil of dogs has been determined using DNA probes raised against human periodontopathogens (Rober et al. 2008). Significant variation was evident in the microbial composition at different locations, again highlighting the importance of oral site in affecting the species present. Interestingly, in this study' bacteria that hybridised to DNA probes targeting the human periodontal pathogens *P. gingivalis*, *Tannerella forsythensis*, *Treponema denticola* (i.e., the so-called red complex of periodontal pathogens; Socransky et al. 1998) and *Aggregatibacter actinomycetemcomitans* were detected in high proportions in all dogs examined. High numbers of *Fusobacterium*-like species were also found and *Streptococcus mitis* was reported as the most predominant streptococcal species. Although the actual species present in the oral cavity of

Table 1 Comparison of specific bacteria frequently found in the mouths of animals and humans

	Host animal					
Bacterial species/Genus	Dog	Cat	Non-human primate	Horse	Swine	Human
Viridans streptococci	+	+	+	+	+	+
β-hemolytic streptococci	+	nd	nd	nd	nd	+
Streptococcus pneumoniae	+	nd	+	+	nd	+
Streptococcus canis	+	+	−	−	−	−
Streptococcus equi	−	−	−	+	−	−
Enterococcus	+	+	+	+	+	+
Staphylococcus aureus	+	+	+	+	+	+
Actinobacillus	−	nd	+	+	nd	+
Neisseria	+	+	+	+	nd	+
Campylobacter	+	+	+	nd	nd	+
Escherichia coli	+	nd	nd	+	nd	+
Haemophilus	+	+	+	nd	nd	+
Capnocytophaga	+	+	+	nd	nd	+
Pasteurella	+	+	nd	+	nd	−
Peptostreptococcus	+	+	+	+	+	+
Actinomyces	+	+	+	+	+	+
Lactobacillus	+	nd	+	nd	nd	+
Eubacterium	+	+	+	+	nd	+
Clostridium	+	+	+	+	+	+
Veillonella	+	+	+	+	nd	+
Megasphera	nd	nd	nd	+	nd	+
Bacteroides	+	+	+	+	nd	+
Prevotella	+	+	+	nd	nd	+
Porphyromonas	+	+	+	nd	nd	+
Fusobacterium	+	+	+	+	nd	+
Treponema	+	+	+	nd	+	+

nd not determined

animals and humans varies, the microbial composition would appear to be generally similar with respect to microbial genera (Table 1). It has therefore been proposed that the oral cavities of animals and humans are equivalent in terms of selection pressures which leads to the presence of core plaque components in all oral cavities with representatives of the genera *Actinomyces*, *Bacteroides*, *Fusobacterium*, *Neisseria*, *Streptococcus* and *Veillonella* consistently found in animals and humans (Dent and Marsh 1981). In terms of plaque-mediated diseases of animals, similarities in the type of causative organisms involved are often made.

Biofilms have several distinct properties when compared with equivalent planktonic growth. Differences between these two modes of growth can manifest as early as the initial attachment of microorganisms to a solid surface, which can result in the activation of signalling molecules on the microbial cell surface leading to the expression of 'biofilm-specific' genes. As the biofilm develops, the three-dimensional biofilm structure will exhibit regional variation with respect to pH, nutrient and oxygen availability. These factors will, in turn, have an effect on the type of microorganism present, their growth rate and gene expression.

The benefits to microorganisms growing within a biofilm are that the resulting community tends to have a collective resistance to host defences and administered antimicrobial agents. To understand the mechanisms behind this 'protective nature' of the biofilm requires an awareness of biofilm structure and development, and this is described later. Of additional concern is that a biofilm may also demonstrate enhanced pathogenicity compared with planktonic growth, a phenomenon referred to as pathogenic synergy (van Steenbergen et al. 1984).

1.2 Formation of Dental Plaque Biofilms

Dental plaque is a normal occurrence on the surface of teeth and particularly so where its mechanical removal by salivary flow is reduced, as encountered at stagnant sites such as in the interproximal regions or fissures of teeth.

The formation of plaque begins with the development of an acquired pellicle on the enamel of the teeth. This pellicle consists of glycoproteins, proline-rich peptides (PRPs), lipids and phosphoproteins and other components originating largely from saliva, although there may also be involvement of gingival crevicular fluid molecules. Pellicle formation is a rapid process, occurring within seconds of a clean tooth surface being exposed to a salivary conditioning film and results in the generation of a surface offering receptors suitable for initial attachment of primary colonising bacteria. These 'early colonisers' make contact with pellicle receptors following either passive transport in saliva or by active motility of the microorganisms to the tooth surface.

Microbial adherence initially involves relatively weak and reversible electrostatic and hydrophobic interactions. However, through the interaction of specific adhesins on the microbial cell surface with target receptors, an irreversible adherence is achieved. Subsequently, 'late colonisers' are recruited to the plaque community, often by aggregation to the primary colonising organisms (Kolenbrander et al. 2000) or to components of the EPS matrix produced by the developing biofilm. The effect of this recruitment to the plaque biofilm is that a complex and diverse population of microorganisms develops with interaction occurring between its members in terms of both metabolic co-operation and competition. The end result of all these interactions is that a relatively stable, climax microbial community occurs within the mature biofilm, where certain combinations of bacterial species will occupy the same locality or alternatively grow in separate locations within the biofilm.

In addition to its 'microbial structure', mature plaque biofilms also have an organised physical structure consisting of stable lower regions of relatively densely packed microbial cells to more open and loosely organised cells in the upper biofilm regions (Fig. 1). Within the biofilm structure, movement of fluid and gases can

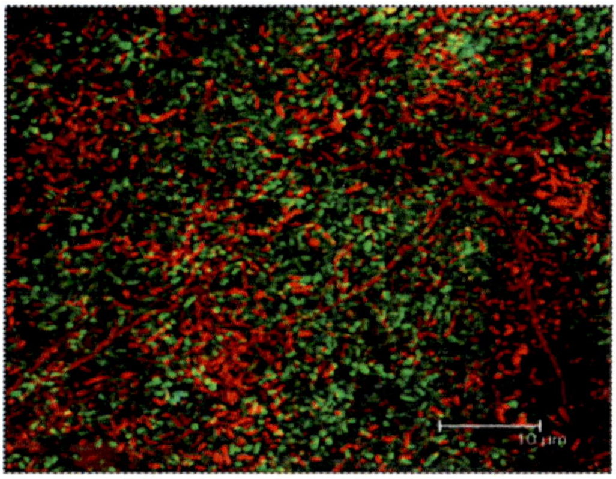

Fig. 1 Microbial structure: mature plaque biofilm

occur through so-called water channels which provide a system of nutrient delivery and means of removal for waste products from within the biofilm structure.

1.3 Opportunistic Oral Pathogens

Generally, the oral microflora of humans and animals exists in a healthy balanced state with the host due to environmental constraints. However, alteration in the local environment may lead to an imbalance in the oral microflora leading to infection. These changes may relate to diet, a reduction in host immune responses or the effects of an administered therapeutic drug, such as an antibiotic. Within a biofilm community, environmental changes lead to different selection pressures, causing shifts in the balance of microbial populations. This can lead to an increase in the numbers of potential pathogens with the result that infection may follow. This so-called ecological plaque hypothesis, first described by Marsh (2003), is used to explain the dynamic nature of plaque-mediated diseases, as discussed later. Since the microorganisms that are involved in oral infection are usually normally harmless and only cause disease when conditions allow, they are often referred to as opportunistic pathogens.

1.4 Plaque-Mediated Diseases in Animals

As with humans, oral infections in animals are highly prevalent with periodontal disease, soft tissue infections and dental caries all encountered to varying degrees. In addition, certain oral infections are also thought to present a risk for potentially more serious systemic disease, although much of the evidence for this relates to studies in humans.

1.5 Periodontal Diseases

Periodontal diseases are a collection of infections involving the degradation of the supporting tissues of the teeth, including the gingiva, periodontal ligament, alveolar bone and root cementum of the tooth. Periodontal diseases are regarded as the most prevalent of all infections encountered in cats and dogs, and have a similar clinical presentation to corresponding human infections. However, in animals, periodontal diseases tend to have an accelerated rate of progression compared to humans, a feature that could relate to the relative difficulty of improving and maintaining oral hygiene in animals.

Gingivitis is often considered to be the first stage of periodontal infection and is clinically characterised by mild inflammation at the gingival margin. Importantly, gingivitis is recognised as a reversible form of periodontitis. Given the many common features between human and animal gingivitis, it is not surprising that researchers have proposed a similarity in the causative microorganisms involved (Cutler and Ghaffar 1997; Syed et al. 1980).

At infected sites, the numbers of Gram negative and proteolytic bacterial genera such as *Porphyromonas* and *Tannerella* increase and these become the predominant members of the biofilm. Studies have indeed demonstrated higher numbers of these 'periodontopathogens' in periodontal infections compared with disease-free counterparts in animals, including dogs (Watson 1994; Gorrel and Rawlings 1996; Isogai et al. 1999; Nordhoff et al. 2008), cats (Mallonee et al. 1988; Norris and Love 1999; Valdez et al. 2000) and sheep (Frisken et al. 1986, 1987; McCourtie et al. 1990; Dreyer and Basson 1992). Some differences have, however, been identified between the *Porphyromonas* isolates of humans and animals, with those originating from humans generally being catalase negative and those from animals often being catalase positive (Harvey et al. 1995; Isogai et al. 1999; Booij-Vrieling et al. 2010). This has led to the suggestion that the catalase-positive *P. gingivalis*-like isolates are actually *Porphyromonas gulae* (Booij-Vrieling et al. 2010).

Periodontopathogens are asaccharolytic, which means they obtain their energy from the breakdown of protein as opposed to carbohydrates. In the case of periodontal disease, the host periodontium provides the source of this protein. Proteolytic activity tends to increase the pH of the local environment, which in turn further promotes the growth of the periodontopathogens. The presence of periodontopathogens and their virulence factors serve to stimulate the animal's cellular and humoral immune defences. As well as combating the causative bacteria, the immune response will also contribute to damage to the periodontium. Immunoglobulins produced against bacterial antigens will activate complement pathways, leading to inflammation and this causes the production of prostaglandins which, in turn, stimulate bone resorption. Unless the numbers of the periodontal pathogens can be reduced, either by the host's immune response or through improved oral hygiene, then progression of the disease from gingivitis to periodontitis is a possibility.

Periodontitis develops from gingivitis and is characterised by the degradation of the actual supporting structures of the teeth, including the alveolar bone.

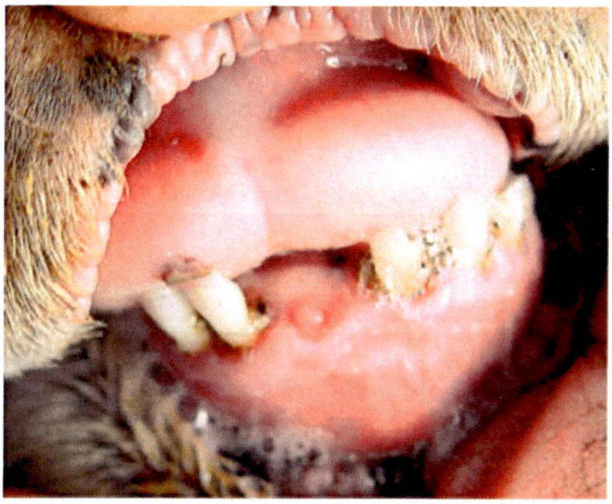

Fig. 2 Ovine 'broken mouth' periodontitis

Periodontitis is an irreversible form of periodontal disease and is graded based on clinical parameters including the extent of gingival recession, erythema, oedema, loss of bone support (as determined by radiographical measurement) and also the number of teeth that may have been lost.

Periodontitis can have serious consequences for affected animals beyond the loss of teeth and the condition has been implicated with dissemination of bacteria to other body sites leading to kidney infection, and heart and liver disease (DeBowes et al. 1996; Scannapieco et al. 2003; Okuda et al. 2004).

Sheep suffer from a naturally occurring form of periodontitis known as 'broken mouth' (Fig. 2) which is a condition affecting sheep grazed on rough pasture, and involves periodontal infection of the incisor teeth, their subsequent loosening and progressive loss. Abattoir surveys in the United Kingdom have found occurrence of incisor loosening or loss in 60–70% of culled ewes (Aitchison and Spence 1984). The prevalence of this infection in any one flock varies between 5% and 70%, with over 50% of flocks in the United Kingdom being affected (West and Spence 2000). 'Broken mouth' periodontitis, in addition to being a painful condition, will reduce the efficiency of grazing of sheep, which contributes to malnutrition, weight loss and systemic health problems (Anderson and Bulgin 1984; Baker and Britt 1990). The implications of broken mouth periodontitis to animal welfare are thus obvious.

Dental problems are the main cause of premature culling of sheep long before the end of their natural reproductive life and are universally acknowledged as an important source of economic loss in the sheep industry, mostly due to high flock replacement costs and the low sale value of broken mouth ewes (up to 30% less than sound-mouthed ewes). Loss of the incisor teeth reduces the efficiency of grazing and results in loss of condition and productivity if the animal is left on rough grazings. It is a major cause of drafting sheep from upland pastures, thus reducing the sustainability of upland enterprises, and results in the sale of older, experienced hefted ewes to lowland systems where their edentulous state can be overcome by using softer and more nutritious feeds. Consequently, this condition is a major economic problem to sheep farmers across the globe, for which no treatment or control methods are available.

1.6 Dental Caries

Dental caries is an infection of teeth caused by plaque bacteria. The infection leads to the dissolution of tooth enamel, largely due to the action of acidic products of carbohydrate fermentation by microorganisms. In order for caries to occur a number of factors have to be present. Firstly, the animal must have an accumulation of plaque that contains the necessary bacteria capable of both surviving in an acidic environment (aciduric bacteria) and also producing acid (acidogenic) under these conditions. Secondly, the diet of the animal has to contain carbohydrates of sufficient quantity and type to yield the necessary acid to degrade the enamel. Given these requirements it is perhaps not surprising that dental caries is not a spontaneous infection in many animals and is particularly uncommon amongst companion animals. Often for caries to occur in such animals, some form of dental trauma has to have preceded the disease. In dogs, naturally occurring caries has an estimated incidence of approximately 5% (Hale 1998). The reasons for this relatively low incidence are many, but particularly relate to the type of oral microflora present (which tends to have relatively low numbers of streptococci), the intake of a diet largely free of fermentable carbohydrates and the normally alkaline pH of canine saliva. When canine caries does occur, it usually develops within pits on the occlusal surfaces or on the necks of the molar teeth. As with dogs, caries is a rare occurrence in cats and it is important to note that in cats' dental caries is not the underlying reason for the majority of dental cavities seen, which are often referred to as feline odontoclastic resorptive lesions. In such lesions saccharolytic, acid-producing bacteria are not involved.

1.7 Systemic Diseases Related to Oral Infection

In recent years, there has been much interest in the potential link between periodontal diseases in humans and systemic conditions such as heart disease, stroke, respiratory diseases, diabetes, osteoporosis and a greater risk of pre-term low birth weight babies. The exact mechanisms for these associations have yet to be determined although elevated bacteraemia with plaque biofilm organisms would be a potential means for the spread of infectious microorganisms. Dissemination of oral bacteria to other body sites has also been associated with systemic disease in animals, with both kidney and heart disease being linked to the spread of oral bacteria (DeBowes et al. 1996). Also, an elevated immune response with increased levels of circulating cytokines could contribute to systemic inflammatory diseases.

1.8 Prevention of Oral Disease in Animals

As mentioned earlier, the loss of teeth and dental function can have serious consequences for an animal, leading to malnutrition and potential systemic disease. Given that gingivitis is a reversible infection caused by plaque biofilm, it follows that effective control of plaque at early stages of infection will prevent progression of the disease. Indeed, it has been suggested that 90% of human periodontitis follows on from gingivitis and that nearly all such cases can be prevented by plaque control. Given the similarities in diseases between humans and animals it is probable that plaque control in animals will have a similar effect in reducing the incidence of periodontitis. However, compared with humans, plaque control in animals can prove problematic.

1.9 Plaque Control in Companion Animals

Proper nutrition and effective oral hygiene are necessary components of oral health and should be jointly promoted in the management of oral disease in dogs and cats. The majority of pet owners do not regularly employ tooth brushing on their animals to remove dental plaque and so plaque removal based on high-fibre chew sticks is often used, which serves to physically scrape plaque off the surfaces of teeth. The effectiveness of such a strategy was demonstrated in a study of dental plaque removal from the teeth of Beagle dogs provided with soft rawhide chew products, where statistically significant reductions in the formation of dental calculus (28.0%), dental plaque (19.0%) and gingivitis (46.0%) occurred (Stookey 2009).

2 Conclusion

Amongst the most prevalent infections of humans are the plaque-mediated diseases, dental caries and periodontal disease. It is not surprising therefore that significant research into the microbiology of human dental plaque has been undertaken. However, despite animals also suffering from equivalent infections, our knowledge of the oral microbiology of animals is comparatively limited. Since animal oral microbiology can, in addition to impacting on animal health, have detrimental effects on human health (in the case of both animal bites and microbial transmission) and an economic impact, there is a clear need to promote research in this area.

The advent of new molecular microbiology methods will clearly be of benefit in this regard with recent comparative studies already revealing the similarities between the oral plaque of certain animals and humans. It is becoming clear that whilst the oral microbiology of animals and humans share similar features, significant differences do exist in respect of the microbial species and the relative proportions of these species in the oral cavity. However, it has been shown that the bacteria within the oral biofilm (dental plaque) responsible for periodontal infection in humans and animals are similar, and hopefully this information can be exploited in our efforts to manage and treat these infections.

References

Aitchison GU, Spence TA (1984) Dental disease in hill sheep: an abattoir survey. J Comp Pathol 94:285–300

Albert TJ, Stevens DL (2010) The first case of *Pasteurella canis* bacteremia: a cirrhotic patient with an open leg wound. Infection 38:483–485

Anderson BC, Bulgin MS (1984) Starvation associated with dental disease in range ewes. J Am Vet Med Assoc 184:737–738

Arons MS, Fernando L, Polayes IM (1982) *Pasteurella multocida* – the major cause of hand infections following domestic animal bites. J Hand Surg Am 7:47–52

Baker JR, Britt DP (1990) Dental calculus and periodontal disease in sheep. Vet Rec 115:411–412

Booij-Vrieling HE, van der Reijden WA, Houwers DJ, de Wit WE, Bosch-Tijhof CJ, Penning LC, van Winkelhoff AJ, Hazewinkel HA (2010) Comparison of periodontal pathogens between cats and their owners. Vet Microbiol 144:147–152

Bygott JM, Malnick H, Shah JJ, Chattaway MA, Karas JA (2008) First clinical case of *Corynebacterium auriscanis* isolated from localized dog bite infection. J Med Microbiol 57:899–900

Colmery B 3rd, Frost P (1986) Periodontal disease. Etiology and pathogenesis. Vet Clin North Am Small Anim Pract 16:817–833

Costerton JW, Geesey GG, Cheng GK (1978) How bacteria stick. Sci Am 238:86–95

Cutler CW, Ghaffar KA (1997) A short-term study of the effects of SBHAN, a novel compound, on gingival inflammation in the beagle dog. J Periodontol 68:448–455

DeBowes LJ, Mosier D, Logan E, Harvey CE, Lowry S Richardson DC (1996), association of periodontal disease and histologic lesions in multiple organs from 45 dogs. J Vet Dent 13:57–60

Dendle C, Looke D (2008) Review article: animal bites: an update for management with a focus on infections. Emerg Med Australas 20:458–467

Dent VE, Marsh PD (1981) Evidence for a basic plaque microbial community on the tooth surface in animals. Arch Oral Biol 26:171–179

Donlan RM, Costerton JW (2002) Biofilms: survival mechanisms of clinically relevant microorganisms. Clin Microbiol Rev 15:167–193

Dreyer WP, Basson NJ (1992) *Porphyromonas gingivalis* as putative pathogen in ovine periodontitis. J Dent Assoc S Afr 47:513–516

Easow JM, Tuladhar R (2007) *Aeromonas hydrophila* wound infection following a tiger bite in Nepal. Southeast Asian J Trop Med Public Health 38:867–870

Escande F, Vallee E, Aubart F (1997) *Pasteurella caballi* infection following a horse bite. Zentralbl Bakteriol 285:440–444

Fine DH (2009) Of mice and men: animal models of human periodontal disease. J Clin Periodontol 36:913–914

Frisken KW, Laws AJ, Tagg JR, Orr MB (1986) The characterization of trypsin-positive black-pigmented *Bacteroides* isolated from sheep periodont disease. Lett Appl Microbiol 3:57–60

Frisken KW, Tagg JR, Laws AJ, Orr MB (1987) Black-pigmented bacteroides associated with broken-mouth periodontitis in sheep. J Periodontal Res 22:156–159

Gest H (2004) The discovery of microorganisms by Robert Hooke and Antoni van Leeuwenhoek, fellows of the Royal Society. Notes Res R Soc Lond 58:187–201

Goldstein EJ, Citron DM, Merkin TE, Pickett MJ (1990) Recovery of an unusual *Flavobacterium* group IIb-like isolate from a hand infection following pig bite. J Clin Microbiol 28:1079–1081

Goldstein EJ (1992) Bite wounds and infection. Clin Infect Dis 14:633–638
Gorrel C, Rawlings JM (1996) The role of a 'dental hygiene chew' in maintaining periodontal health in dogs. J Vet Dent 13:31–34
Haddad V Jr, Duarte MR, Neto DG (2008) Tegu (teiu) bite: report of human injury caused by a *Teiidae* lizard. Wilderness Environ Med 19:111–113
Hale FA (1998) Dental caries in the dog. J Vet Dent 15:79–83
Harvey CE, Emily PP (1993) Restorative dentistry. Small animal dentistry. Mosby Yearbook, St. Louis, pp 213–265
Harvey CE, Thornsberry C, Miller BR (1995) Subgingival bacteria – comparison of culture results in dogs and cats with gingivitis. J Vet Dent 12:147–150
Isogai H, Kosako Y, Benno Y, Isogai E (1999) Ecology of genus *Porphyromonas* in canine periodontal disease. Zentralbl Veterinärmed B 46:467–473
Kolenbrander PE, Andersen RN, Kazmerak KM, Palmer RJ (2000) Coaggregation and coadhesion in oral biofilms. In: Allison DG, Gilbert P, Lappin-Scott HM, Wilson M (eds) Community structure and co-operation in biofilms (Society for General Microbiology symposium no. 59). Cambridge University Press, Cambridge, pp 65–85
Król J, Florek M, Pliszczak-Król A, Staroniewicz Z (2006) Microbiologic analysis of bite wounds in humans caused by dogs and cats. Medycyna Wet 62:498–501
Love DN, Vekselstein R, Collings S (1990) The obligate and facultatively anaerobic bacterial flora of the normal feline gingival margin. Vet Microbiol 22:267–275
Mallonee DH, Harvey CE, Venner M, Hammond BF (1988) Bacteriology of periodontal disease in the cat. Arch Oral Biol 33:677–683
Marsh PD (2003) Are dental diseases examples of ecological catastrophes? Microbiology 149:279–294
Marsh PD (2006) Dental plaque as a biofilm and a microbial community-implications for health and disease. BMC Oral Health 6(Suppl 1):S14
McCourtie J, Poxton IR, Brown R, Whittaker CR, Spence JA, Aitchison GU (1990) A longitudinal study of the cultivable subgingival anaerobic bacteria isolated from sheep during the development of broken mouth periodontitis. J Med Microbiol 31:275–283
Nordhoff M, Rühe B, Kellermeier C, Moter A, Schmitz R, Brunnberg L, Wieler LH (2008) Association of *Treponema* spp. with canine periodontitis. Vet Microbiol 127:334–342
Norris JM, Love DN (1999) Associations amongst three feline *Porphyromonas* species from the gingival margin of cats during periodontal health and disease. Vet Microbiol 65:195–207
Okuda K, Kato T, Ishihara K (2004) Involvement of periodontopathic biofilm in vascular diseases. Oral Dis 10:5–12
Osterlund A, Nordlund E (1997) Wound infection caused by *Staphylococcus hyicus* subspecies hyicus after a donkey bite. Scand J Infect Dis 29:95
Rober M, Quirynen M, Haffajee AD, Schepers E, Teughels W (2008) Intra-oral microbial profiles of beagle dogs assessed by checkerboard DNA–DNA hybridization using human probes. Vet Microbiol 127:79–88
Scannapieco FA, Bush RB, Paju S (2003) Associations between periodontal disease and risk for atherosclerosis, cardiovascular disease, and stroke. A systematic review. Ann Periodontol 8:38–53
Sin Fai Lam KN, Mah PK, Chuah SC, Chew LS (1993) *Pasteurella multocida* septicaemia following a dog bite. Singapore Med J 34:271–273
Socransky SS, Haffajee AD, Cugini MA, Smith C, Kent RL Jr (1998) Microbial complexes in subgingival plaque. J Clin Periodontol 25:134–144
Stookey GK (2009) Soft rawhide reduces calculus formation in dogs. J Vet Dent 26:82–85
Syed SA, Svanberg M, Svanberg G (1980) The predominant cultivable dental plaque flora of beagle dogs with gingivitis. J Periodont Res 15:123–136
Takada K, Hayashi K, Sasaki K, Sato T, Hirasawa M (2006) Selectivity of Mitis Salivarius agar and a new selective medium for oral streptococci in dogs. J Microbiol Methods 66:460–465

Tindall JP, Harrison CM (1972) *Pasteurella multocida* infections following animal injuries, especially cat bites. Arch Dermatol 105:412–416

Valdez M, Haines R, Riviere KH, Riviere GR, Thomas DD (2000) Isolation of oral spirochetes from dogs and cats and provisional identification using polymerase chain reaction (PCR) analysis specific for human plaque *Treponema* spp. J Vet Dent 17:23–26

van Steenbergen TJM, van Winkelhoff AJ, de Graaff J (1984) Pathogenic synergy: mixed infections in the oral cavity. Antonie Leeuwenhoek 50:789–798

Watson ADJ (1994) Diet and periodontal disease in dogs and cats. Aust Vet J 71:313–318

Weber DJ, Wolfson JS, Swartz MN, Hooper DC (1984) *Pasteurella multocida* infections. Report of 34 cases and review of the literature. Medicine (Baltimore) 63:133–154

West DM, Spence JA (2000) Diseases of the oral cavity. In: Martin WB, Aitken ID (eds) Diseases of sheep, 3rd edn. Blackwell Science Ltd., Oxford, pp 125–131

Westling K, Farra A, Cars B, Ekblom AG, Sandstedt K, Settergren B, Wretlind B, Jorup C (2006) Cat bite wound infections: a prospective clinical and microbiological study at three emergency wards in Stockholm, Sweden. J Infect 53:403–407

Wunder JA, Briner WW, Calkins GP (1976) Identification of the cultivable bacteria in dental plaque from the beagle dog. J Dent Res 55:1097–1102

Yaqub S, Bjørnholt JV, Hellum KB, Steinbakk M, Enger AE (2004) Bite wound infections. Tidsskr Nor Laegeforen 124:3194–3196

Evidence and Significance of Biofilms in Chronic Wounds in Horses

Samantha J. Westgate, Steven L. Percival, Peter D. Clegg, Derek C. Knottenbelt, and Christine A. Cochrane

Abstract Equine wounds have a high risk of becoming infected due to their environment. Infected wounds encompass diverse populations of microorganisms that fail to respond to antibiotic treatment, resulting in chronic non-healing wounds. In human wounds this has been attributed to the ability of bacteria to survive in a biofilm phenotypic state. Biofilms are known to delay wound healing, principally due to their recalcitrance towards antimicrobial therapies and components of the innate immune response. The presence of biofilms in equine wounds partly explains the reluctance of many lower limb wounds to heal. Non-healing limb wounds in horses are a well documented welfare and economic concern. Therefore, there is a need to develop future treatments in order to increase the healing rate, decrease the cost of treatment and reduce suffering associate with equine wounds.

S.J. Westgate (✉) • D.C. Knottenbelt
Faculty of Health and Life Sciences, Department of musculoskeletal Biology, University of Liverpool, Leahurst, Neston, South Wirral CH64 7TE, UK
e-mail: knotty@liverpool.ac.uk

S.L. Percival
Department of Pathology, Medical School, West Virginia University, Morgantown, WV, USA
e-mail: biofilms@hotmail.co.uk

P.D. Clegg
Department of Musculoskeletal Biology, University of Liverpool, Leahurst Campus, Neston, Cheshire CH64 5SD, UK
e-mail: p.d.clegg@liverpool.ac.uk

C.A. Cochrane
Faculty of Health and Life Sciences, Institute of Ageing and Chronic Disease, University of Liverpool, Leahurst, Neston. CH64 7TE, United Kingdom
e-mail: c.a.cochrane@liverpool.ac.uk

S.L. Percival et al. (eds.), *Biofilms and Veterinary Medicine*,
Springer Series on Biofilms 6, DOI 10.1007/978-3-642-21289-5_6,

1 Introduction

Trauma and surgical wounds are highly prevalent in horses and represent a significant management challenge to veterinary surgeons (Carter et al. 2003). Horses are prone to trauma wounds, such as grazes, lacerations and puncture wounds. A large percentage of trauma and surgical wounds fail to heal despite extensive veterinary intervention. These wounds are described as chronic.

Surgical and traumatic wounds can become infected. In colic surgery, 16% of complications are a consequence of the surgical wound breaking down due to infection (Mair and Smith 2005; Proudman et al. 2002). The prevalence of traumatic wounds in horses is high (Collins et al. 2000; Hernandez and Hawkins 2001; Singer et al. 2003) and a high percentage of these wounds progress to a chronic state. Equine lower limb wounds are particularly susceptible to colonisation from microorganisms and are notoriously problematic to heal (Cochrane et al. 2003). Maintaining a wound microenvironment that helps to suppress microbial proliferation, particularly at problematic anatomical sites, is key to successful wound treatment (Knottenbelt 2003).

Chronic wounds in horses have a similar pathophysiology to human chronic wounds (Cochrane 2003). This is of particular interest as laboratory rats do not commonly suffer from chronic wounds, thus there is currently no adequate animal model for human chronic wounds (Dorsett-Martin and Wysocki 2008). Management practices utilised for human chronic wounds are being employed successfully in the management of chronic equine wounds (Wilmink et al. 2006). In human studies, it was estimated that as many as 80% of human infections are in some way biofilm related (Kolenbrander et al. 2002; Wolcott and Rhoads 2008), with wounds being just one area in which biofilms are suspected to hinder treatment. Research has successfully identified biofilms in equine wounds (Cochrane et al. 2009) and chronic wounds are being re-addressed with a focus on the role of bacterial biofilms (Westgate et al. 2010). Appropriate treatment methods resulting from biofilm research would significantly decrease the duration of chronic wounds. This knowledge base can also be employed by veterinary practitioners to implement treatment strategies that prevent equine wounds from developing biofilms and from entering into a chronic non-healing state. Data collected from equine cases can also be compared to and extrapolated back to human studies.

2 Equine Wound Healing

The equine wound healing process is fundamentally the same as all mammalian wound healing; however, there are a number of anomalies in the horse that are worthy of discussion. It has been known for some time that wounds in the equine lower limb heal significantly slower than wounds to the thoracic region (Theoret et al. 2001). Proposed anatomical reasons for the failure of distal limb healing include a relatively poor blood supply and reduced tissue oxygenation in the limb compared to the thoracic region. These factors make wound bed conditions more favourable for microbes (Theoret et al. 2001).

When skin is damaged an inflammatory response is immediately initiated leading to the formation of a localised fibrin/fibronectin clot. The clot is multifunctional. It provides a temporary barrier against infection and a matrix for cellular migration, and it contains the growth factors, responsible for triggering fibroblast proliferation, migration and subsequent granulation tissue formation (Greiling and Clark 1997). Clotting and inflammation within the wound bed are transient mechanisms. The closed wound forms a scar and creates a permanent barrier to microbial ingress. Interestingly, wound closure is more rapid in ponies than horses. Ponies have a better local defence mechanism compared to horses. This means the inflammatory phase is shorter, resulting in a shorter time frame for the completed healing process (Wilmink et al. 2002).

Secondary intention healing involves re-epithelialisation (Dahm et al. 2002) and concurrent tissue contraction (Cochrane et al. 2003). Re-epithelialisation involves keratinocyte migration across the wound (Dahm et al. 2002). This centripetal contraction begins at wound margins on the surface of the wound and gradually progresses deeper into the wound. Keratinocyte migration involves cellular migration rather than cellular proliferation, whereas the re-development of the lower stratification layers requires standard differentiation of the supra basal cells (Clark 1985). Inflammatory cells, fibroblasts and capillaries combine to form contractile tissue, including granulation tissue. The formation of granulation tissue triggers the change of fibroblasts into myofibroblasts, which adhere firmly to surrounding granulation tissue (Singer et al. 1984). The strength by which the wound resists contraction controls the trigger for this change, where increased resistance triggers the switch (Darby and Gabbiani 1990). Fibroblast cells show phenotypic differences and the number and density of fibroblast cells effect wound contraction where an increase in density and number significantly increase wound contraction, and subsequent wound healing rate (Cochrane et al. 2003; Wilmink et al. 2001). The final stage of wound healing requires angiogenesis. This is triggered by growth factors (Cochrane 1997) including fibroblast growth factor 2 (FGF2) (Broadley et al. 1989; Frank et al. 1995) and vascular endothelial growth factor (VEGF) (Carter et al. 2003).

In acute wounds such as surgical incision wounds, the healing process is conventionally taken to have been completed within 2 weeks with restoration of tissue tensile strength occurring within a few weeks thereafter. However, the type of wound and associated complications can extend this time frame (Cochrane et al. 2003; Schwartz et al. 2002). The chronic wound does not progress sequentially through the "normal" stages of wound healing (Wilmink and Van Weeren 2005). Inhibited healing stages form the principal pathological processes.

3 Types of Equine Wound

The cause of equine wounds can be categorised as traumatic or surgical, and the healing process can be further classified as acute or chronic. Over 20% of trauma wounds have been reported to result from a bite or kick from another horse

Table 1 A description of open wounds commonly seen in horses. Information from (Knottenbelt 2007)

Wound type	Description
Puncture	A very small deep wound that can prove fatal if it penetrates a vital organ
Graze	Superficial removal of the epidermis. Healing is usually uncomplicated
Incised	A clean cut with a sharp defined margin (includes surgical wounds)
Laceration	Traumatic skin tear that can be accompanied by bruising
Complicated	Involves multiple structures and commonly displays healing difficulties
Burn	Can result from heat, scalding, friction, chemicals, freezing and sun. The extent of a burn is described as the percentage of the body it affects

(Knubben et al. 2008), either as a result of aggressive behaviour or exuberant playful behaviour. This kind of injury is more regularly seen when herds are newly mixed or unstable (Waring 2003). Other injuries result from the horse's immediate environment such as damaged fencing or sharp objects in the field or stable, and can occur when the horse is ridden. Deep injuries resulting from cross country jumping and racing are common. Horses are owned mainly for the competition industry or as pleasure animals. In both cases, relatively large amounts of money are available for individual equine treatments compared to farmed animals such as cattle and sheep. For this reason, surgical wounds, for ailments such as surgical colic are common place in the equine industry. Due to the wide array of causes of injury, there are a number of wound types that can present to the veterinary surgeon (Table 1)

When assessing a wound, the whole animal should be assessed initially. Tissue viability, oedema, wound location and patient age should all be considered when deciding on appropriate wound treatment strategies (Adam and Southwood 2006).

4 Factors Known to Delay Equine Wound Healing

A number of factors can delay the healing process (Knottenbelt 2007). These are discussed below.

4.1 Wound Location

Healed wounds have a low tensile strength, which equates to only 50% of the original strength 6-month post-injury. This means that wounds located over joints can experience delayed healing in response to a continued high tensile pressure (Rivera and Spencer 2007). However, in some cases, repeated controlled movement can in some circumstances improve joint healing and not detrimentally affect the wound healing process (Johnson 1990).

4.2 Poor or Impaired Blood and Oxygen Supply

Wound regeneration requires healthy granulation tissue containing new blood vessels and an adequate oxygen supply (Nakada et al. 2006; Sen 2009). An adequate oxygen supply is vital for efficient cellular respiration, a prerequisite for the acceptance of skin grafts (Sen 2009), and to prevent the establishment of anaerobic bacterial colonies (Bakker 2000). In horses, hyperbaric medicine is currently used to treat ischemic injuries such as laminitis (Slovis 2008).

4.3 Poor Nutritional State of the Horse

A poor nutritional state prior to and throughout the duration of the injury is detrimental to wound healing (Dorsett-Martin and Wysocki 2008). In particular, burn wound healing can be significantly hindered by inadequate nutrition (Nguyen et al. 1996).

4.4 Wound pH

Bacterial infection can alter the pH of the wound thus inhibiting enzymes and growth factors required for successful wound healing (Schreml et al. 2009). Humans who have a high skin pH harbour bacterial flora such as *Staphylococcus aureus*, whereas *Staphylococcus epidermidis* isolates were more commonly cultured from humans with a low skin pH (Lambers et al. 2006).

4.5 Iatrogenic Factors

Inappropriate use of dressings, sutures and surgical equipment can detrimentally affect wounded tissue thus hindering the wound healing process (Bertone 1996). Maggot therapy they can pose an additional iatrogenic risk resulting from ammonia being absorbed into the blood stream (Hall and Wall 1995).

4.6 Elevated Cortisone Levels

High levels of cortisol resulting from prolonged stress or Equine Cushing's disease can prevent the wound from progressing successfully through the wound healing process (Elliot 2001; Van der Kolk 1997).

4.7 Necrotic Tissue

Necrotic tissue provides a breeding site for microorganisms (Johnson et al. 2007; Zyl et al. 2010) and prevents the wound from being supplied with a new vascular system in response to healthy granulation tissue (Knottenbelt 1997).

4.8 An Imbalance in Matrix Metalloproteinases and Tissue Inhibitors of Metalloproteinases

Imbalances in the relative concentrations of matrix metalloproteinases (MMPs) and tissue inhibitors of metalloproteinases (TIMPs) are known to delay wound healing significantly (Percival and Cochrane 2010). In chronic wounds the expression of TIMPs is reduced when compared to the expression of MMPs. This results in an elevation in overall proteinase activity, leading to elevated protein degradation within the wound bed that results in a delay in wound healing (Lefebvre-Lavoie et al. 2005).

The above scenarios can lead to a wound being classified as chronic. Chronic wounds are characterised by a stalled healing process that persistently refuses to progress through the normal wound healing stages (Cochrane et al. 2003; Grinnell 1992). Visually, chronic wounds appear red and inflamed (White and Cutting 2008), and often produce fetid wound exudates (Cochrane 2003). Particular concern exists in equine wounds that are at risk from infection. The risk of infection is heightened in horses compared to humans. The horses' local environment provides a reservoir of microorganisms capable of colonising and infecting open wounds (Galuppo et al. 1999).

5 Bacteria and Their Role in Equine Wound Healing

Infection with bacteria causes a prolonged inflammatory phase which delays wound closure (Grinnell 1992). Historically the term "critical colonisation" was used to determine if a wound was at risk of becoming infected. The traditional practice of reporting a wound as infected purely based on a bacterial number greater than 1×10^5 viable organisms per gram is now considered largely outdated (Bowler 2003; Hendrickson and Virgin 2005). A more meaningful classification of the microbial status is the point when the host's immune response is triggered (Kingsley 2001).

Traumatic wounds are an obvious infection risk as the foreign body responsible for causing the wound will introduce environmental bacteria directly into the wound (Adam and Southwood 2006). When the host's immune response is impaired or the wound is not adequately cleaned, these wounds can become chronically infected (Kingsley 2001).

Post-surgical infections are a serious concern for veterinary surgeons (Griffiths et al. 2003). Indigenous microbiota have the ability to penetrate and colonise the hosts tissue (Galuppo et al. 1999). The infecting isolates can be environmental pathogens such as *Actinobacillus equuli,* a species usually associated with septicaemia in foals (Smith and Ross 2002). More often, however, wound infections result from commensal flora acting as opportunistic pathogens. *Streptococcus zooepidemicus*, *S. aureus*, *alpha haemolytic Streptococcus* sp., *E. coli* and *P. aeruginosa* have been isolated from a variety of clean and clean-contaminated orthopaedic procedures (MacDonald et al. 1994). Post-operative infections in the horse colonised with bacteria such as methicillin-resistant *Staphylococcus aureus* (MRSA) are still on the increase. Based on growing concerns with MRSA treatment of these cases requires continued monitoring and investigation (Trostle et al. 2001).

The percentage of horses that develop infected surgical wounds is significantly higher than the figure reported in human (Cruse and Foord 1980) and small animal studies (Vasseur et al. 1988). Horses are at an increased risk of infection as a result of their heavily colonised local environment. However, variations in the definitions employed to denote an "infected wound" could also partly explain the differences in reported infection rates (MacDonald et al. 1994).

The strategies used by bacteria to cause and induce persistent infections in wounds are not fully understood; however, they are likely to be multifactorial. Bacteria can alter wound pH to make it more favourable for them, and concurrently less appropriate for successful wound healing (Schreml et al. 2009). In human studies, *S. aureus* and *E. coli* have been reported to act via specific adherence proteins to suppress the influx of inflammatory cells (Chavakis et al. 2002) and to enhance bacterial attachment to human cells (Fexby et al. 2007), respectively. *P. aeruginosa* has been widely reported as an opportunistic wound pathogen in human studies (Chincholikar and Pal 2002). By residing deep within the wound, *P. aeruginosa* manages to keep the wound in a prolonged inflammatory state (Fazli et al. 2009). Synergistic interactions with other bacteria such as *Burkholderia cepacia* result in a total increase in the pathogenicity of both organisms (Bjarnsholt et al. 2009). This synergy has yet to be reported in equine wounds but it is more than likely that synergistic encounters occur on a regular basis in wounds.

Reduced effectiveness of antibiotic may be indicative of the presence of a bacterial biofilm (Clutterbuck et al. 2007a; Phillips et al. 2008). It has been estimated that 99% of natural bacteria exist within a stable biofilm (Rhoads et al. 2008). That is to say they are attached to a surface rather than as independent planktonic isolates (Hausner and Wuertz 1999), and that biofilms can be linked with more than 80% of persistent infections (National Institute of Health, USA, 1997). Despite the high prevalence of natural biofilms, research investigating the significance to this lifestyle particularly in equine veterinary medicine is very limited.

6 Biofilms

It has been proposed that chronic, non-healing wounds may be a result of bacteria inhabiting the wound and forming an antimicrobial barrier known as a biofilm (Graham 1999; Serralta et al. 2001). Biofilms can be defined as a group of micro-organisms that live in a self-synthesised matrix (Mertz 2003), and they are diverse, dynamic and unique, reflecting the environment and inhabiting species (Serralta et al. 2001). Single species bacteria can form biofilms particularly on medical devices (Harrison-Balastra et al. 2003); however, more commonly in nature, biofilms are complex, containing multispecies colonies of aerobic and anaerobic microorganisms (Mertz 2003). When present within a biofilm microorganisms communicate with each other and manipulate the hosts response to infection via intracellular communication known as quorum sensing (Nadell et al. 2008).

Biofilms in humans have been linked to a number of diseases such as cystic fibrosis (Kipnis et al. 2006; Singh et al. 2000), endocarditis (Benn et al. 1997) and catheter-related infections (Marrie and Costerton 1984; Morris et al. 1999). Despite their strong pathogenic effects, some biofilms are advantageous. For example, on hair follicles, biofilms protect against invading pathogens (Bais et al. 2004; Darveau et al. 2003; Mertz 2003) and in the human gut, commensal bacteria produce bacteria that protect against food borne pathogens (Lee et al. 2000) a process known as colonisation resistance. In wounds, however, biofilms at present are considered unwanted with their presence considered to cause a delay in the healing process. This concept has been reported both in humans (Kirker et al. 2009) and now equine wounds (Clutterbuck et al. 2007b; Cochrane et al. 2009).

7 Biofilm Development

The formation of a biofilm is a multistage process. The initiation of a biofilm requires attachment of planktonic bacteria to a surface. In the case of equine wounds the wound bed. A high population density and low nutritional environment have been considered by many researchers to provide cues that trigger bacteria to settle and attach. Attachment has been aided by bacterial surface appendages fimbriae or pili (O'Toole et al. 2000).

The process of bacterial attachment to a surface alters the bacterial phenotype (Watnick and Kolter 2000) and marks an end to the planktonic phenotype (Serralta et al. 2001). Phenotype characteristics expressed by attached bacteria can vary. Such phenotypic changes are reflected by the physical and chemical characteristics of the surface to which the bacteria become attached (Landry et al. 2006). The initial attachment of bacteria to a surface is considered to be reversible process (Golovlev 2002). Consequently, the attached bacteria can be easily washed off from a surface. Following reversible adhesion, irreversible adhesion occurs where the bacteria become more firmly attached on the surface. At this stage in the biofilm formation process, it is proposed that a genetic involvement occurs (O'Toole and Kolter 1998). This has been demonstrated in *P. aeruginosa*. For example, it has been found that removal of the genes required for *P. aeruginosa* attachment to a surface has been shown to render the bacteria incapable of biofilm formation (O'Toole and Kolter 1998).

To produce a mature biofilm, surface attachment must become irreversible. A number of mechanisms are known to act concurrently to achieve this. Reversible attachment triggers the production of the EPS, which in turn acts to cement the bacteria in place leading to irreversible attachment (Hall-Stoodley et al. 2004). Type IV pili (O'Toole et al. 2000) and the *lasR* and *lasI* genes also play a role in irreversible attachment prior to biofilm formation (Sauer et al. 2002). The settlements produce chemical cues which attract increasing numbers of bacteria to the attached cluster. The attracted bacteria may be the same as the primary species or represent novel species including protozoa and fungi (Serralta et al. 2001).

Production of an exopolysaccharide (EPS) layer follows bacterial attachment. This creates a hydrated barrier between the settled bacterial species and the external environment. The bacteria produce chemical signals in response to nutrient availability. These chemical signals shape the development of the mature biofilm. Genes including *lasB* and rhlA are partly responsible for this cell-to-cell signalling (Davies et al. 1998).

A mature biofilm contains "mushrooms" and "pillars" (Lewandowski et al. 1995; Serralta et al. 2001). These structures contain fluid channels which transport nutrients, waste products and chemical messages throughout the biofilm (Nadell et al. 2009). These chemical messages, known as quorum sensing molecules are responsible for interspecies communication within the biofilm (Brenner et al. 2007; Williams 2007). Biofilms remodel their structure in response to a changing environment. Increased differentiation of the biofilm is seen in low nutrient conditions (Rice et al. 2005). Bacterial genes pool in the fluid channels and can be exchanged via horizontal transfer (Wuertz et al. 2004), causing the change in phenotype seen within biofilm residing bacteria (Prigent-Combaret et al. 1999).

Biofilms are dynamic and cells are constantly reproduced and lost from the biofilm. Cells are lost by erosion and sloughing, which causes a gradual and massive cell loss, respectively (Stoodley et al. 2001). The final stage of biofilm formation involves the dispersal of individual planktonic cells from the biofilm. These cells establish new colonies elsewhere (Stoodley et al. 2002). Different methods have been proposed for bacterial dispersion. *Pseudomonas aeruginosa* uses active dispersal. This involves the production of a matrix dissolving enzyme that releases sections of biofilm encased bacteria into the local environment (Hall-Stoodley et al. 2004). *Proteus mirabilis* achieves dispersal in response to their swarming ability (Rather 2005), and *Myxococcus xanthus* produces a slim trail that allows for the transfer of planktonic isolates along the attached surface (Mertz and Forest 2002). Other species have demonstrated the ability to alter surface components (Neu 1996) and to use quorum sensing signals (Parsek and Greenberg 2005) in order to facilitate bacterial dispersion. It is likely that combinations of these methods are used throughout a multispecies biofilm. Dispersion allows planktonic isolates to maintain their biofilm phenotype (O'Toole and Kolter 1998).

The time scale for the production of a mature single species, *P. aeruginosa* biofilm in vitro is 10 h. Attachment is seen at 3 h, the first indications of biofilm production can be observed by light microscopy at 5 h and a distinct biofilm complete with nutrient channels is present by 10 h (O'Toole and Kolter 1998). This time scale is likely to vary between species. In vivo, the host immune response can retard biofilm production.

Multiple aerobic species reside near the surface of in vivo biofilms, whilst bacteria that reside deep within the biofilm are often anaerobic. This reflects the observation that oxygen levels can be almost entirely depleted in the deep biofilm layers (Stewart and Franklin 2008).

Evolutionary biologists proposed that life within a biofilm would be detrimental to some isolates, raising the question of altruism within a biofilm (Nadell et al. 2009). If intra-specific competition is considered, it seems unlikely that a biofilm forming phenotype would benefit from producing EPS substances that were of benefit to non-EPS producing species. Investing energy into the production of biofilm materials results in a decrease in available energy for growth and reproduction, thus leaving this method open to "evolutionary cheaters" (isolates that do not produce EPS, but benefit from residing within it) (Xavier and Foster 2007). However, biofilm production also provides a successful method of intra-specific competition (Xavier and Foster 2007). Non-biofilm phenotypes residing close to biofilm forming phenotypes can be smothered by the EPS thus demonstrating a selective advantage to being an EPS producer. Concurrently, isolates of the same species are shunted further up the biofilm as the fluid channels are formed. This provides the shunted isolates with improved access to nutrients and oxygen (Xavier and Foster 2007), whilst simultaneously out competing neighbouring non-EPS producing species. As such, a biofilm lifestyle demonstrates altruism, and an opportunity for intra-specific competition.

8 Evidence of Biofilms in Horse Wounds

8.1 Clinical In Vivo Identification of Equine Biofilms

Clinically, biofilms are visualised as a thick, slimy shiny layer on the wound surface (Goodrich 2006). This layer can be difficult to remove with standard cleansing techniques. Typically non-healing equine limb wounds can show a number of concurrent healing complications (Fig. 1). When considering the presence of a bacterial biofilm, it is often necessary to consider the biofilm secondary to a current identified wound healing inhibitor such as exuberant granulation tissue or necrotic wound tissue. Healing abnormalities such as the production of exuberant granulation tissue and the presence of necrotic tissue will delay healing independently whilst simultaneously providing an ideal environment to house a persistent bacterial infection. Typically the wound exudates from a biofilm infected wound are thick, yellow in colour and have a fetid odour (Figs. 2 and 3). When this slime is examined microscopically it can confirm the presence of a bacterial biofilm (Schierle et al. 2009).

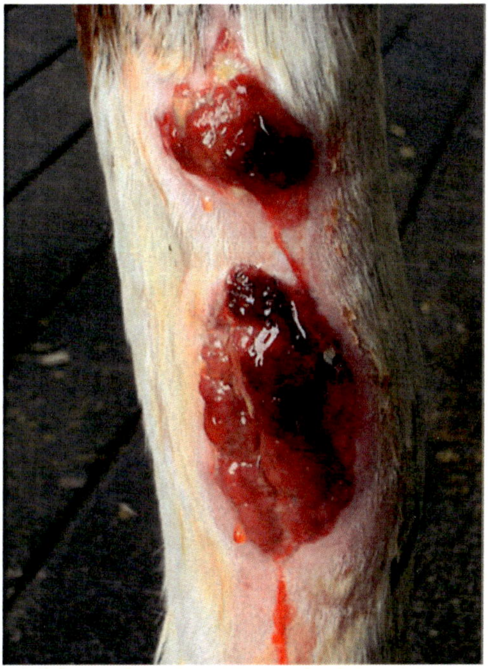

Fig. 1 A 2-week-old non-healing equine wound producing a translucent wound exudate. This wound had multiple factors known to inhibit efficient healing. It was infected by multiple bacterial species including *Staphylococcus aureus* and *Pseudomonas aeruginosa*, it was producing exuberant granulation tissue, and it contained necrotic tissue

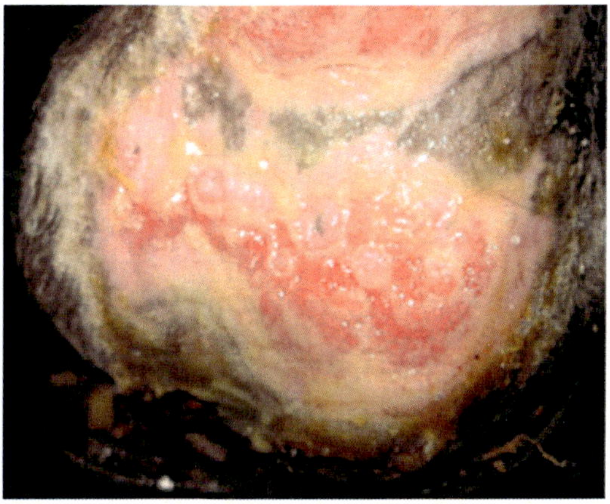

Fig. 2 A 140-day-old equine wound with a slimy surface layer, characteristic of an inherent biofilm. The slime layer was produced throughout the whole surface of the wound. Despite repeated cleansing with saline at 3-day intervals (to coincide with the dressing change), the slime layer was persistently present each time the wound was undressed

8.2 *In Vitro Visualisation of Biofilms*

There are a wide number of methods available for the visualisation of in vitro biofilms. These include indirect methods that produce viable but non-culturable bacteria and direct viable methods that allow for direct observations of growing biofilms (Lindsay and Von Holy 2006). Indirect methods involve manual removal of the biofilm or removal via sonication, followed by a subsequent step that allows for quantification of the removed biofilm. Direct methods involve microscopy of the

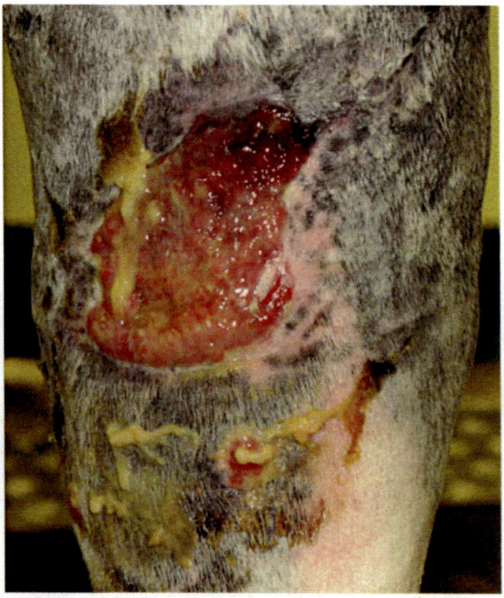

Fig. 3 A 3-week-old non-healing equine wound caused by barbed wire. The wound produced exuberant granulation tissue, which was removed by shape debridement. The *left* of the wound shows the persistent *yellow* discharge produced by the wound, whilst the *top right corner* shows healthy granulation tissue. The treatment aims involved removal of the biofilm and preventing the spread of the biofilm into the healthy granulation tissue

intact biofilm via confocal microscopy, scanning electron microscopy (SEM) and atomic force microscopy. Biofilms are usually visualised as clusters of cells which can grow in pillar and mushroom shapes (Lewandowski et al. 1995), and are encased by a visible or identifiable layer of EPS (Davis et al. 2008).

9 Visualisation of Equine Biofilms on Debrided Tissue

9.1 Gram Stains

Histological stains have been used extensively to investigate the presence of biofilms in wound tissues (Davis et al. 2008). Staining with hematoxylin and eosin (H and E) and using Brown and Brenn staining method allows for the visualisation of large colonies at 200× magnification and individual bacterial isolates at 1,000× magnification. Other stains such as Gram stain and Mason's Trichrome can be equally effective.

Gram stains distinguish Gram positive from Gram-negative bacterial infections. Gram-positive bacteria can be visualised as individual isolates spread evenly throughout the wound tissue (Fig. 4). At a low magnification, biofilms appear as dark purple (Gram positive) and pink (Gram negative) patches in the wound tissue (Fig. 6). When the magnification is increased, individual colonies can be seen within these patches, demonstrating the presence of large bacterial clusters (Fig. 5b). The clusters can be exclusively Gram positive, exclusively Gram negative or contain a mixture of both, demonstrating the presence of multispecies biofilms.

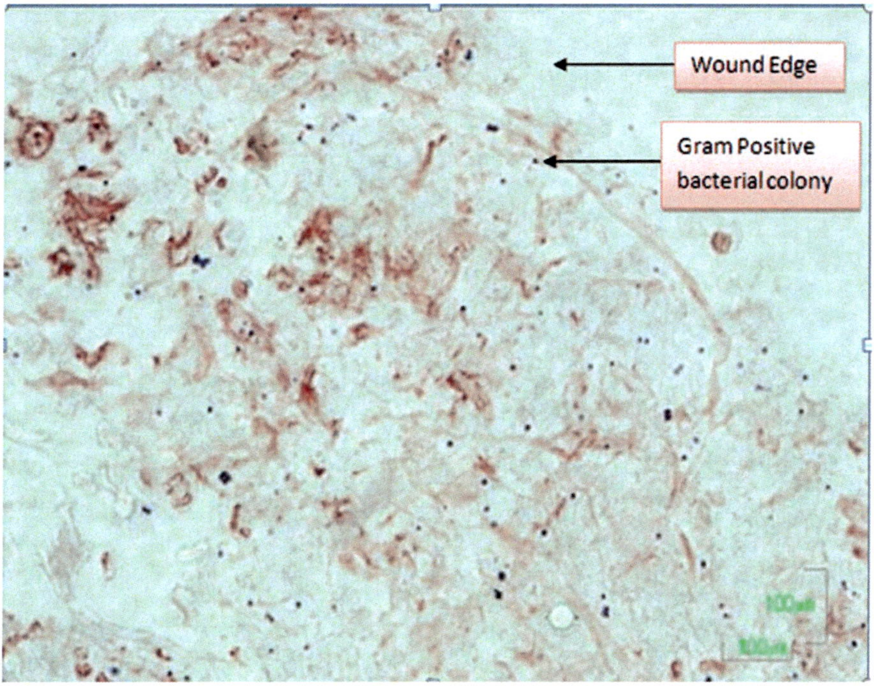

Fig. 4 Wound tissue colonised with non-biofilm forming bacteria. The Gram-positive colonies are regularly spaced throughout the wound tissue (×1,000 magnification)

Distinguishing Gram-negative isolates within a Gram-positive cluster can be difficult due to the dark nature of the Gram-positive stain.

By being located close to the wound edge, the bacteria have access to a constant oxygen supply. Anaerobic colonies can be found deep within wound tissue; however, it requires specialised sample and culture techniques to isolate and identify anaerobic sections of the biofilm (James et al. 2008).

10 Visualisation of In Vitro Biofilms Grown Using Equine Wound Bacterial Isolates

10.1 *Stains Used to Visualise Bacteria and Their Biofilms*

10.1.1 Coomassie Brilliant Blue Stain

Coomassie brilliant blue stain was used to stain a *P. aeruginosa* biofilm grown over night on a microscope slide. Coomassie blue stains undenatured proteins and can be used to stain both bacteria and the protein within the EPS matrix. The aim was to differentiate a non-biofilm forming bacterial phenotype (Fig. 6a) from a biofilm

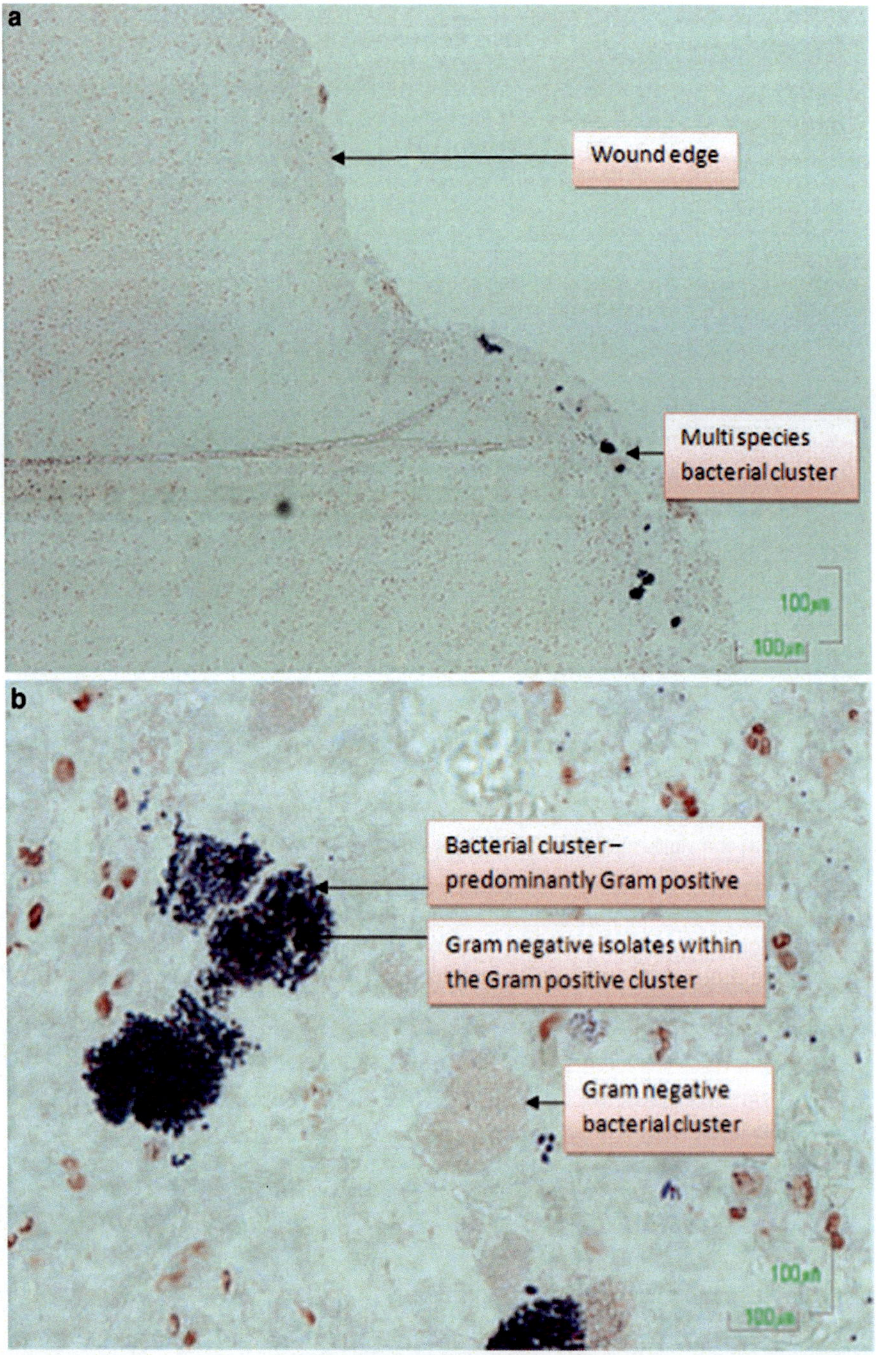

Fig. 5 (**a**) Chronic wound tissue displaying Gram-positive (*purple*) and Gram-negative (*pink*) bacterial clusters, close to the wound edge (×10 magnification). (**b**) This is the same section as (**a**), but visualised at ×1,000 magnification. Individual colonies can be seen clustering together. Very few planktonic colonies are present surrounding the clusters. The clusters contain Gram positive, Gram negative and mixed species colonies

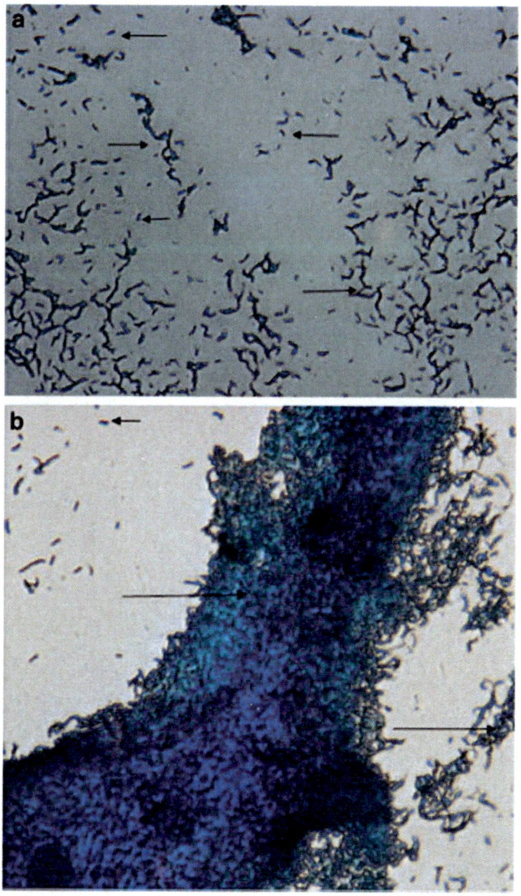

Fig. 6 (**a**) A non-biofilm forming *Pseudomonas* isolate. Isolates are evenly spaced throughout the slide and many isolates are not touching other bacteria (*left pointing arrows*). Some isolates show slight aggregation (*right pointing arrows*). There is minimal evidence of EPS matrix between the bacteria (×1,000 magnification). (**b**) A biofilm forming *Pseudomonas* isolate. Isolates are tightly clustered together and encased in a protein based cover (*right pointing arrows*). There is minimal evidence of individual isolates (*left pointing arrow*) (×1,000 magnification)

forming phenotype (Fig. 6b). Figure 6b clearly shows the bacteria living closely together and within a matrix material. The matrix is thinner on the edge of the biofilm. Thick EPS layers can be seen as light blue areas on the right-hand side of the picture.

10.2 Counter Staining Bacteria and Bacterial Biofilms

In vivo biofilm forming bacteria can be stained with propidium iodide solution and then counter stained with Concanavalin A – Fluorochrome fluorescein isothiocyanate (ConA-FiTC). Propidium iodide stains nucleic acid, and therefore bacteria. It fluoresces red when excited with light at a wavelength of 488 nm (Fig. 7a). ConA-FiTC attaches to carbohydrate residues on glycoproteins and fluoresces bright green (Fig. 7b). A positive result in response to ConA-FiTC demonstrates the presence of the EPS matrix (Ivnitsky et al. 2007; Kania et al. 2008).

The presence of the EPS can be seen clearly in areas where the bacteria are in thick clusters (Fig. 7b). The individual isolates that can be seen surrounding the main bacterial cluster (Fig. 7a) cannot be seen surrounding the main bacterial

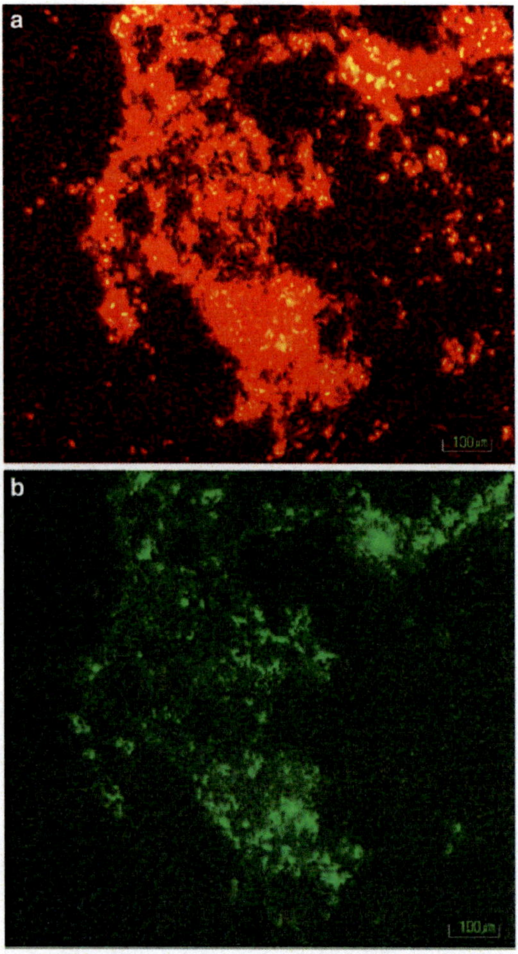

Fig. 7 (**a**) *Staphylococcus aureus* bacteria grown overnight on a microscope slide and stained with propidium iodide. Stained bacteria were visualised on a fluorescent microscope. Large aggregates of bacteria cover the centre and top right of the slide. Individual bacteria can be visualised on the periphery of the main bacterial cluster. (**b**) *Staphylococcus aureus* biofilm matrix counter stained with ConA-FiTC and visualised on a fluorescent microscope. There are strong areas of florescence throughout the attached bacteria, but little evidence of carbohydrate residues surrounding the main colonised area

cluster on the ConA-FiTC stained slide. This demonstrates that the non-clustered bacteria are not EPS producers.

10.3 Scanning Electron Microscopy

SEM is a useful tool for identifying and visualising medical biofilms. SEM allows for a high resolution and a high magnification of the biofilm landscape (Schaudinn et al. 2009). When SEM shows both large aggregates of bacteria and evidence of an EPS matrix, then the image can be taken as showing evidence of a bacterial biofilm (Davis et al. 2008; James et al. 2008; Sun et al. 2008). Figure 8 shows a *P. aeruginosa* biofilm grown on a Calgary 96-well plate pin lid. The rods on the left of the picture are clustered but are not EPS producing, whilst those on the right of the picture have produced a thick EPS layer that is covering the *Pseudamonas* isolates.

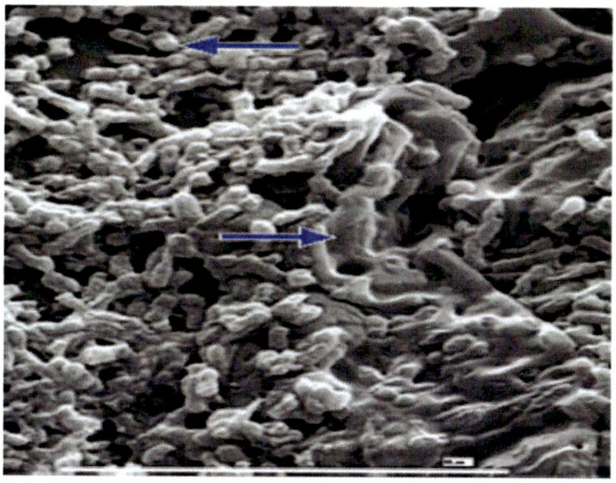

Fig. 8 A 24-h *Pseudomonas aeruginosa* biofilm grown in vitro on a polystyrene pin lid and visualised by scanning electron microscopy. The image shows that the isolates are clustered tightly together (*left pointing arrow*) and are beginning to produce EPS material (*right pointing arrow*) (*scale bar* = 10μm, magnification = ×3500)

Concern surrounds the treatment process required for SEM microscopy. Early research showed that SEM treatment procedures can remove EPS material, dehydrate the matrix and uncover individual bacteria, giving the illusion of a monolayer of bacteria (Little et al. 1991). Different treatment procedures cause different changes to the biofilm architecture (Araujo et al. 2003), where some techniques effectively preserve the bacteria whilst stripping the EPS and vice versa (Fratesi et al. 2004). Furthermore different treatment techniques have different effects when the bacteria are attached to different surfaces. In vivo surfaces add an extra level of complication as the tissue and the biofilm can be dehydrated by the preparation techniques. Where the EPS is not removed, a limitation of SEM is that individual bacteria cannot be examined in detail as SEM cannot penetrate the EPS matrix (Marrie et al. 1982).

Visualising bacteria within a biofilm is also made more difficult because the phenotypic size and shape of biofilm encased bacteria can change dramatically, making them unidentifiable via standard techniques (Costerton 1999; Webster et al. 2004).

10.4 Other Methods for Visualising Biofilms

The combined use of fluorescent in situ hybridisation (FISH) and confocal microscopy allow for a detailed image of biofilm forming bacteria within the EPS matrix (Malic et al. 2009). As with SEM microscopy, confocal images provide a high resolution and high magnification image; however, confocal microscopy can also penetrate different levels of the biofilm. This allows for the visualisation of the biofilm's EPS surface, the bacteria beneath the surface and the tissues to which the bacteria are attached. This method has been used to visualise biofilms on equine wound dressings (Parsek and Greenberg 2005).

There are a number of novel biofilm visualisation methods such as labelling antigens to MRSA biofilms (Brady et al. 2007) and the use of genetic techniques that can differentially tag Gram-negative and Gram-positive isolates (Lagendijk et al. 2009). In order to overcome the limitations of any single visualisation method, authors have endeavoured to combine multiple methods such as FISH/cLSM – SEM (Schaudinn et al. 2009), PNA FISH (Malic et al. 2009) and atomic force microscopy with epifluorescence microscopy (Mangold et al. 2008). Biofilm visualisation methods that destruct host tissue are not suitable for the visualisation of wound biofilms.

10.5 In Vitro Evidence of Biofilms

A number of models have been proposed which aim to mimic biofilm growth, and allow wound bacteria to be grouped as biofilm forming, non-biofilm forming or somewhere in between. Most of these methods identify the isolates in response to the EPS layer they produce or in response to bacterial attachment (the initial stage of biofilm formation). When possible, microscopic images can also be incorporated into these methods.

10.6 Congo Red Agar

Congo red agar can be used to distinguish biofilm forming *S. aureus* isolates from non-biofilm forming isolates (Stevens et al. 2008; Ymele-Leki and Ross 2007). A colour change from red to black is seen in response to the EPS slime layer produced by the biofilm forming isolates (Fig. 9) (Baselga et al. 1993). This colour change was supported by the presence of the *icaA* and *icaD* genes that are strongly

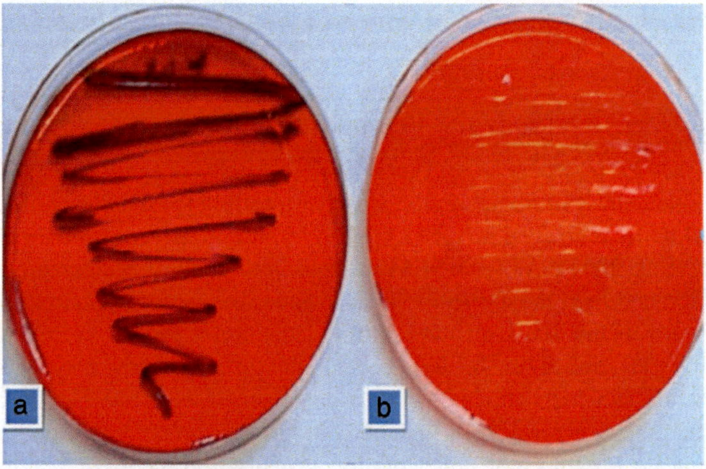

Fig. 9 Equine wound isolates of *Staphylococcus aureus* grown on Congo red agar for 72 h at 37°C. The *black coloration* of the bacteria on plate (**a**) indicates a biofilm forming *Staphylococcus aureus* phenotype. The *red coloration* of plate (**b**) is indicative of a non-biofilm forming phenotype

correlated to the production of slime production (Arciola et al. 2002). Initially, CRA was only validated for *S. aureus*; however, it has also been used to assess biofilm formation in *S. epidermidis* cultured from medical implants (Arciola et al. 2002; Mateo et al. 2008).

11 Crystal Violet Assay

The crystal violet assay provides a quantitative index of biofilm mass (Hadi et al. 2010; Izano et al. 2007). Bacterial attachment can be graded according to the optical density of the wells. Wells that have more attached bacteria will attract more crystal violet and therefore show a greater optical density. Figure 10 shows a crystal violet assay staining 27 bacteria isolated from equine wounds. There is large variation in the biofilm forming ability of the tested isolates, demonstrating that different genus, species and even bacterial strains isolated from the same wound can differ greatly in their ability to form biofilms.

A limitation of the crystal violet assay as an indicator of biofilm formation is that the 96-well plate is made of polystyrene. Different bacteria vary in their ability to adhere to different surfaces and polystyrene may not represent the ability of all bacteria to adhere to a wound surface. Crystal violet dyes bind more appropriately to some EPS components than others (Burton et al. 2007), thus some bacteria may be incorrectly recorded as weak biofilm formers.

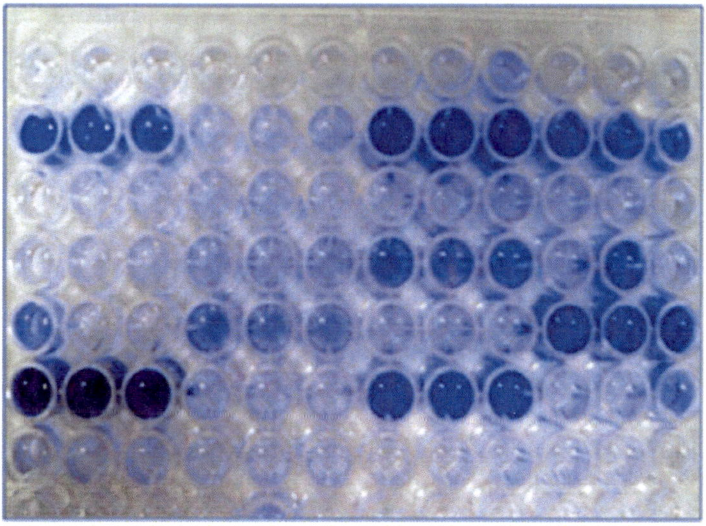

Fig. 10 A 96-well plate, Crystal Violet assay. Isolates were tested in triplicate from *left* to *right*. Darker wells indicate a greater amount of bacteria attached to the well surface. The first three wells on the *top line* were control wells containing broth only

12 Significance of Wound Biofilms

Biofilms cause a significant delay to equine wound healing (Cochrane et al. 2009). Non-healing wounds are a welfare issue for the horse, a management issue for the veterinary practitioner and an economic issue for equine owners and businesses (DeRossi et al. 2009; Merckoll et al. 2009). Biofilm infected wounds can be difficult to diagnose, since wound biofilms are not well documented in the veterinary field.

An equine wound infected with bacterial biofilms can be mistakenly reported as if it were uninfected, leading to inappropriate treatment. Biofilms can exist deep in the wound bed making bacterial isolation difficult using standard techniques such as wound swabs (Kaeberlein et al. 2002). This is because, when bacteria are protected by a biofilm, they exist in a viable but "non-culturable" state (Graham 1999). In some cases isolates are retrievable from the wound, but do not produce successful cultures (Leigh et al. 1974). False negative cultures occur as a result of inappropriate temperature, agar and aerobic/anaerobic conditions.

12.1 Biofilms and Resistance

Compared to planktonic bacteria, biofilm residing isolates notoriously demonstrate resistance against antimicrobial agents and the host's immune response (Table 2). One thousand times increased resistance has been documented for *S. aureus* within a biofilm compared to in their planktonic state (Leid et al. 2002). The host's immune response is less effective at removing a biofilm-related infection than a planktonic infection (Clutterbuck et al. 2007b). Time lapse video microscopy demonstrated that leukocytes can become trapped within a biofilm, rendering them incapable of phagocytosis. These leukocytes have been termed antiphagocytic and the process termed "frustrated phagocytosis" (Hoiby et al. 2001; Hyde et al. 1998; Leid et al. 2002). These findings were contrary to older studies that reported an inability of leukocytes to penetrate the physical barrier of the biofilm (Bolister et al. 1991; DeBeer et al. 1994; Hoyle et al. 1990; Nichols et al. 1988). The apparent contradictions have been ascribed to the bacterial species responsible for biofilm formation (Anderl et al. 2000) and to the age of the biofilm. Older mature biofilms develop fluid channels, which provide transport routes for leukocytes into the biofilm, whereas immature biofilms, such as the 2-day biofilm used in Hoyle's study (1990) may not yet have become complex enough to acquire such systems

Table 2 Examples of antibiotic resistance in post-operative wound infection in horses. Adapted from (Clutterbuck et al. 2007b)

Bacterial organism	Antibiotic resistance	References
Actinobacillus spp.	Penicillin	Smith and Ross (2002)
Staphylococcus epidermis	Methicillin	Trostle et al. (2001)
Actinobacillus equuli	Penicillin	Gay and Lording (1980)

(Leid et al. 2002). These fluid channels can also transport antimicrobial agents into the biofilm; however, antimicrobial agents are also less effective against biofilm residing isolates than planktonic isolates (Costerton et al. 1999; Serralta et al. 2001). Once inside the biofilm the inactivity of antibacterial agents such as penicillin can be explained by a slower growing bacterial phenotype (Tuomanen et al. 1986). Sessile biofilm residing bacteria display an increased ability to transfer plasmids (including genes that code for multidrug resistance) via conjugation (Hausner and Wuertz 1999). The number of chance encounters of individual bacteria is higher in biofilms than in planktonic bacteria (Hausner and Wuertz 1999). These findings suggest that the ever increasing immergence of multidrug resistant strains could be linked to their ability to form and live within a polymicrobial biofilm. Furthermore, bacteria within a biofilm can demonstrate resistance in the absence of standard mechanisms such as efflux pumps and target mutations (De Kievit et al. 2001).

A decrease in nutrient availability and/or high levels of waste products can also lower the pH and oxygen levels of the micro-environment. This inhibits the effectiveness of antimicrobial agents such as aminoglycosides (Zhang and Bishop 1996), whilst biofilm living bacteria simultaneously experience protection from pH fluctuations (Roberts 1996). Bacteria persist within the biofilm throughout antimicrobial treatment and re-establish the infection once conditions become more favourable (Costerton et al. 1999; Loo et al. 2000; Wimpenny and Colasanti 1997). Planktoninc bacteria resulting from erosion and sloughing may also augment antibiotic resistance since cells removed from the biofilm via sloughing maintain their biofilm phenotypes (Donlan 2002). A persistent bacterial infection triggers the wound to remain in the inflammatory phase therefore leading to the development of a chronic equine wound.

12.2 *In Vitro Assessment of Antibiotic Resistance in Biofilm Isolates*

Traditionally, bacterial susceptibility to antibiotic agents was reported as the minimum inhibitory concentration (MIC). Since the observation that biofilm residing bacteria show increased antibiotic resistance compared to their planktonic counterparts, the MIC value provides limited practical use. In order to address this, it is necessary to report the minimum biofilm eradication concentration (MBEC) as well as their MIC for biofilm forming isolates (Ceri et al. 1999; Sepandj et al. 2007).

A 96-well plate can be used to test bacteria at a range of serial dilutions against a chosen antibiotic. Biofilms are grown on a polystyrene pin lid by incubating the pins in wells containing inoculated broth. Plates are shaken throughout the incubation period to encourage bacterial attachment to the pins. The MIC of the chosen isolate can be read directly from the 96-well plate. The MBEC is determined by sonicating the biofilm covered pins into fresh broth. At the point where the antibiotic is no longer active against the bacteria, the wells appear cloudy due to the proliferative bacterial growth (Fig. 11) (Sepandj et al. 2007). Quantitative data can be obtained from the 96-well plates by reading the optical density of the wells. MIC values and MBEC values

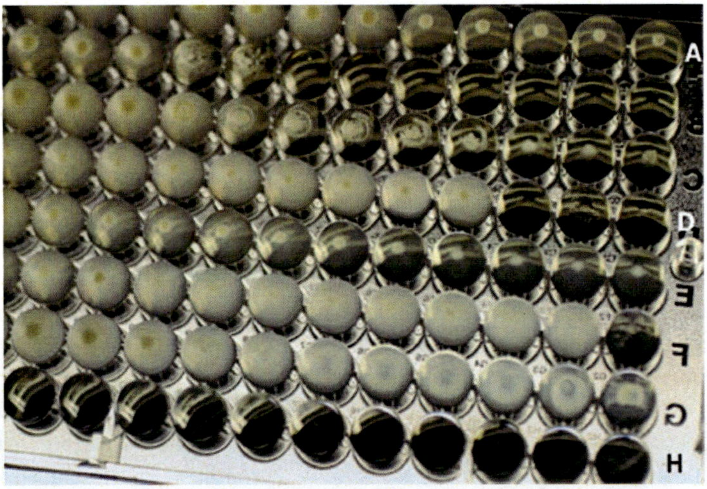

Fig. 11 A 96-well plate showing the MBEC results for seven *Staphylococcus* isolates. Isolates were tested against Penicillin G (1 unit) at 12 serial dilutions ranging from 1,000 to 0.49 μg/ml. Isolate A showed only a decreased bacterial growth, even at the highest Penicillin concentrations. Isolate D demonstrated a clear MBEC value of 250 μg/ml. Row H contained broth only

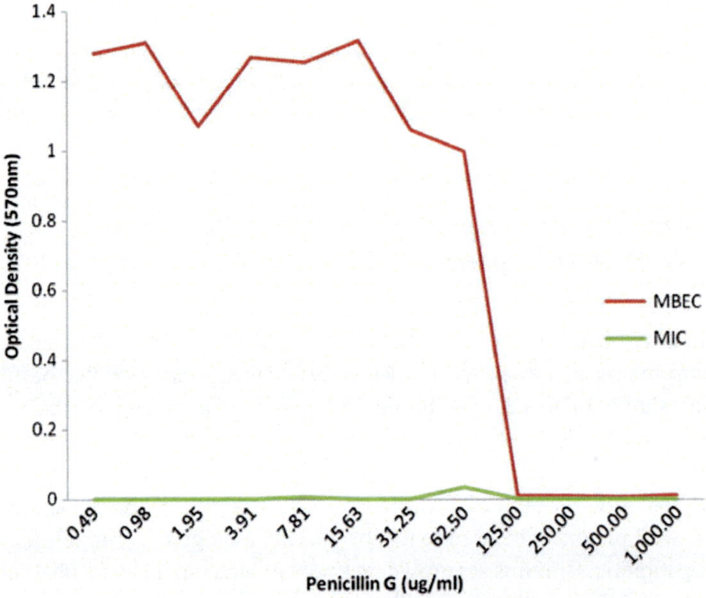

Fig. 12 MIC and MBEC concentrations of *S. warneri* cultured from a chronic equine wound

can be compared for individual isolates in response to the chosen antimicrobial agent. Low levels of antibiotics such as Penicillin G are effective against planktonic isolates, but much higher levels of the same antibiotic are required to inhibit bacterial growth in their biofilm state (Fig. 12).

13 Treatment to Remove Biofilms and Biofilm Residing Bacteria

Chronic wounds have individual causes and are infected with unique bacterial communities (Bjarnsholt et al. 2008). They can contain non-bacterial microbes, occur at varying locations and vary in their level of chronicity in response to the nutritional state of the animal. As such, each case requires an individual treatment strategy (Rhoads et al. 2008).

Wound treatment involves four principle management steps: wound assessment, wound preparation, definitive treatment and aftercare (Kumar and Leaper 2005). Treatment methods employed in the management of horse wounds focus on rapid and efficient wound evaluation, scrupulous aseptic surgical techniques and conscientious and prolonged aftercare. Appropriate antibiotic treatment regimes are employed when the wound is at risk of, or known to be infected (Griffiths et al. 2003). When a biofilm is thought to be hindering healing, these treatment regimes should be adjusted accordingly. Treatment with low level inappropriate antibiotic agents can enhance the spread on antibiotic resistance genes and select for antibiotic resistant populations.

Currently, the most effective management of a wound that is complicated by biofilm formation involves sharp surgical debridement of the infected tissue (Apelqvist et al. 1993; Cochrane et al. 2009; Wolcott and Rhoads 2008). Following biofilm removal, topical antimicrobials can be applied to the wound to prevent bacterial re-attachment and therefore limit the risk of re-infection (Lipsky and Hoey 2009). Topical treatment without debridement is unlikely to remove the biofilm (Davis et al. 2008). Where infection is suspected primary closure of the wound should not be attempted until the biofilm has been removed. Following biofilm removal, if the vascular supply and healthy tissue are plentiful primary closure can be reconsidered (Hendrickson and Virgin 2005; Zyl and Rayner 2008).

Novel treatment methods including chemical agents (Chen and Stewart 2000), bacteriophages (Donlan 2009; Lu and Collins 2007), dairy products (Busscher et al. 2000) and quorum sensing inhibitors (Balaban et al. 2007) are currently being investigated for their ability to breakdown biofilms. Unfortunately, wound biofilms are particularly difficult to treat. Treatment must disrupt and remove the biofilm without demonstrating cytotoxic effects on the newly granulating tissue. Alternative treatments such as Maggot therapies are being used to remove necrotic tissue in equine wounds (Phillips et al. 2008; Sherman et al. 2007). Where the bacterial biofilm is within necrotic tissue, maggot therapy could be a useful tool. Medical grade honey has also demonstrated bactericidal effects against biofilm encased isolates (Merckoll et al. 2009).

14 Conclusion

The process of equine wound healing follows the stages seen in all mammalian wound healing. There are many causes of equine wounds, but appropriate treatment should reflect not only the cause of the wound, but also how the wound is responding to treatment.

There are a number of methods now available to assess whether microorganisms within a wound bed are existing within a polymicrobial biofilm. There are limitations to most of the currently available methods. The application of multiple simultaneous methods provides useful data; however, it can be costly and is not always applicable to the in vivo situation. Other less invasive and less costly methods can be time consuming and of limited use within the clinical setting.

Removal of microorganisms is essential for the successful application of any wound treatment procedure. In addition to the traditional laboratory methods that report the MIC of infecting isolates, the MBEC provides useful data that should also be considered, particularly in situations where microorganisms demonstrate a decreased susceptibility to antimicrobial agents. In this way, more appropriate doses of antimicrobial drugs can be employed (Brooun et al. 2000). More appropriate doses of antimicrobials used at the correct time throughout treatment will help hinder the current spread of multiple resistance genes amongst clinical bacterial.

Appropriate treatment of biofilm infected wounds would benefit equine welfare and decrease the overall cost of treating wounds. Equine wound data can be extrapolated to aid human wound healing, particularly in the case of diabetic ulcers that have been infected by bacterial biofilms. Despite the high level of biofilm residing bacteria in the environment, and their likely prevalence in mammalian wounds, there are still many questions surrounding their identification and treatment methods. Continued research into equine biofilms is essential in order to further understand equine chronic wounds.

References

Adam EN, Southwood LL (2006) Surgical and traumatic wound infections, cellulitis, and myositis in horses. Vet Clin North Am Equine Pract 22:335–361

Anderl JN, Franklin MJ, Stewart PS (2000) Role of antibiotic penetration limitation in *Klebsiella pneumoniae* biofilm resistance to ampicillin and ciprofloxacin. Antimicrob Agents Chemother 44:1818–1824

Apelqvist J, Larsson J, Agardh CD (1993) Long-term prognosis for diabetic patients with foot ulcers. J Intern Med 233:485–491

Araujo JC, Teran FC, Oliveira RA, Nour EAA, Montenegro MAP, Campos JR, Vazoller RF (2003) Comparison of hexamethyldisilazane and critical point drying treatments for SEM analysis of anaerobic biofilms and granular sludge. J Electron Microsc 52:429–433

Arciola CR, Campoccia D, Gamberini S, Cervellati M, Donati E, Montanaro L (2002) Detection of slime production by means of an optimised Congo red agar plate test based on a colourimetric scale in *Staphylococcus epidermidis* clinical isolates genotyped for ica locus. Biomaterials 23:4233–4239

Bais HP, Fall R, Vivanco JM (2004) Biocontrol of *Bacillus subtilis* against infection of arabidopsis roots by *Pseudomonas syringae is* facilitated by biofilm formation and surfactin production1. Plant Physiol 134:307–319

Bakker DJ (2000) Hyperbaric oxygen therapy and the diabetic foot. Diab/Metab Res Rev 16: S55–S58

Balaban N, Cirioni O, Giacometti A, Ghiselli R, Braunstein JB, Silvestri C, Mocchegiani F, Saba V, Scalise G (2007) Treatment of staphylococcus aureus biofilm infection by the quorum-sensing inhibitor RIP. Antimicrob Agents Chemother 51:2226–2229

Baselga R, Albizu I, De La Cruz M, Del Cacho E, Barberan M, Amorena B (1993) Phase variation of slime production in *Staphylococcus aureus*: implications in colonization and virulence. Infect Immun 61:4857–4862

Benn M, Hagelskjaer LH, Tvede M (1997) Infective endocarditis, 1984 through 1993: a clinical and microbiological survey. J Intern Med 242:15–22

Bertone AL (1996) Infectious arthritis. In: McIlraith CW, Trotter G (eds) Joint disease in the horse. WB Saunders, Philadelphia, pp 397–409

Bjarnsholt T, Kirketerp-Møller K, Jensen PØ, Madsen KG, Phipps R, Krogfelt K, Høiby N, Givskov M (2008) Why chronic wounds will not heal: a novel hypothesis. Wound Repair Regen 16:2–10

Bjarnsholt T, Jenson PO, Fiandaca MJ, Pedersen J, Hansen CR, Andersen CB, Pressier T, Givskov M, Hoiby N (2009) *Pseudomonas aeruginosa* biofilms in the respiratory tract of cystic fibrosis patients. Pediatr Pulmonol 44:547–558

Bolister N, Basker M, Hodges NA, Marriott C (1991) The diffusion of beta-lactam antibiotics through mixed gels of cystic fibrosis-derived mucin and *Pseudomonas aeruginosa* alginate. J Antimicrob Chemother 27:285–293

Bowler PG (2003) The 105 bacterial growth guidelines: reassessing its clinical relevance in wound healing. Ost Wound Manag 49:44–53

Brady RA, Leid JG, Kofonow J, Costerton JW, Shirtliff ME (2007) Immunoglobulins to surface-associated biofilm immunogens provide a novel means of visualization of methicillin-resistant *Staphylococcus aureus* biofilms. Appl Environ Microbiol 73:6612–6619

Brenner K, Karig DK, Weiss R, Arnold FH (2007) Engineered bidirectional communication mediates a consensus in a microbial biofilm consortium. Proc Natl Acad Sci USA 104: 17300–17304

Broadley KN, Aquino AM, Woodward SC, Buckley-Sturrock A, Sato Y, Rifkin DB, Davidson JM (1989) Monospecific antibodies implicate basic fibroblast growth factor in normal wound repair. Lab Investig 61:571–575

Brooun A, Songhua L, Lewis K (2000) A dose-response study of antibiotic resistance in *Pseudamonas aeruginosa* biofilms. Antimicrob Agents Chemother 44:640–646

Burton E, Yakandawala N, LoVetri K, Madhyastha MS (2007) A microplate spectrofluorometric assay for bacterial biofilms. J Ind Microbiol Biotechnol 34:1–4

Busscher HJ, Free RH, Van Weissenbruch R, Albers FWJ, Van Der Mei HC (2000) Preliminary observations on influence of dairy products on biofilm removal from silicone rubber voice prostheses in vitro. J Dairy Sci 83:641–647

Carter CA, Jolly DG, Worden CE, Hendren DG, Kane CJM (2003) Platelet-rich plasma gel promotes differentiation and regeneration during equine wound healing. Exp Mol Pathol 74:224–255

Ceri H, Olson ME, Stremick C, Read RR, Morck D, Buret A (1999) The calgary biofilm device: new technology for rapid determination of antibiotic susceptibilities of bacterial biofilms. J Clin Microbiol 37:1771–1776

Chavakis T, Hussain M, Kanse SM (2002) *Staphylococcus aureus* extracellular adherence protein serves as an anti-inflammatory factor by inhibiting the recruitment of host leukocytes. Nat Med 8:687–693

Chen X, Stewart PS (2000) Biofilm removal caused by chemical treatments. Water Res 34: 4229–4233

Chincholikar DA, Pal RB (2002) Study of fungal and bacterial infections of the diabetic foot. Indian J Pathol Microbiol 45:15–22

Clark RAF (1985) Cutaneous tissue repair: basic biological considerations. J Am Acad Dermatol 13:701–725

Clutterbuck AL, Cochrane CA, Dolman J, Percival SL (2007a) Evaluating antibiotics for use in medicine using a poloxamer biofilm model. Ann Clin Microbiol Antimicrob 6:2

Clutterbuck AL, Woods EJ, Knottenbelt DC, Clegg PD, Cochrane CA, Percival SL (2007b) Biofilms and their relevance to veterinary medicine. Vet Microbiol 121:1–17

Cochrane CA (1997) Models *in vivo* of wound healing in the horse and the role of growth factors. Vet Dermatol 8:259–272

Cochrane CA, Pain R, Knottenbelt DK (2003) *In-vitro* wound contraction in the horse: differences between body and limb wounds. Wounds 15:175–181

Cochrane CA, Freeman K, Woods E, Welsby S, Percival SL (2009) Biofilm evidence and the microbial diversity of horse wounds. Can J Microbiol 55:197–202

Collins MN, Friend TH, Jousan FD, Chen SC (2000) Effect of density on displacement, falls, injuries and orientation during horse transportation. Appl Anim Behav Sci 67:169–179

Costerton JW (1999) Introduction to biofilm. Int J Antimicrob Agents 11:217–221

Costerton JW, Steward PS, Greenberg EP (1999) Bacterial biofilms: a common cause of persistent infections. Science 284(5418):1318–1322

Cruse PJ, Foord R (1980) The epidemiology of wound infection. A 10-year prospective study of 62,939 wounds. Surg Clin North Am 60:27–40

Dahm AM, De Bruin A, Limat A, VON Tscharner C, Wyder M, Suter MM (2002) Cultivation and characterisation of primary and subcultured equine keratinocytes. Equine Vet J 34:114–120

Darby I, Gabbiani G (1990) Alpha-smooth muscle actin is transiently expressed by myofibroblasts during experimental wound healing. Lab Investig 63:21–29

Darveau RP, McFall-Ngai M, Ruby E, Miller S, Mangan DF (2003) Host tissues may actively respond to beneficial microbes. ASM News 69:86–191

Davies DG, Parsek MR, Pearson JP, Iglewski BH, Costerton JW, Greenberg EP (1998) The involvement of cell-to-cell signals in the development of a bacterial biofilm. Science 280:295–298

Davis SC, Ricotti C, Cazzaniga A, Welsh E, Eaglstein WH, Mertz PM (2008) Microscopic and physiological evidence forbiofilm-associated wound colonisation *in vivo*. Wound Repair Regen 16:23–29

De Kievit TR, Parkins MD, Gilllis RJ, Srikumar R, Ceri H, Poole K, Iglewski BH, Storey DG (2001) Multidrug efflux pumps: expression patterns and contribution to antibiotic resistance in *Pseudomonas aeruginosa* biofilms. Antimicrob Agents Chemother 45:1761–1770

DeBeer D, Srinivasan R, Stewart PS (1994) Direct measurement of chlorine penetration into biofilms during disinfection. Appl Environ Microbiol 60:4339–4344

DeRossi R, Coelho ACAO, Mello GS, Frazílio FO, Leal CRB, Facco GG, Brum KB (2009) Effects of platelet-rich plasma gel on skin healing in surgical wound in horses. Acta Cir Bras 24:276–281

Donlan RM (2002) Biofilms: microbial life on surfaces. Emerg Infect Dis 8:881–890

Donlan RM (2009) Preventing biofilms of clinically relevant organisms using bacteriophage. Trends Microbiol 17:66–72

Dorsett-Martin WA, Wysocki AB (2008) Rat models of skin wound healing. In: Conn PM (ed) Sourcebook of models for biomedical research. Humana, Totowa, NJ, pp 631–638

Elliot M (2001) Cushing's disease: a new approach to therapy in equine and canine patients. Br Homeopath J 90:33–36

Fazli M, Bjarnsholt T, Kirketerp-Møller K, Jørgensen B, Andersen AS, Krogfelt K, Givskov M, Tolker-Nielsen T (2009) Non-RNAdon distribution of *Pseudamonas aeruginoa* and *Staphylococcus aureus* in chronic wounds. J Clin Microbiol. doi:10.1128/JCM.01395-09

Fexby S, Bjarnsholt T, Ostrup JP (2007) Biological Trojan horse: antigen 43 provides specific bacterial uptake and survival in human neutrophils. Infect Immun 75:30–34

Frank S, Hubner G, Breier G, Longaker MT, Greenhalgh DG, Werner S (1995) Regulation of vascular endothelial growth factor expression in cultured keratonocytes, implications for normal and impaired wound healing. Am Soc Biochem Mol Biol 270:12607–12613
Fratesi SE, Lynch FL, Kirkland BL, Brown LR (2004) Effects of SEM preparation techniques on the appearance of bacteria and biofilms in the Carter Sandstone. J Sed Res 74:858–867
Galuppo LD, Pascoe JR, Jang SS, Willits NH, Greenman SL (1999) Evaluation of iodophor skin preparation techniques and factors influencing drainage from ventral midline incisions in horses. J Am Vet Med Assoc 215:969
Gay CC, Lording PM (1980) Peritonitis in horses associated with *Actinobacillus equuli*. Aust Vet J 56:296–300
Golovlev EL (2002) The mechanism of formation of *Pseudomonas aeruginosa* biofilm, a type of structured population. Mikrobiologiia 71:293–300
Goodrich LR (2006) Osteomyelitis in horses. Vet Clin North Am Equine Pract 22:389–417
Graham DY (1999) Antibiotic resistance in *Helicobacter pylori*: implications for therapy. Gastroenterology 117:1032–1033
Greiling D, Clark RAF (1997) Fibronectin provides a conduit for fibroblast transmigration from collagenous stroma into fibrin clot provisional matrix. J Cell Sci 110:861–870
Griffiths DA, Simpson RA, Shorey BA, Speller DCE, Williams NB (2003) Single-dose preoperative antibiotic prophylaxis in gastrointestinal surgery. Lancet 308:325–328
Grinnell F (1992) Wound repair, keratinocyte activation and integrin modulation. J Cell Sci 101:1–5
Hadi R, Vickery K, Deva A, Charlton T (2010) Biofilm removal by medical device cleaners: comparison of two bioreactor detection assays. J Hosp Infect 74:160–167
Hall MJR, Wall R (1995) Myiasis of humans and domestic animals. Adv Parasitol 35:257–334
Hall-Stoodley L, Costerton JW, Stoodley P (2004) Bacterial biofilms: from the natural environment to infectious diseases. Nat Rev Microbiol 2:95–108
Harrison-Balastra C, Cazzaniga AL, Davis SC, Mertz PM (2003) A wound-isolated *Pseudomonas aeruginosa* grows a biofilm in vitro within 10 hours and is visualized by light microscopy. J Dermatol Surg 29:1–5
Hausner M, Wuertz S (1999) High rates of conjugation in bacterial biofilms as determined by quantitative in situ analysis. Appl Environ Miocrobiol 65:3710–3713
Hendrickson D, Virgin J (2005) Factors that affect equine wound repair. Vet Clin Equine 21:33–44
Hernandez J, Hawkins DL (2001) Training failure among yearling horses. Am J Vet Res 62:1418–1422
Hoiby N, Krogh JH, Moser C, Song Z, Ciofu O, Kharazmi A (2001) *Pseudomonas aeruginosa* and the in vitro and in vivo biofilm mode of growth. Microbes Infect 3:23–35
Hoyle BD, Jass J, Costerton JW (1990) The biofilm glycocalyx as a resistance factor. J Antimicrob Chemother 26:1–5
Hyde JA, Darouiche RO, Costerton LW (1998) Strategies for prophylaxis against prosthetic valve endocarditis: a review article. J Heart Valve Dis 7:316–326
Ivnitsky H, Katz I, Minz D, Volvovic G, Shimoni E, Kesselman E, Semiat R, Dosoretz CG (2007) Bacterial community composition and structure of biofilms developing on nanofiltration membranes applied to wastewater treatment. Water Res 41:3924–3935
Izano EA, Wang H, Ragunath C, Ramasubbu N, Kaplan JB (2007) Detachment and Killing of *Aggregatibacter actinomycetemcomitans* biofilms by dispersin B and SDS. J Dent Res 86:618–622
James GA, Swogger E, Wolcott R, Pulcini ED, Secor P, Sestrich J, Costerton JW, Stewart PS (2008) Biofilms in chronic wounds. Wound Repair Regen 16:37–44
Johnson D (1990) The effect of continuous passive motion on wound-healing and joint mobility after knee arthroplasty. J Bone Joint Surg Am 72:421–426
Johnson P, Hayman J, Quek T (2007) Consensus recommendations for the diagnosis, treatment and control of *Mycobacterium ulcerans* infection (Bairnsdale or Buruli ulcer) in Victoria, Australia. Med J Aust 186:64–68

Kaeberlein T, Lewis K, Epstein SS (2002) "Uncultivable" microorganisms in pure culture in simulated natural environment. Science 296:1127–1129

Kania RE, Lamers GEM, Vonk JM, Dorpmans E, Struik J, Huy PT, Hiemstra P, Bloemberg GV, Grote JJ (2008) Characterization of mucosal biofilms on human adenoid tissues. Laryngoscope 118:128–134

Kingsley A (2001) A reactive approach to wound infection. Nurs Stand 15:50–58

Kipnis E, Sawa T, Wiener-Kronish J (2006) Targeting mechanisms of *Pseudomonas aeruginosa* pathogenesis. Me´ Malad Infect 36:78–91

Kirker KR, Secor PR, James GA, Fleckman P, Olerud JE, Stewart PS (2009) Loss of viability and induction of apoptosis in human keratinocytes exposed to *Staphylococcus aureus* biofilms in vitro. Wound Repair Regen 17:690–699

Knottenbelt DC (1997) Equine wound management: are there significant differences in healing at different sites on the body? Vet Dermatol 8:273–290

Knottenbelt DC (2003) Handbook of equine wound management. WB Saunders, Liverpool

Knottenbelt DC (2007) Handbook of equine wound management. WB Saunders, China

Knubben JM, Furst A, Gygax L, Stauffacher M (2008) Bite and kick injuries in horses: prevalence, risk factors and prevention. Equine Vet J 40:219–223

Kolenbrander PE, Andersen RN, Blehert DS, Egland PG, Foster JS, Palmer RJ (2002) Communication among oral bacteria. Microb Mol Biol Rev 66:486–505

Kumar S, Leaper DJ (2005) Classification and management of acute wounds. Surgery 23:47–51

Lagendijk EL, Validov S, Lamers GEM, de Weert S, Bloemberg GV (2009) Genetic tools for tagging Gram-negative bacteria with mCherry for visualization *in vitro* and in natural habitats, biofilm and pathogenicity studies. FEMS Microbiol Lett 305:81–90

Lambers H, Piessens S, Bloem A (2006) Natural skin surface pH is on average below 5, which is beneficial for its resident flora. Int J Cosmet Sci 28:359–370

Landry RM, An D, Hupp JT, Singh PK, Parsek MR (2006) Musin-*Pseudomonas aeruginosa* interactions promote biofilm formation and antibiotic resistance. Mol Microbiol 59:142–151

Lee YK, Lim CY, Teng WL, Ouwehand AC, Tuomola EM, Salminen S (2000) Quantitative approach in the study of adhesion of lactic acid bacteria to intestinal cells and their competition with *Enterobacteria*. Appl Environ Microbiol 66:3692–3697

Lefebvre-Lavoie J, Lussier JG, Theoret CL (2005) Profiling of differentially expressed genes in wound margin biopsies of horses using suppression subtractive hybridization. Physiol Genomics 22:157–170

Leid JG, Shirtliff ME, Costerton JW, Stoodley AP (2002) Human leukocytes adhere to, penetrate, and respond to *Staphylococcus aureus* biofilms. Infect Immun 70:6339–6345

Leigh DA, Simmons K, Norman E (1974) Bacterial flora of the appendix fossa in appendicitis and postoperative wound infection. J Clin Path 27:997–1000

Lewandowski L, Stoodley AP, Roe F, (1995) Internal mass transport in heterogeneous biofilms: recent advances. In: NACE International Annual Conference and Corrosion Show, Houston

Lindsay D, Von Holy A (2006) Bacterial biofilms within the clinical setting. What healthcare professionals should know. J Hosp Infect 64:313–325

Lipsky BenjaminÂ A, Hoey C (2009) Clinical practice: topical antimicrobial therapy for treating chronic wounds. Clin Infect Dis 49:1541–1549

Little B, Wagner P, Ray R, Pope R, Scheetz R (1991) Biofilms: an ESEM evaluation of artifacts introduced during SEM preparation. J Ind Microbiol Biotechnol 8:213–221

Loo CY, Corliss DA, Ganeshkumar N (2000) *Streptococcus gordonii* biofilm formation: identification of genes that code for biofilm phenotypes. Am J Microbiol 182:1374–1382

Lu TK, Collins JJ (2007) Dispersing biofilms with engineered enzymatic bacteriophage. Proc Natl Acad Sci USA 104:111

MacDonald DG, Morley PS, Bailey JV, Barber SM, Fretz PB (1994) An examination of the occurrence of surgical wound infection following equine orthopaedic surgery (1981–1990). Equine Vet J 6:323–326

Mair TS, Smith LJ (2005) Survival and complication rates in 300 horses undergoing surgical treatment of colic. Part 2: Short-term complications. Equine Vet J 37:303–309

Malic S, Hill KE, Hayes A, Percival SL, Thomas DW, Williams DW (2009) Detection and identification of specific bacteria in wound biofilms using peptide nucleic acid fluorescent *in situ* hybridization (PNA FISH). Microbiology 155:2603–2611

Mangold S, Harneit K, Rohwerder T, Claus G, Sand W (2008) Novel combination of atomic force microscopy and epifluorescence microscopy for visualization of leaching bacteria on pyrite. Appl Environ Microbiol 74:410–415

Marrie TJ, Costerton JW (1984) Scanning and transmission electron microscopy of in situ bacterial colonization of intravenous and intraarterial catheters. J Clin Microbiol 19(5): 687–693

Marrie TJ, Nelligan J, Costerton JW (1982) A scanning and transmission electron microscopic study of an infected endocardial pacemaker lead. Circulation 66:1339–1341

Mateo M, Maestre J, Aguilar L, Giménez M, Granizo J, Prieto J (2008) Strong slime production is a marker of clinical significance in *Staphylococcus epidermidis* isolated from intravascular catheters. Eur J Clin Microbiol Infect Dis 27:311–314

Merckoll P, Jonassen TØ, Vad ME, Jeansson SL, Melby KK (2009) Bacteria, biofilm and honey: a study of the effects of honey on 'planktonic' and biofilm-embedded chronic wound bacteria. Scand J Infect Dis 41:341–347

Mertz PM (2003) Cutaneous biofilms: friend or foe? Wounds 15:129–132

Mertz AJ, Forest KT (2002) Bacterial surphase motility: slime trails, grappling hooks and nozzles. Curr Microbiol 12:297–303

Morris NS, Stickler DJ, McLean RJ (1999) The development of bacterial biofilms on indwelling urethral catheters. World J Urol 17:345–350

Nadell CD, Xavier JB, Levin SA, Foster KR (2008) The evolution of quorum sensing in bacterial biofilms. PLoS Biol 6:14

Nadell CD, Xavier JB, Foster KR (2009) The sociobiology of biofilms. FEMS Microbiol Rev 33:206–224

Nakada T, Saito Y, Chikenji M, Koda S, Higuchi M, Kawata K, Ishida S, Takahashi S, Kondo S, Kubota Y, Kubota I, Shimizu Y (2006) Therapeutic outcome of hyperbaric oxygen and basic fibroblast growth factor on intractable skin ulcer in legs: preliminary report. Plast Reconstr Surg 117:646–651

Neu TR (1996) Significance of bacterial surface-active compounds in interactions of bacteria with interfaces. Microbiol Rev 60:151–166

Nguyen TT, Gilpin DA, Meyer DA (1996) Current treatment of severely burned patients. Ann Surg 223:14–25

Nichols WW, Dorrington SM, Slack MP, Walmsley HL (1988) Inhibition of tobramycin diffusion by binding to alginate. Antimicrob Agents Chemother 32:518–523

O'Toole GA, Kolter R (1998) Initiation of biofilm formation in *Pseudomonas fluorescens* WCS365 proceeds via multiple, convergent signalling pathways: a genetic analysis. Mol Microbiol 28:449–461

O'Toole G, Kaplan HB, Kolter R (2000) Biofilm formation as microbial development. Annu Rev Microbiol 54:49–79

Parsek MR, Greenberg EP (2005) Sociomicrobiology: the connection between quorum sensing and biofilms. Trends Microbiol 13:27–33

Percival SL, Cochrane CA (2010) MMP and microbial enzymes. In: Percival SL, Cutting K (eds) Microbiology of wound. CRC, New York

Phillips P, Sampson E, Yang O, Antonelli P, Progulske-Fox A, Schultz G (2008) Bacterial biofilms in wounds. Wound Heal South Afr 1:10–12

Prigent-Combaret C, Vidal O, Dorel C, Lejeune P (1999) Abiotic surface sensing and biofilm-dependent regulation of gene expression in *Escherichia coli*. J Bacteriol 181:5993–6002

Proudman CJ, Smith JE, Edwards GB, French NP (2002) Long-term survival of equine surgical colic cases. Part 1. Patterns of mortality and morbidity. Equine Vet J 34:432–437

Rather PN (2005) Swarmer cell differentiation in *Proteus mirabilis*. Environ Microbiol 7: 1065–1073

Rhoads DD, Wolcott RD, Percival SL (2008) Biofilms in wounds: management strategies. J Woundcare 17:502–508

Rice SA, Koh KS, Queck SY, Labbate M, Lam KW, Kjelleberg S (2005) Biofilm formation and sloughing in *Serratia marcescens* are controlled by quorum sensing and nutrient cues. J Bacteriol 187:3477–3485

Rivera AE, Spencer JM (2007) Clinical aspects of full-thickness wound healing. Clin Dermatol 25:39–48

Roberts IS (1996) The biochemistry and genetics of capsular polysacharide production in bacteria. Annu Rev Microbiol 50:285–315

Sauer K, Camper AK, Ehrlich GD, Costerton JW, Davies DG (2002) *Pseudomonas aeruginosa* displays multiple phenotypes during development as a biofilm. J Bacteriol 184:1140–1154

Schaudinn C, Carr G, Gorur A, Jaramillo D, Costerton JW, Webster P (2009) Imaging of endodontic biofilms by combined microscopy (FISH/cLSM – SEM). J Microsc 235:124–127

Schierle CF, Garza MDl, Mustoe TA, Galiano RD (2009) Staphylococcal biofilms impair wound healing by delaying reepithelialization in a murine cutaneous wound model. Wound Repair Regen 17:354–359

Schreml S, Szeimies RM, Karrer S, Heinlin J, Landthaler M, Babilas P (2009) The impact of the pH value on skin integrity and cutaneous wound healing. J Eur Acad Dermatol Venereol 24:373–378

Schwartz AJ, Wilson DA, Keegan KG (2002) Factors regulating collagen synthesis and degredation during second-intension healing of wounds in the thoracic region and distal aspect of the forelimb of horses. Am J Vet Clin 63:1564–1570

Sen CK (2009) Wound healing essentials: let there be oxygen. Wound Repair Regen 17:1–18

Sepandj F, Ceri H, Gibb A, Read R, Olson M (2007) Minimum inhibitory concentration versus minimum biofilm eliminating concentration in evaluation of antibiotic sensitivity of enterococci causing peritonitis. Perit Dial Int 27:464–468

Serralta VW, Harrison-Balestra C, Cazzaniga AL (2001) Lifestyles of bacteria in wounds: presence of biofilms? Wounds 13:29–34

Sherman RA, Morrison S, Ng D (2007) Maggot debridement therapy for serious horse wounds – a survey of practitioners. Vet J 174:86–91

Singer II, Kawka DM, Kazazis DM, Clark RAF (1984) The *in vivo* codistribution of fibronectin and actin firers in granulation tissue: immunofluorescence and electron microscopic studies of the fibronexus at the myofibroblast surface. J Cell Biol 98:2091

Singer ER, Saxby F, French NP (2003) A retrospective case-control study of horse falls in the sport of horse trials and three-day eventing. Equine Vet J 35:139–145

Singh PK, Schaefer AL, Parsek MR, Moninger TO, Welsh MJ, Greenberg EP (2000) Quorum-sensing signals indicate that cystic fibrosis lungs are infected with bacterial biofilms. Nature 407:762–764

Slovis N (2008) Review of equine hyperbaric medicine. J Equine Vet Sci 28:760–767

Smith MA, Ross MW (2002) Postoperative infection with *Actinobacillus* spp. in horses: 10 cases (1995–2000). J Am Vet Med Assoc 221:1306–1310

Stevens N, Tharmabala M, Dillane T, Greene CM, O'Gara JP, Humphreys H (2008) Biofilm and the role of the *ica* operon and *aap* in *Staphylococcus epidermidis* isolates causing neurosurgical meningitis. Clin Microbiol Infect 14:719–722

Stewart PS, Franklin MJ (2008) Physiological heterogeneity in biofilms. Nat Rev Micro 6:199–210

Stoodley P, Wilson S, Hall-Stoodley L, Boyle JD, Lappin-Scott HM, Costerton JW (2001) Growth and detachment of cell clusters from mature mixed-species biofilms. Appl Environ Microbiol 67:5608–5613

Stoodley P, Sauer K, Davies DG, Costerton JW (2002) Biofilms as complex differentiated communities. Annu Rev Microbiol 56:187–209

Sun Y, Dowd SE, Smith E, Rhoads DD, Wolcott RD (2008) *In vitro* multispecies Lubbock chronic wound biofilm model. Wound Repair Regen 16:805–813

Theoret CL, Barber SM, Moyana TN, Gordon JR (2001) Expression of transforming growth factor β_1 and β_3, and basic fibroblast growth factor in full-thickness skin wounds of equine limbs and thorax. Vet Surg 30(3):269–277

Trostle SS, Peavey CL, King DS, Hartmann FA (2001) Treatment of methicillin-resistant Staphylococcus epidermidis infection following repair of an ulnar fracture and humeroradial joint luxation in a horse. J Am Vet Med Assoc 4:554–559, 527
Tuomanen E, Cozens R, Tosch W, Zak O, Tomasz A (1986) The rate of killing of *Escherichia coli* by β-lactam antibiotics is strictly proportional to the rate of bacterial growth. J Gen Microbiol 132:1297–1304
Van der Kolk JH (1997) Equine Cushing's disease. Equine Vet Educ 9:209–214
Vasseur PB, Levy J, Dowd E (1988) Surgical wound infection rates in dogs and cats: data from a teaching hospital. Vet Surg Clin N Am 17:60–64
Waring GH (2003) Agonistic behaviour. In: Waring GH (ed) Horse behaviour. Noyes/William Andrew, New York, pp 253–269
Watnick P, Kolter R (2000) Biofilm: city of microbes. J Bacteriol 182:2675–2679
Webster P, Wu S, Webster S, Rich KA, McDonald K (2004) Ultrastructural preservation of biofilms formed by non-typeable *Haemophilus influenzae*. Biofilms 1:165–182
Westgate SJ, Percival SL, Knottenbelt DC, Clegg PD, Cochrane CA (2010) Chronic equine wounds: what is the role of infection and biofilms? Wounds 22:138–145
White R, Cutting K (2008) Critical colonisation of chronic wounds: microbial mechanisms. Wounds 4:70–78
Williams P (2007) Quorum sensing, communication and cross-kingdom signalling in the bacterial world. Microbiology 153:3923–3938
Wilmink JM, Van Weeren PR (2005) Second-intention repair in the horse and pony and management of exuberant granulation tissue. Vet Clin North Am Equine Pract 21:15–32
Wilmink JM, Nederbragt H, Van Weeren PR, Stilk PW, Barneveld A (2001) Differences in wound contraction between horses and ponies: the in vitro contraction capability of fibroblasts. Equine Vet J 33:499–505
Wilmink JM, Herten J, Weeren PR, Barneveld A (2002) Retrospective study of primary intention healing and sequestrum formation in horses compared to ponies under clinical circumstances. Equine Vet J 34:270–273
Wilmink JM, Van Den Boom R, Van Weeren PR, Barneveld A (2006) The modified Meek technique as a novel method for skin grafting in horses: evaluation of acceptance, wound contraction and closure in chronic wounds. Equine Vet J 38:324–329
Wimpenny JWT, Colasanti RA (1997) A unifying hypothesis for the structure of microbial films based on cellular automation models. FEMS Microb Ecol 22:1–16
Wolcott RD, Rhoads DD (2008) A study of biofilm-based wound management in subjects with critical limb ischaemia. J Wound Care 17:145–155
Wuertz S, Okabe S, Hausner M (2004) Microbial communities and their interactions in biofilm systems: an overview. Water Sci Technol 49:327–336
Xavier JB, Foster KR (2007) Cooperation and conflict in microbial biofilms. Proc Natl Acad Sci USA 104:876–881
Ymele-Leki P, Ross JM (2007) Erosion from *Staphylococcus aureus* biofilms grown under physiologically relevant fluid shear forces yields bacterial cells with reduced avidity to collagen. Appl Environ Microbiol 73:1834–1841
Zhang TC, Bishop PL (1996) Evaluation of substrate and pH effects in a nitrifying biofilm. Water Environ Res 68:1107–1115
Zyl N, Rayner SG (2008) Penetrating thoracic injury with associated abdominal visceral involvement in a mare. Equine Vet Educ 20:414–417
Zyl AV, Daniel J, Wayne J, McCowan C, Malik R, Jelfs P, Lavender CJ, Fyfe JA (2010) *Mycobacterium ulcerans* infections in two horses in south-eastern Australia. Aust Vet J 88:101–106

Osteomyelitis in the Veterinary Species

Peter D. Clegg

Abstract Osteomyelitis has been defined as an inflammatory process of bone that is accompanied by bone destruction, and is caused by infecting microorganisms. Osteomyelitis has a number of aetiologies in the veterinary species, but haematogenous, traumatic and iatrogenic causes are the most frequent. Osteomyelitis is frequently challenging to manage and treat, due to the ability of bacteria to evade and overcome both host defences and antimicrobial agents. Prior to availability of antimicrobial agents, bacterial osteomyelitis had a 50% mortality rate in man and whilst the prognosis has improved considerably with the advent of such drugs, treatment can often be extremely problematic. Treatment usually requires prolonged antibiosis frequently administered in conjunction with surgical debridement of the affected bone.

1 Introduction

Osteomyelitis has been defined as an inflammatory process of bone that is accompanied by bone destruction, and is caused by infecting microorganisms. If the infection is limited to bone only, the disease is termed osteitis, but where extension occurs into the bone marrow, osteomyelitis is the relevant terminology. Osteomyelitis is frequently challenging to manage and treat, due to the ability of bacteria to evade and overcome both host defences and antimicrobial agents (Ciampolini and Harding 2000). Prior to availability of antimicrobial agents, bacterial osteomyelitis had a 50% mortality rate in man (Joyner and Smith 1936), and whilst the prognosis has improved considerably with the advent of such drugs, treatment can often be extremely problematic (Goodrich 2006).

P.D. Clegg (✉)
Department of Musculoskeletal Biology, University of Liverpool, Leahurst Campus, Neston, Cheshire CH64 5SD, UK
e-mail: p.d.clegg@liverpool.ac.uk

S.L. Percival et al. (eds.), *Biofilms and Veterinary Medicine*, Springer Series on Biofilms 6, DOI 10.1007/978-3-642-21289-5_7,

2 Classifications

In most classifications, osteomyelitis is split into three forms dependent on the aetiology: haematogenous infections (which almost invariably occur in neonates secondary to septicaemia), traumatic (secondary to laceration or puncture wounds) and iatrogenic (secondary to surgical intervention and most commonly associated with the presence of orthopaedic implants).

3 Pathophysiology of Osteomyelitis

Osteomyelitis requires initial bacterial contamination and adhesion, followed by infection and subsequent chronicity. Bacteria reach the bone by either haematogenous routes, from direct inoculation or by direct extension from adjacent soft tissue infections. The latter route, whilst common in man, is relatively rare in the veterinary species. In man, a variety of patient predisposing factors have been identified (Ciampolini and Harding 2000), although in the veterinary species there is less understanding of such risk factors other than those relating to immunocompromise.

The role of bone necrosis and ischaemia is a key event in the establishment of osteomyelitis (Ciampolini and Harding 2000). Interestingly, healthy bone is extremely resistant to bacterial infection (Andriole et al. 1973); however, as soon as there is either bone necrosis, contamination or the presence of foreign bodies (for instance the presence of orthopaedic implants), the ability of bone to become infected is considerably increased. The importance of the presence of foreign material in potentiating infection was demonstrated in a rabbit model of orthopaedic implant infection, where as few as 50 *staphylococcal* organisms would lead to establishment of infection in a rabbit model of hip arthroplasty, whereas approximately 10,000 organisms were required to produce an infection in the absence of a foreign body (Southwood et al. 1985).

Following initial bone infection, purulent material can spread widely through the medullary canal. On occasions the infection can exit the cortical bone, burst through the periosteum and lead to infection in the surrounding soft tissues, or drain externally. Purulent material results in increase in pressure in the medullary cavity, and thrombosis in blood vessels, which can lead to bone necrosis. Such necrotic bone can act as a sequestrum and continue to act as a focus of infection. Small sequestra can be resorbed by osteoclasts; however, more frequently, sequestra become walled off by a lining of compact new bone, known as an involucrum. The involucrum frequently is perforated by an opening, which allows drainage of purulent material to the outside (Catto 1980).

There is considerable evidence that biofilms are a consistent feature of microbial colonisation in osteomyelitis and in implant-related bone infections (Brady et al. 2008). It has been frequently postulated that the presence of the biofilm and the adherent mode of growth of bacteria in biofilms will reduce the susceptibility of these organisms to host clearance mechanisms and antibiotic therapy, and thus may be a fundamental factor in acute and chronic osteomyelitis (Costerton 2005; Gristina and Costerton 1984; Gristina et al. 1985; Sedghizadeh et al. 2009). Once bacteria are established by adherence on a bone surface, and surrounded by a biofilm, it makes them relatively resistant to evade both host defences, and from antimicrobial therapy. Furthermore, the biofilm allows establishment of an intercellular signalling network and horizontal transfer of genetic material, all of which improves the survival of bacteria in osteomyelitis, and limits our ability to manage infection therapeutically (Davies et al. 1998; Mah and O'Toole 2001; Mah et al. 2003).

4 Osteomyelitis and Bacteriology

There is a diverse range of species of bacteria identified in osteomyelitis, and the bacterial species identified may vary in different regions. However, a number of studies have determined the bacteriology in orthopaedic infections and osteomyelitis. Whilst such studies are useful, it is important that in all cases of suspected cases of osteomyelitis appropriate culture and sensitivity are undertaken to determine the most appropriate antimicrobial treatment.

In adult horses, a study of 233 horses with osteomyelitis and septic arthritis and tenosynovitis determined that 91% of bacterial isolates were aerobic or facultatively anaerobic and only 9% were anaerobic. The most common bacteria isolated were *Enterobacteriaceae* (29%) *non-β-haemolytic streptococci* (13%), *Coagulase positive staphylococci* (12%), *β-haemolytic streptococci* (9.4%) *and coagulase negative staphylococci* (7.3%) (Moore et al. 1992). A more recent report of infections subsequent to long bone fracture repair and arthrodesis in the horse identified that 32% of infections were Gram-positive, 28% were Gram-negative and 40% were mixed infections. *Enterobacter cloacae* (24.5%) was the most commonly cultured Gram-negative bacteria (Ahern et al. 2010), whilst coagulase-negative *staphylococcus* (21%) was the most common Gram-positive isolate. Positive culture was associated with a lower rate of discharge from hospital. In foals, enteric Gram-negative organisms are most commonly isolated (Goodrich 2006). In foals past the immediate neonatal period, osteomyelitis has been associated with *rhodococcus* infection, particularly on farms where such infection is endemic (Firth et al. 1993).

In dogs *staphylococcus* and *streptococcus species* have been identified as the most frequent bacterial isolates from canine osteomyelitis (Hodgin et al. 1992; Johnson et al. 1984; Muir and Johnson 1992; Walker et al. 1983).

5 Osteomyelitis in the Horse

5.1 Haematogenous Osteomyelitis

Haematogenous osteomyelitis is almost exclusively seen in the neonatal foal, and is associated with immunocompromise subsequent to failure of passive transfer of immunity. Affected animals may have had a history of septicaemia, and frequently there will be a focus of infection associated at a site such as at the umbilicus. In many cases there will be multiple sites affected. There is a very close relationship between neonatal osteomyelitis and articular sepsis, and whilst both articular sepsis can progress on to cause osteomyelitis in the adjacent epiphysis, and similarly epiphyseal osteomyelitis can burst into the adjacent joint, most frequently they are separate conditions. A classification for neonatal sepsis has been proposed in the horse (Firth 1983; Firth and Goedegebuure 1988). S-type osteomyelitis derives from the synovium and results in articular sepsis. This is most commonly seen in the early neonate in the first few days of life. E-type osteomyelitis occurs in the epiphyseal bone beneath the articular cartilage, and whilst can extend into the joint cavity in the early stages is limited to the epiphyseal bone. This is seen most frequently in foals at a slightly older age group than the S-type disease, with the condition being often identified in foals up to a few weeks old. There are distinct predilection sites for E-type osteomyelitis (Firth and Goedegebuure 1988). P-type osteomyelitis is associated with the physeal growth plate, and can extend into the adjacent epiphysis or metaphysis. Again this is seen in a slightly older age group of foals again, with the condition being identified in animals often a few weeks to a few months of age. In addition, specific syndromes of osteomyelitis in the neonate affecting the small tarsal bones (T-type) and carpal bones (C-type) have been described (Firth et al. 1985). Other rare presentations of osteomyelitis in the foal include disease affecting the vertebrae or ribs (Boswinkel et al. 2006; Neil et al. 2010).

Haematogenous osteomyelitis is rare in the adult horse, although a syndrome of proximal sesamoid bone osteomyelitis subsequent to catheterisation of the dorsal metatarsal artery for purposes of blood pressure monitoring under general anaesthesia has been reported (Barr et al. 2005).

Treatment of osteomyelitis in the horse, no matter the age, will involve long-term antimicrobial therapy for at least 4 weeks, with the selection of the antibiotic being chosen on the basis of culture and sensitivity testing. Commonly, treatment will involve surgical lavage and debridement of the affected area. As in the neonate there is a close relationship between osteomyelitis and septic arthritis, therapy of the infected synovial structures is vital, usually involving arthroscopic debridement and lavage of the affected joint. Data have indicated that thoroughbred foals affected with neonatal sepsis are less likely to start a race than unaffected control animals and were also older when they first started racing compared to controls (Smith et al. 2004).

5.2 *Osteomyelitis Secondary to Trauma*

Horses are at a high risk of sustaining orthopaedic trauma resulting in osteomyelitis. Horses frequently sustain traumatic injuries as a result of wounds from external objects, for instance collision with fencing material. In particular horses frequently sustain kick injuries, from other horses, which have the potential to lead to osseous trauma, bacterial contamination of bone and resulting in osteitis or osteomyelitis

(Harrison et al. 1991) (Figs. 1–3). The commonest finding is that of a bone sequestration subsequent to either cortical fracture, or periosteal disruption and cortical necrosis. Clinical signs usually relate most obviously to failure of a wound to heal, and continuing purulent drainage from the site of an earlier wound. Initially radiographic imaging may show either no abnormality or may identify cortical bone fragmentation. Over a number of weeks, classic radiographic changes of a sequestrum formation appears (obvious fragments of radiopaque bone), surrounded by an involucrum (Figs. 4–6). Frequently, a tract or sinus will be identified connecting the sequestrum with the outside. Certain bones, in particular the pedal bone, may not show such obvious changes as a sequestrum, and will generally show signs of evidence of bone resorption and loss (Cauvin and Munroe 1998). Rarely do such traumatic injuries progress onto osteomyelitis, although every case has this potential (Fig. 7).

Treatment usually requires initial antimicrobial therapy to treat any secondary soft tissue infection. Some sequestration has the ability to resolve without veterinary input, subsequent to osteoclastic resorption of any sequestrum. More commonly, surgical removal of the sequestered bone is required to resolve the issue, followed by

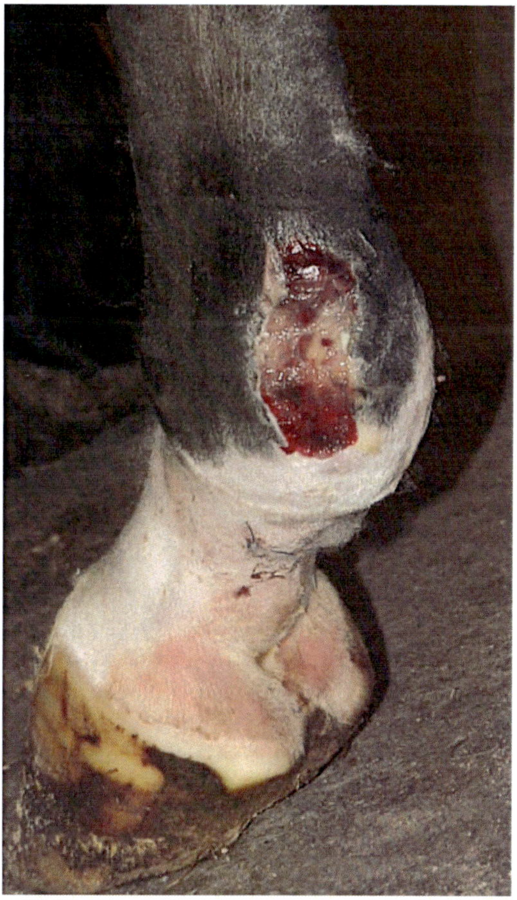

Fig. 1 A on-healing wound on the plantar aspect of a horses fetlock as a result of septic tenosynovitis of the digital flexor tendon sheath and osteomyelitis of the lateral proximal sesamoid bone

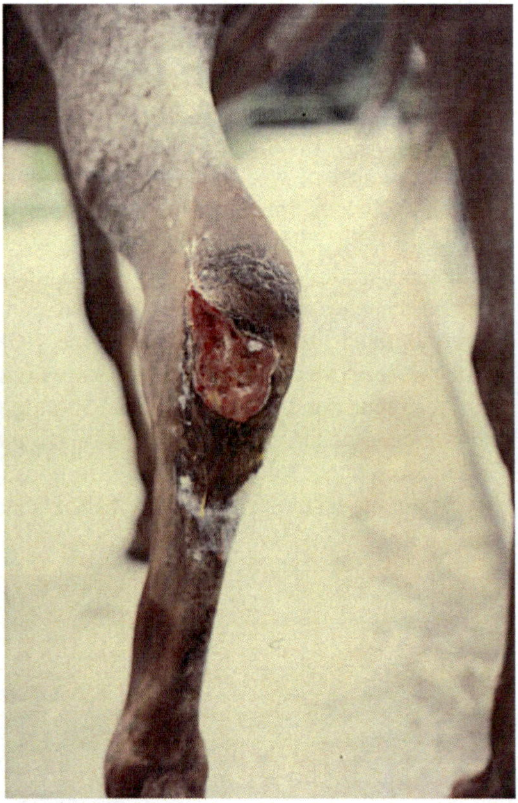

Fig. 2 A large non-healing wound on the plantar aspect of a horses tarsus as a result of severe osteomyelitis of the sustentaculum tali of the calcaneal bone

debridement and lavage of any involucrum. Long-term antimicrobial therapy is usually not required to effect recovery unless there is extension of the infection into the medullary cavity. Prognosis for full recover is often excellent in the case of bone sequestration. However, if osteomyelitis occurs the prognosis will often be guarded.

5.3 *Iatrogenic Causes of Osteomyelitis*

Osteomyelitis in the horse is a major issue subsequent to fracture repair, especially subsequent to internal fixation of fractures. Otherwise, osteomyelitis is a relatively rare occurrence in the horse. Postoperative infection (POI) has been reported to occur in 8% of clean equine orthopaedic surgeries and 52% of clean-contaminated surgeries (MacDonald et al. 1994). Fractures involving long bones had a fivefold chance of becoming infected, in comparison to simple articular fractures. Simple arthroscopic surgery has been shown to have a POI rate of approximately 1.5% in the horse (Figs. 8 and 9). Risk factors for POI in the horse have been shown to include increasing age, female gender, contaminated surgical procedures, surgery lasting >90 min and administration of preoperative antimicrobial drugs (MacDonald

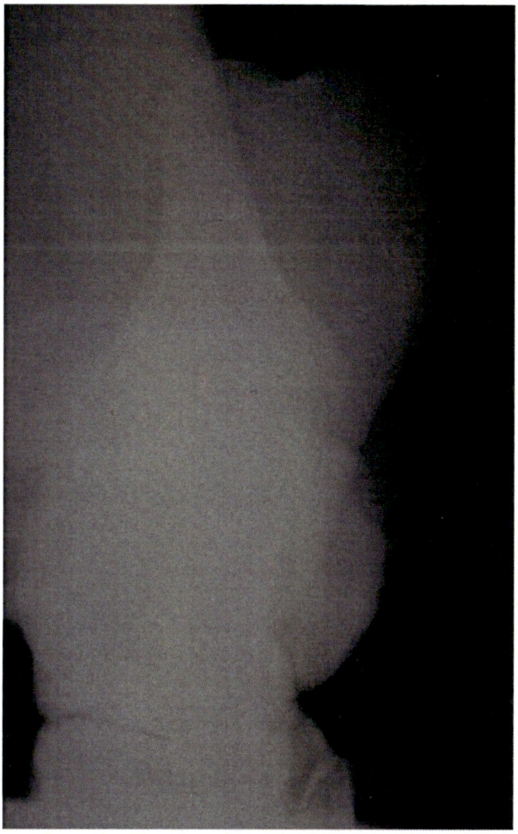

Fig. 3 A Radiograph of the previous horse demonstrating bone lysis and production on the sustentaculum tali as a consequence of osteomyelitis

et al. 1994). A recent report detailing risk factors for POI subsequent to surgical repair of long bone fractures or arthrodesis identified that POI occurred in 28% of cases, and of the cases that became infected, 59% survived to be discharged from hospital. Repairs where POI did not occur were 7.25 times more likely to be discharged from hospital. Closed fractures were 4.23 times more likely to remain uninfected and 4.59 times more likely to be discharged from hospital than open fractures. If a fracture was managed by closed reduction followed by internal fixation, there was a 2.5-fold reduction in rate of POI and a 5.9 times greater chance of being discharged from hospital compared to fractures managed by open reduction and internal fixation (Ahern et al. 2010).

Whilst prevention has to be the most important intervention to manage osteomyelitis, by appropriate antibiosis and optimal surgical technique, it is always likely that cases of osteomyelitis will occur if ones practice performs internal fixation of major fractures in the horse. Treatment in established cases includes long-term antibiosis, as well as removal of any metallic implants once osseous stability has been achieved. Frequently, the prognosis is guarded in cases where osteomyelitis occurs in the face of internal fixation of a fracture (Ahern et al. 2010).

In recent years, much emphasis has been placed on using techniques to obtain high levels of antibiosis into the relevant osseous structures. This has led to the

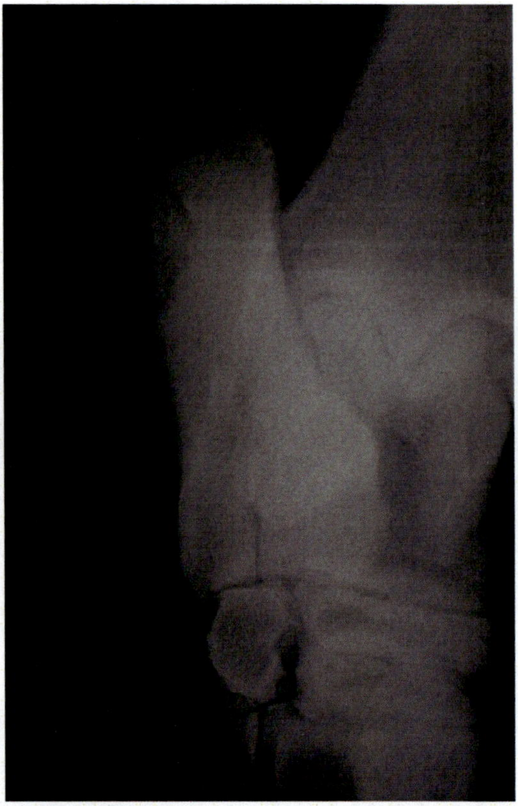

Fig. 4 A fracture of the calcaneus in a horse which is beginning to sequestra due to septic osteitis

use of techniques such as implantation of polymethylmethacrylate antibiotic impregnated beads into the fracture site (Booth et al. 2001; Butson et al. 1996), intravenous regional perfusion (Pille et al. 2005; Rubio-Martinez and Cruz 2006; Scheuch et al. 2002; Werner et al. 2003) and intraosseous regional perfusion (Kettner et al. 2003; Rubio-Martinez et al. 2006; Scheuch et al. 2002). All techniques have the ability to sustain large concentration of antibiotics into a targeted site, and may result in a more rapid resolution of sepsis.

6 Osteomyelitis in Companion Animals (Dogs and Cats)

As in other species, osteomyelitis in dogs and cats is a relatively uncommon condition, which has the potential to be serious and difficult to treat. Similar to the other veterinary species, osteomyelitis can occur through either haematogenous spread of infection in the neonate, due to traumatic injury or through iatrogenic routes, most commonly secondary to surgical intervention. As in other animals, osteomyelitis is most common as a result of bacterial infection, although fungal osteomyelitis is occasionally reported in these animals (Caywood 1983; Caywood et al. 1978; Langley-Hobbs 2006).

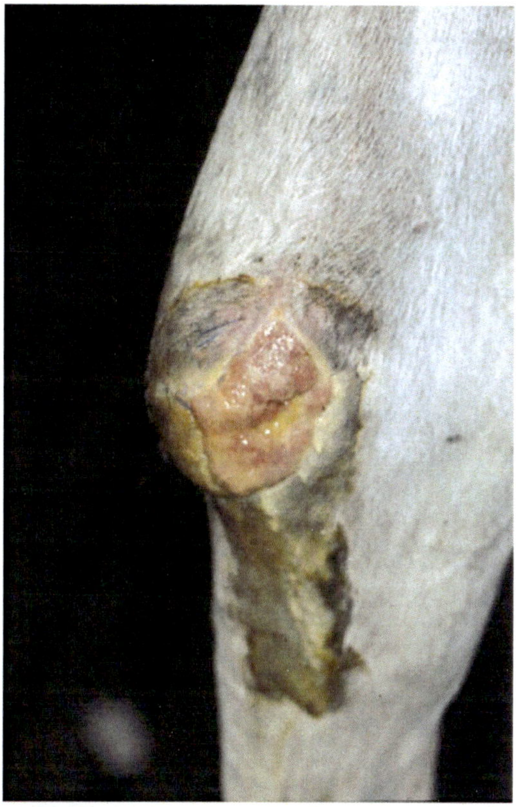

Fig. 5 The appearance of the horse 2 weeks, later. There is an obvious draining tract affecting the calcaneus due to the presence of a sequestrum and septic osteitis

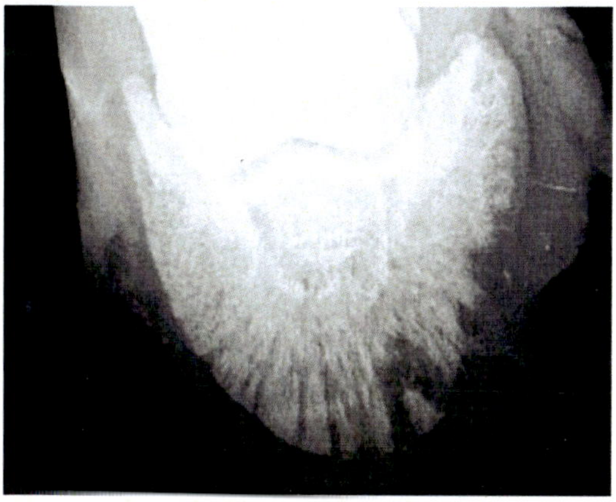

Fig. 6 A radiograph demonstrating a sequestrum in a pedal bone in a horse

6.1 Haematogenous Osteomyelitis

Haematogenous forms of osteomyelitis, as in other species, occur most commonly in the neonate, when circulating bacterial emboli as a consequence of septicaemia lodge in the metaphysis region of bones. Septicaemia can occur through a number

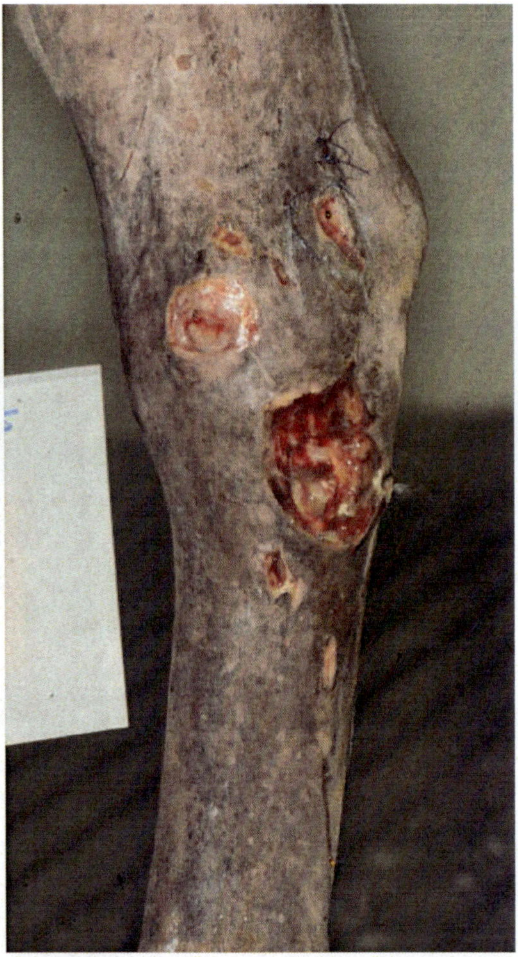

Fig. 7 Trauma is common in the horse. This horse has sustained multiple wounds to the hind limb and has led to damage to the second metatarsal bone and osseous sequestration

of causes, although umbilical infections are a frequent primary source of infection. Metaphyseal infection will result in necrosis, hyperaemia and leukocytosis at the site, and ultimately bone destruction. There is a frequent association with articular sepsis ("joint ill") in such cases either through concomitant synovial sepsis, or from infection breaking out of the bone and contaminating adjacent articular structures. Haematogenous forms of osteomyelitis in the neonate dog and cat are almost certainly encountered less frequently than the condition seen in farm animals or horses.

In most cases, the animal will be pyrexic and show other signs of septicaemia. Usually the animal is lame, and there will be swelling overlying the affected bones, and these sites are painful on palpation. Frequently multiple sites and limbs will be affected, and in cases where articular sepsis is occurring, there will be swelling and effusion associated with the adjacent joints.

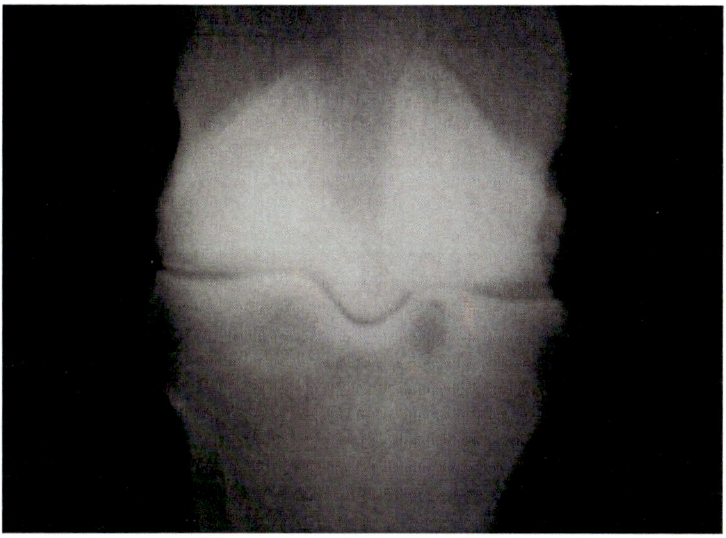

Fig. 8 A Radiograph of a horse with septic arthritis of the fetlock joint and septic osteomyelitis of the pastern

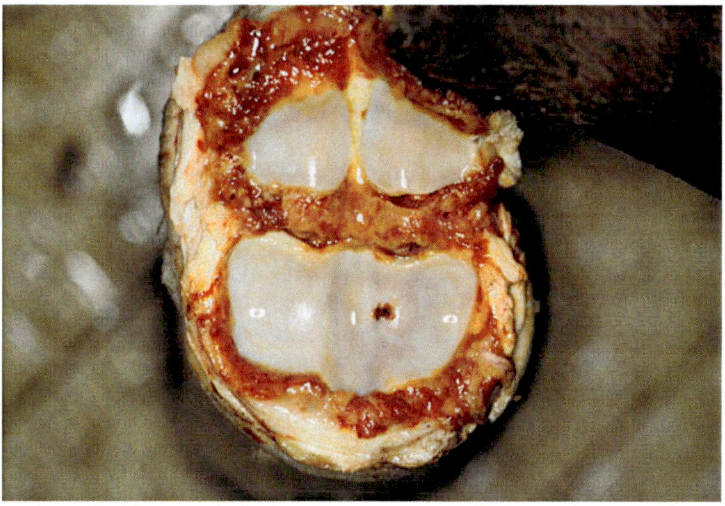

Fig. 9 Postmortem view of the proximal pastern demonstrating an erosion through the articular cartilage into an osteomyelitic area in the subchondral bone

Haematogenous forms of osteomyelitis are relatively rare in the adult dog, although there are sporadic case reports describing such diseases (Cabassu and Moissonnier 2007; Emmerson and Pead 1999; Rabillard et al. 2011).

Diagnosis is usually confirmed by radiology. Frequently early in the course of the disease, no bone destruction may be evident, and soft tissue swelling over the affected site is all that can be determined. Later in the course of the disease, bone destruction is evident and can be identified as irregular lysis within the metaphysis. Eventually, this lysis can extend and lead to destruction of overlying cortical bone. Definitive diagnosis relies on needle aspiration or surgical biopsy of infected material, and relevant culture and sensitivity. Smears of the aspirated material can be examined, with appropriate staining, to identify bacteria in the sample prior to culture results becoming available. As most of these animals will be septicaemic, blood cultures should be taken, prior to antibiotic administration, for appropriate culture and sensitivity testing. Furthermore, the umbilical region should always be carefully assessed to determine whether this is the source of any septic foci. In some cases, ultrasonographic examination of the umbilical remnants can be useful in assessing any sites of potential sepsis.

In dogs and cats, such cases are generally managed through medical routes in the first instance, with high doses of bactericidal antibiotics being administered. Antibiotics should ideally not be administered until all samples have been collected for appropriate bacterial identification, and their administration should be prolonged, with generally administration continuing for at least 2–4 weeks after clinical signs have resolved. Surgical treatment for haematogenous osteomyelitis is rarely required in the neonatal dog or cat and the prognosis is generally favourable if treatment is commenced promptly.

6.2 Osteomyelitis Secondary to Trauma

Trauma as a cause of osteomyelitis in dogs and cats is relatively rare and usually occurs as a consequence of a penetrating wound (frequently as a consequence of bites). The most frequent cause of traumatic osteomyelitis is as a consequence of either open fractures, or as a result of surgical intervention (see below).

Affected animals usually demonstrate lameness, and pain and swelling associated with the traumatic wound. There is frequently continuing purulent drainage from the site and failure of any wound to heal. Initially radiology can be unrewarding, but radiographs should be repeated if suspicious of osteomyelitis every 7–10 days, in order to identify any bone lysis that may be developing. Treatment is again prolonged antibiosis, with antibiotic choice being determined following appropriate culture and sensitivity testing, with treatment continuing for 2–4 weeks after cessation of any clinical signs. Surgical treatment through debridement and lavage is often indicated in order to remove infected and necrotic material. If surgical treatment is undertaken, a decision will have to be made whether the wound is left open to heal by secondary intention, or whether primary skin closure can be achieved. If primary closure is performed, it may be necessary to place some form of drain to allow removal of any inflammatory exudate.

6.3 Iatrogenic Causes of Osteomyelitis

Iatrogenic osteomyelitis is most frequently seen after surgical treatment of fractures, and the risk is obviously the greatest in open fractures that are already contaminated. However there is a risk of osteomyelitis occurring subsequent to any surgery being performed which involves bone, for instance following joint arthroplasty surgery. Osteomyelitis can become evident months or years after surgical intervention, particularly in cases where orthopaedic implants have been placed. In cases of chronic osteomyelitis lameness may or may not be present, and the animal may present simply with swelling and pain associated with palpation of the limb. Frequently there are draining tracts, which drain a purulent exudate intermittently. Commonly such draining tracts heal and appear to respond well to short-term antibiotic therapy, only to recur weeks to months following cessation of antibiosis. It is normal for most affected animals to be well and show no systemic illness as a consequence of such osteomyelitis.

Diagnosis of acute iatrogenic osteomyelitis is frequently made on clinical signs, and whilst radiology can be useful, in acute cases the classic signs of osteomyelitis of bone lysis may yet not be apparent and may require repeat radiographs at 10- to 14-day intervals, until there has been sufficient bone loss to be identified. Obviously, the best approach to iatrogenic osteomyelitis is by prevention through attention to good surgical principles and techniques. If there is an open fracture, surgical debridement and lavage are essential, and attention to removal of all necrotic and contaminating material is perquisite. The removal of exudate via closed-suction drainage is often required. Fracture stability has a major influence on controlling osteomyelitis, and rigid external or internal fixation of the fracture is extremely beneficial in the control of osteomyelitis. In acute cases, appropriate antibiosis should be provided following culture and sensitivity testing and once again, treatment should be prolonged (for 4–6 weeks).

In chronic cases, radiological examination is usually diagnostic, and it is important to identify what is the focus of infection. Frequently areas of bone lysis will be identified around orthopaedic implants such as plates, screws or pins. Bone sequestration is not an uncommon feature in chronic osteomyelitis. Treatment is usually though surgical removal of the implants and debridement of any affected bone. If a sequestrum is present, it is important that the sequestered piece of dead bone is removed and the remaining involucrum in the bone is curetted. It is important to maintain bone stability, if the bone has yet to achieve full healing, and different methods of rigid internal fixation may have to be considered, for instance, placement of an external fixator. Achieving rigid stabilization of the bone is key to resolving osteomyelitis subsequent to fracture. If there is loss of bone material, the placement of an autologous cancellous bone graft can assist considerably in bone formation. Appropriate long-term antibiosis dependent on culture and sensitivity is always necessary. Again, where wounds can be closed, it is usually prudent to place some form of closed-suction drainage.

Osteomyelitis consequent to surgical intervention and fracture repair is a major complication, and its treatment can frequently be expensive and prolonged. The prognosis is not always favourable. Whilst with appropriate interventions, many cases can be resolved, some cases can be extremely problematic to manage successfully. In certain cases, management by either amputation of the affected limb, or euthanasia of the animal may be the most appropriate course of action.

6.4 Fungal Osteomyelitis in Dogs and Cats

Fungal bone infections are extremely rare in Northern temperate regions such as the United Kingdom, but are not infrequently seen in tropical and hot regions of the world. Such diseases are more frequently seen in either young or immunosuppressed animals. Frequently the animals will show signs of systemic illness, with signs of depression, anorexia and weight loss. The animal will often show signs of other body systems, including respiratory, gastrointestinal and ocular manifestations of fungal infection. There are a number of fungal species involved in such mycotic diseases, and they can manifest with a variety of musculoskeletal signs. Frequently such animals will be lame, with swelling and nodule formation associated with the bones. They can have a variety of radiological manifestations that affect the bones. It is important in such cases to define other systemic manifestations of the disease through appropriate imaging and clinical pathological diagnostic techniques (Langley-Hobbs 2006; Oxenford and Middleton 1986).

7 Osteomyelitis in Production Animals

Osteomyelitis is rare in farm animals and is sporadically seen either secondary to trauma, or occasionally seen with haematogenous spread in neonates from umbilical infections, or from trauma such as tail biting in pigs and castration wounds in any species. In cows infection with actinomyces and (historically) brucellosis can be a cause, whilst in pigs brucellosis, atrophic and necrotic rhinitis have been identified as causes. In cows, large sequestra are a possibility, especially where long bones have been affected by trauma. In particular, adult cows appear to produce an extremely florid bone reaction, in an attempt to wall off sequestra. Cows are frequently affected by sepsis of the pedal bones, subsequent to extension of infection from the claw.

Treatment is as in other species, with prolonged antibiosis, in combination with appropriate surgical drainage and debridement. The economics of food animal production mean that treatment may not be attempted, with euthanasia being the most appropriate management method in many cases.

References

Ahern BJ, Richardson DW, Boston RC, Schaer TP (2010) Orthopedic infections in equine long bone fractures and arthrodeses treated by internal fixation: 192 cases (1990–2006). Vet Surg 39:588–593

Andriole VT, Nagel DA, Southwick WO (1973) A paradigm for human chronic osteomyelitis. J Bone Joint Surg Am 55:1511–1515

Barr ED, Clegg PD, Mark Senior J, Singer ER (2005) Destructive lesions of the proximal sesamoid bones as a complication of dorsal metatarsal artery catheterization in three horses. Vet Surg 34:159–166

Booth TM, Butson RJ, Clegg PD, Schramme MC, Smith RK (2001) Treatment of sepsis in the small tarsal joints of 11 horses with gentamicin-impregnated polymethylmethacrylate beads. Vet Rec 148:376–380

Boswinkel M, van der Lugt JJ, Sloet van Oldruitenborgh-Oosterbaan MM (2006) Vertebral osteomyelitis caused by *Rhodococcus equi* in a three-and-half-month-old Dutch Warmblood foal. Tijdschr Diergeneeskd 131:612–616

Brady RA, Leid JG, Calhoun JH, Costerton JW, Shirtliff ME (2008) Osteomyelitis and the role of biofilms in chronic infection. FEMS Immunol Med Microbiol 52:13–22

Butson RJ, Schramme MC, Garlick MH, Davies JV (1996) Treatment of intrasynovial infection with gentamicin-impregnated polymethylmethacrylate beads. Vet Rec 138:460–464

Cabassu J, Moissonnier P (2007) Surgical treatment of a vertebral fracture associated with a haematogenous osteomyelitis in a dog. Vet Comp Orthop Traumatol 20:227–230

Catto ME (1980) Locomotor system. In: Anderson JR (ed) Muir's textbook of pathology. Edward Arnold, Ltd, London, UK, pp 874–941

Cauvin ER, Munroe GA (1998) Septic osteitis of the distal phalanx: findings and surgical treatment in 18 cases. Equine Vet J 30:512–519

Caywood DD (1983) Osteomyelitis. Vet Clin North Am Small Anim Pract 13:43–53

Caywood DD, Wallace LJ, Braden TD (1978) Osteomyelitis in the dog: a review of 67 cases. J Am Vet Med Assoc 172:943–946

Ciampolini J, Harding KG (2000) Pathophysiology of chronic bacterial osteomyelitis. Why do antibiotics fail so often? Postgrad Med J76:479–483

Costerton JW (2005) Biofilm theory can guide the treatment of device-related orthopaedic infections. Clin Orthop Relat Res 437:7–11

Davies DG, Parsek MR, Pearson JP, Iglewski BH, Costerton JW, Greenberg EP (1998) The involvement of cell-to-cell signals in the development of a bacterial biofilm. Science 280:295–298

Emmerson TD, Pead MJ (1999) Pathological fracture of the femur secondary to haematogenous osteomyelitis in a weimaraner. J Small Anim Pract 40:233–235

Firth EC (1983) Current concepts of infectious polyarthritis in foals. Equine Vet J 15:5–9

Firth EC, Alley MR, Hodge H (1993) *Rhodococcus equi*-associated osteomyelitis in foals. Aust Vet J 70:304–307

Firth EC, Goedegebuure SA (1988) The site of focal osteomyelitis lesions in foals. Vet Q 10:99–108

Firth EC, Goedegebuure SA, Dik KJ, Poulos PW (1985) Tarsal osteomyelitis in foals. Vet Rec 116:261–266

Goodrich LR (2006) Osteomyelitis in horses. Vet Clin North Am Equine Pract 22:389–417, viii-ix

Gristina AG, Costerton JW (1984) Bacterial adherence and the glycocalyx and their role in musculoskeletal infection. Orthop Clin North Am 15:517–535

Gristina AG, Oga M, Webb LX, Hobgood CD (1985) Adherent bacterial colonization in the pathogenesis of osteomyelitis. Science 228:990–993

Harrison LJ, May SA, Edwards GB (1991) Surgical treatment of open splint bone fractures in 26 horses. Vet Rec 128:606–610

Hodgin EC, Michaelson F, Howerth EW, Austin F, Davis F, Haase AS (1992) Anaerobic bacterial infections causing osteomyelitis/arthritis in a dog. J Am Vet Med Assoc 201:886–888

Johnson KA, Lomas GR, Wood AK (1984) Osteomyelitis in dogs and cats caused by anaerobic bacteria. Aust Vet J 61:57–61

Joyner AL, Smith DT (1936) Acute staphylococcus osteomyelitis. Surg Gynecol Obstet 63:1–6

Kettner NU, Parker JE, Watrous BJ (2003) Intraosseous regional perfusion for treatment of septic physitis in a two-week-old foal. J Am Vet Med Assoc 222(346–350):316

Langley-Hobbs S (2006) Diseases and disorders of bone. In: Houlton J, Cook J, Innes J, Langley-Hobbs S (eds) BSAVA manual of canine and feline musculoskeletal disorders. BSAVA, Quedgeley, pp 34–49

MacDonald DG, Morley PS, Bailey JV, Barber SM, Fretz PB (1994) An examination of the occurrence of surgical wound infection following equine orthopaedic surgery (1981–1990). Equine Vet J J26:323–326

Mah TF, O'Toole GA (2001) Mechanisms of biofilm resistance to antimicrobial agents. Trends Microbiol 9:34–39

Mah TF, Pitts B, Pellock B, Walker GC, Stewart PS, O'Toole GA (2003) A genetic basis for *Pseudomonas aeruginosa* biofilm antibiotic resistance. Nature 426:306–310

Moore RM, Schneider RK, Kowalski J, Bramlage LR, Mecklenburg LM, Kohn CW (1992) Antimicrobial susceptibility of bacterial isolates from 233 horses with musculoskeletal infection during 1979–1989. Equine Vet J J24:450–456

Muir P, Johnson KA (1992) Anaerobic bacteria isolated from osteomyelitis in dogs and cats. Vet Surg 21:463–466

Neil KM, Charman RE, Vasey JR (2010) Rib osteomyelitis in three foals. Aust Vet J 88:96–100

Oxenford CJ, Middleton DJ (1986) Osteomyelitis and arthritis associated with *Aspergillus fumigatus* in a dog. Aust Vet J 63:59–60

Pille F, De Baere S, Ceelen L, Dewulf J, Croubels S, Gasthuys F, De Backer P, Martens A (2005) Synovial fluid and plasma concentrations of ceftiofur after regional intravenous perfusion in the horse. Vet Surg 34:610–617

Rabillard M, Souchu L, Niebauer GW, Gauthier O (2011) Haematogenous osteomyelitis: clinical presentation and outcome in three dogs. Vet Comp Orthop Traumatol 24:146–150

Rubio-Martinez LM, Cruz AM (2006) Antimicrobial regional limb perfusion in horses. J Am Vet Med Assoc 228(706–712):655

Rubio-Martinez LM, Lopez-Sanroman J, Cruz AM, Tendillo F, Rioja E, San Roman F (2006) Evaluation of safety and pharmacokinetics of vancomycin after intraosseous regional limb perfusion and comparison of results with those obtained after intravenous regional limb perfusion in horses. Am J Vet Res 67:1701–1707

Scheuch BC, Van Hoogmoed LM, Wilson WD, Snyder JR, MacDonald MH, Watson ZE, Steffey EP (2002) Comparison of intraosseous or intravenous infusion for delivery of amikacin sulfate to the tibiotarsal joint of horses. Am J Vet Res 63:374–380

Sedghizadeh PP, Kumar SK, Gorur A, Schaudinn C, Shuler CF, Costerton JW (2009) Microbial biofilms in osteomyelitis of the jaw and osteonecrosis of the jaw secondary to bisphosphonate therapy. J Am Dent Assoc 140:1259–1265

Smith LJ, Marr CM, Payne RJ, Stoneham SJ, Reid SW (2004) What is the likelihood that Thoroughbred foals treated for septic arthritis will race? Equine Vet J 36:452–456

Southwood RT, Rice JL, McDonald PJ, Hakendorf PH, Rozenbilds MA (1985) Infection in experimental hip arthroplasties. J Bone Joint Surg Br 67:229–231

Walker RD, Richardson DC, Bryant MJ, Draper CS (1983) Anaerobic bacteria associated with osteomyelitis in domestic animals. J Am Vet Med Assoc 182:814–816

Werner LA, Hardy J, Bertone AL (2003) Bone gentamicin concentration after intra-articular injection or regional intravenous perfusion in the horse. Vet Surg 32:559–565

Biofilms and Implication in Medical Devices in Humans and Animals

Samuel J. Hooper, Steven L. Percival, Christine A. Cochrane, and David W. Williams

Abstract The increasing use of medical devices in modern medicine, from surgical sutures to indwelling tubes for feeding or catheterization, has been accompanied by a rise in device-related infections. Many such devices provide an ideal surface to which microorganisms can attach and form biofilms. Biofilm infections are extremely difficult to eradicate and, therefore, are a serious concern, often associated with increased mortality and morbidity.

This chapter aims to describe some of the processes behind the development of biofilms on medical devices, and provide examples of biofilm-related infections from human and veterinary medicine.

1 Introduction

Microbial biofilms occur when microorganisms adhere to a surface and produce an array of extracellular polymers that both provides a structural matrix and facilitates further adherence of microorganisms (Costerton et al. 2005). Biofilm communities are generally polymicrobial in composition, with complex interactions occurring between the species that are present (Jakubovics 2010).

The surface for microbial attachment and biofilm development may be that of a living tissue, or an inert, non-living material such as environmental substrata,

S.J. Hooper (✉) • D.W. Williams
Oral Microbiology Group, Tissue Engineering and Reparative Dentistry, School of Dentistry, Cardiff University, Heath Park, Cardiff CF14 4XY, UK
e-mail: hoopersj@cardiff.ac.uk

S.L. Percival
Department of Pathology, Medical School, West Virginia University, Morgantown, WV, USA

C.A. Cochrane
Faculty of Health and Life Sciences, Institute of Ageing and Chronic Disease, University of Liverpool, Leahurst, Neston. CH64 7TE, United Kingdom

S.L. Percival et al. (eds.), *Biofilms and Veterinary Medicine*,
Springer Series on Biofilms 6, DOI 10.1007/978-3-642-21289-5_8,

industrial fabrications, or the biomaterials of medical devices. In the case of human infection, medical devices that are frequently associated with biofilm development include central venous catheters (Machado et al. 2009), heart valves (Venditti 2009), orthopaedic implants (Esteban et al. 2010), urinary catheters (Stickler 2008), intraocular lenses (Behlau and Gilmore 2008) and dental implants (Busscher et al. 2010).

Biofilms are considered the principle growth form of microorganisms, and are found in all environments including those associated with industry and in human and veterinary medicine. In clinical situations, the occurrence of biofilms is now recognised as a significant cause of human infection and importantly, certain properties of biofilms mean that the successful treatment of these infections is often difficult. Key amongst these biofilm properties is their resistance to removal by host defence mechanisms (Fedtke et al. 2004), as well as increased tolerance against administered antimicrobials (Anderson and O'Toole 2008). The latter is clearly evident when comparing free-living or planktonic growth susceptibility profiles with those of biofilms (Hill et al. 2005). Furthermore, the ability of microorganisms to form biofilms on host surfaces is also associated with enhanced virulence (Coenye et al. 2007).

As mentioned above, biofilms on medical devices are difficult to eradicate and therefore associated infections often lead to increased mortality and morbidity for infected individuals. Whilst research on biofilms formed by animal pathogens is relatively limited, it is likely that biofilms are involved in many animal diseases, including urinary tract infection, pneumonia, endocarditis and mastitis (Bonifait et al. 2008; Clutterbuck et al. 2007). Based on human studies, biofilm involvement in infection is currently estimated at 65%, with many of these being associated with the presence of an implanted medical device (Hetrick and Schoenfisch 2006).

Biofilm-related infections may be caused by a single microbial species or a mixture of species, depending on the medical device and its duration of use. Indwelling medical devices in humans are particularly susceptible to colonisation by Gram-positive bacteria, including *Staphylococcus aureus*, Gram-negative bacteria and yeasts, such as *Candida*. The source of these biofilm and disease-causing microorganisms may be from the skin of the host, cross-contamination from healthcare workers, tap water to which entry ports are exposed, or other sources within the local environment (Safdar et al. 2002).

In addition to being associated with host infection, the occurrence of a biofilm on a medical device may also be responsible for the failure of function of the device itself (Davis et al. 2002). This in turn can cause deterioration in both the health and quality of life of the individual, together with an increased cost of treatment for the healthcare provider.

1.1 Biofilms on Medical Devices

Adherent microbial populations have been demonstrated on the surfaces of a wide range of medical devices, examples of which are listed in Table 1. Biofilm-related

Table 1 Medical devices associated with biofilm-related infections

Medical devices
Urinary catheters
Intravenous catheters
Endotracheal tubes
Feeding tubes
Pacemakers/automated intracardiac devices
Prosthetic heart valves
Prosthetic joints
Voice prostheses
Penile implants
Breast implants
Intrauterine devices
Contact lenses
Dental implants
Vascular grafts
Biliary stents
Tympanostomy tubes
Orthopedic devices (fixators, nails, screws)

Schinabeck and Ghannoum (2006)

infections have also been reported to be associated with the use of many of these devices, and specific examples of these are discussed below.

1.2 Urinary Tract Infections in Catheterised Individuals

Urinary catheters are widely used in the treatment of humans and animals to facilitate drainage of the bladder in cases where there is urinary incontinence, typically arising from neurological dysfunction or trauma. In humans the standard Foley catheter is widely used, with an estimated 100 million Foley catheters employed worldwide per year. In the case of animals, short-term catheterisation is used to collect sterile urine, remove urinary obstruction and also empty the bladder. Longer-term, indwelling catheters are also used for the removal of urine and these are more likely to lead to biofilm formation with subsequent urinary tract infection (Barsanti et al. 1985).

A catheterised urinary tract provides an ideal environment for a number of different bacteria to grow as biofilms. These biofilms develop on the inner luminal surface of the catheter, but the origin of the microorganisms may be from extra-luminal contamination, occurring either at the time of catheter insertion or subsequently whilst the catheter is being used. Intra-luminal contamination frequently occurs when drainage bags are replaced, temporarily exposing the inner lumen of the catheter to the environment. Regardless of the route of contamination, microbial access will be provided to a vulnerable body cavity, i.e. the bladder, which due to catheterisation has lost the normally protective function of the flushing effect of regular urinary flow, which removes contaminating microorganisms. Once the bladder has become infected, the catheter design, which may incorporate an inflatable balloon for retention of the catheter in the bladder, ensures that complete emptying of the bladder does not occur, with the result that a residual pool of infected urine remains in the bladder. Eradication of such an infection is therefore difficult.

To highlight the problem of catheter-associated urinary tract infection (UTI) in animals, Bubenik et al. (2007) examined the frequency of UTIs in 147 catheterised and 99 non-catheterised dogs. For 66 of the catheterised dogs, UTI was evident, and the risk of infection was found to increase by 27% for each day of continued catheterisation. The biofilms on indwelling urinary catheters may initially be composed of a single microbial species, but longer periods of catheterisation inevitably lead to multispecies biofilms (Donlan 2001a, b). Catheter-associated urinary tract infection in cats and dogs is primarily caused by the bacteria, *Escherichia coli*, *Proteus mirabilis*, *Pseudomonas*, *Enterococcus* and *Klebsiella*. Of particular concern, however, are those infections involving urease-producing bacteria, most notably *P. mirabilis*, which can lead to the formation of crystalline biofilms. Urease production by *P. mirabilis* causes the breakdown of urea with release of ammonia. This in turn elevates the pH of the urine leading to crystallisation of magnesium and calcium salts resulting in catheter encrustation and blockage, as well as the generation of obstructive kidney and bladder stones. Catheter encrustation increases the risk of complications of infection leading to diseases such as pyelonephritis and septicaemia (Kunin 1987). Furthermore, the bacteria within crystalline biofilms have been found to have elevated resistance to antibiotics, so reliable strategies for the clinical prevention of such infections are urgently needed.

1.3 Peritoneal Catheter Infection

Peritoneal dialysis was first performed in dogs in 1946 and is now a frequently used procedure in veterinary medicine, primarily for the treatment of acute and chronic renal failure. In peritoneal dialysis, a catheter is left in place and this passes through the body wall. Infections can subsequently occur from bacteria migrating within or along the outside of the catheter and into the peritoneal cavity. Biofilm formation on peritoneal catheters has been recognised for several decades following scanning and confocal microscopical analysis (Dasgupta and Costerton 1989; Gorman et al. 1993). Experimental studies in rabbits have revealed that extensive biofilm formation on the external catheter surface occurred within 3 weeks and proceeded from the cutaneous exit site of the catheter through to the peritoneal cavity. In these studies, the principle infecting agents were staphylococci most likely originating from the animals' skin (Read et al. 1989). Similar studies have demonstrated *Staphylococcus epidermidis* biofilm progression along peritoneal catheters in pigs (McDermid et al. 1993).

1.4 Catheter-Related Bloodstream Infections

In humans, catheter-related bloodstream infection (CRBSI) is the third-most common nosocomial infection in the intensive care unit (ICU) and is associated with the use of intravascular catheters. Bacterial biofilms have been found to form on central venous catheters (CVCs) within the first 24 h after insertion. In veterinary medicine, intravenous catheters are widely employed in the administration of medications and fluids. In this regard, the primary function of the catheter is to optimise the administration of fluids whilst reducing the risk of complications (Tan et al. 2003). Studies into CRBSI in animals are limited when compared with humans, although research in dogs has shown the occurrence of alterations in renal blood flow related to *Pseudomonas* biofilm formation on intravascular catheters. The extent of biofilm formation and infection could be reduced through employing careful aseptic techniques on catheter insertion together with the use of an antibiotic lock technique (Bach et al. 1998).

Catheter-tract infections have also been studied in experimental animals where they are highly detrimental to the health of the animal and reduce the functional lifespan of surgical implants. In one study of 53 *Macaca mulatta* monkeys that were catheterised, infections were encountered in 30.2% and the mean catheter lifespan in such cases decreased from 354 days for uninfected catheters to 147 days. The main infecting species were found to be coagulase-negative staphylococci (~42%), together with *S. aureus* (~22%). Nearly half of the bacterial isolates recovered were meticillin-resistant Gram-positive bacteria (Taylor and Grady 1998).

1.5 Ventilator-Associated Pneumonia

Ventilator-associated pneumonia (VAP) is one of the most frequently occurring nosocomial infections in humans on intensive care units and is second only to urinary tract infection amongst hospital-acquired infections of humans (Pneumatikos et al. 2009; Ramirez et al. 2007). The actual true incidence of VAP in humans is unclear as no gold-standard diagnostic test exists, but is believed to occur in 8–28% of susceptible patients, with mortality rates ranging between 24 and 76%. Key to the infection is that the process of intubation and mechanical ventilation, which bypasses the normal host defence mechanisms, provides an unprotected conduit between the oral cavity and lungs. The presence of an endotracheal tube is strongly associated with the subsequent occurrence of VAP and the development of a biofilm within the lumen of the endotracheal tube is deemed an important factor in the development of infection.

Compared with humans, veterinary emergency and critical care is still a relatively new and developing discipline. Long-term (>24 h) mechanical ventilation in companion animals is relatively uncommon, but is practiced. On rare occasions rabbits, guinea pigs, chinchillas and many other small exotic mammals may also be intubated, although again, not routinely (Johnson 2010). The practice of mechanical ventilation is more frequently used for large animals and the process is similar to that used for humans. Animal models of VAP have been developed and it has been shown that piglets that were mechanically ventilated for 4 days consistently developed pneumonia (Marquette et al. 1995). Such spontaneous occurrence of pneumonia of mechanically intubated pigs is believed to be most commonly caused by the bacterial species *Pasteurella multocida* and *Streptococcus suis* (Marquette et al. 1996), although *Klebsiella oxitoca*, *P. aeruginosa*, *S. epidermidis* and *S. aureus* may also be responsible (Wermert et al. 1998).

Interestingly, it has been found that in mechanically ventilated humans, dental plaque biofilm is also modified by the additional presence of potential respiratory pathogens such as *S. aureus* and Gram-negative bacteria including *Pseudomonas aeruginosa* and *Enterobacteriaceae*. There is mounting evidence that such oropharyngeal colonization is a prerequisite for the development of VAP and might be the original source of organisms that seed the endotracheal biofilm.

1.6 Gastronomy Tubes

Feeding tubes are frequently used to provide sustenance for both human and animal patients who cannot consume calories by swallowing. A gastronomy tube provides nutrition directly to the stomach either by the nasogastric route or via an incision in the abdomen (Wortinger 2006). These tubes are usually made of polyurethane or silicone, and the formation of microbial biofilms within these tubes is often considered an inevitable consequence of microbial overgrowth in the gut. A range of Gram-positive and Gram-negative bacteria are capable of attaching to and colonising the lumen of feeding tubes. Fungal contamination of gastronomy devices has also been reported and associated with deterioration of the tubes (Dautle et al. 2002, 2003). Many of these microorganisms are considered pathogenic and may pose a threat to the treated individual and particularly to those who are immunocompromised, so infected feeding devices must be removed and replaced. Both gastronomy tubes and nasogastric tubes are frequently used to feed cats suffering from major organ failure, injury or post-surgery and, generally, complications are rare. However, infection following fungal and bacterial colonisation of the tube has been reported.

1.7 Pacemakers

Cardiovascular implants maintain defective tissues and rectify congenital defects. In human and veterinary medicine, the prevalence of non-valvular cardiovascular device-related infection varies depending on the implanted device. Infection of left ventricular assist devices ranges from 25 to 70%, whereas infection of devices used to close congenital defects are considered rare (Bluhm 1985; Cohen et al. 2002; Lemire et al. 1975).

Retrospective studies of pacemaker-related infections in humans show a prevalence of 0–19% (Baddour et al. 2003). In veterinary medicine, the use of cardiovascular devices is limited predominantly to pacemakers and patent ductus arteriosus (PDA) occlusion devices in dogs. The incidence of infection of PDA occlusion devices is unknown, whereas pacemaker infection rates range from 1 to 16% (Wess et al. 2006; Domenech et al. 2005; Oyama et al. 2001; Sisson et al. 1991; Darke et al. 1989).

Clinical signs of cardiovascular device-related infections in dogs are variable, but similar to those seen in humans and include fever, lethargy, immune complex-mediated disease, embolic events, sepsis, and cellulitis or abscess at the site of the implanted device (Karchmer and Longworth 2003).

Three main factors related to the pathogenesis of device-related infections are: microbial virulence factors, host response to the device and the device's physical characteristics (Baddour et al. 2003). For example, microorganisms such as *S. aureus* are able to produce adhesion molecules that facilitate binding to the host. Additionally, contact with a device may strip the host's endothelium and expose proteins to which microorganisms can adhere. Biofilm formation on devices is a key to the success of microbial colonisation. This biofilm limits the effectiveness of the host immune response and the ability of antimicrobials to reach therapeutic concentrations.

In humans, a number of hypotheses have been proposed to explain the pathogenesis of device-related infections. The most common hypothesis is that contaminants are inoculated at the time of implantation. The high percentage of pacemakers infected with cutaneous organisms such as *Staphylococcus* and *Streptococcus* may be explained by this hypothesis (del Rio et al. 2003; Laguno et al. 1998; Arber et al. 1994). The other commonly postulated route of infection is haematogenous spread secondary to bacteraemia not associated with the device.

In a study conducted by Fine and co-workers (2007) using dogs, four pacemaker recipients showed signs of infection, 90 days after implantation. Three of these dogs had *Staphylococcus* infections, and the other had a co-infection with *P. aeruginosa* and *Corynebacterium*. The *Pseudomonas*/*Corynebacterium* infection occurred in a patient when the sterile technique was compromised during pacemaker implantation. Organisms such as *Pseudomonas*, *Corynebacterium* and *Serratia* are primarily opportunistic pathogens and rarely cause infection in the absence of a break in the host's defences. Therefore, the infection of these four pacemakers is most likely to have occurred at the time of surgery.

1.8 Orthopaedic Devices

Musculoskeletal infection is one of the most common complications associated with surgical fixation of bones fractured during trauma. These infections often involve bacterial colonisation and biofilm formation on the fracture fixation device, as well as infection of the surrounding tissues. Generally, infection presents within the first two postoperative months or many months to years post-surgery when a delayed or late developing infection is observed (Gustilo et al. 1987).

As a consequence of infection, fracture healing can be delayed or prevented and implant loosening can be observed. In order to achieve a successful outcome and to allow fracture healing, surgical removal of the device is often required in addition to prolonged courses of antibiotic therapy.

Orthopaedic procedures in humans, including all fracture types, fixation techniques and prostheses, are associated with an average infection rate of 5% in the United States which equates to 100,000 infections per year costing 15,000 US$ per incidence (Darouiche 2004).

Recently, Kuijer et al. (2007) investigated infection incidence of implanted, degradable tissue-engineered (TE) scaffold biomaterials in rabbit knee osteochondral defects. Sterile, polyester copolymer scaffolds of different compositions and cell-accessible pore volumes were surgically inserted into rabbit osteochondral defects for periods of 3 weeks up to 9 months. Infection assessment included observation of pus or abscesses in or near the knee joint and post-mortem histological examination. Of 228 implanted TE scaffolds, ten appeared to be infected: six scaffolds without cell seeding (3.6%) and four cell-seeded scaffolds (6.3%). These infections were evident across all scaffold types, independent of polymer composition or available pore volume, and up to 9 months.

1.9 Suture Infections

The success of any surgical procedure must include accurate closure and stabilization of the wound margins by sutures. However, the presence of foreign materials in a wound significantly enhances the susceptibility of tissue infection in the host (Blomstedt et al. 1977; Österberg and Blomstedt 1979). The ultimate consequence of suturing can be postoperative infection resulting in compromised wound healing.

Inflammatory responses arise from foreign body reactions to sutures in the form of exudates which may decrease resistance to infection and ultimately impair wound healing (Trimbos et al. 1989). It is possible that sutures may also serve as a pathway for bacteria to enter a surgical wound. This is a physical process that is enhanced by the capillary action of the suture material (Chu and Williams 1984).

The physical construction of some suture materials may protect contaminating bacteria and enable microorganisms to multiply (Everett 1970). Relatively inert synthetic suture materials are associated with less inflammation than sutures manufactured from natural materials (Selvig et al. 1998).

1.10 Ophthalmic Infections

Evidence is increasing that biofilms can play a role in ocular infections. Prosthetic materials that come in contact with the eye, including contact lenses, intraoral lenses (IOLs), scleral buckles and suture material, have all been associated with an increased risk of infection and are all demonstrably capable of harbouring bacterial biofilms. For example, *S. epidermidis* has been found to colonize IOLs and as such has been linked to postoperative endophthalmitis following cataract surgery and IOL insertion (Kodjikian et al. 2003, 2004). Intraocular silicone prostheses have been used in the treatment of ophthalmic disease or injury that precludes the salvage of the eye in animals, including horses (Provost et al. 1989) and dogs (Lin et al. 2007).

1.11 Factors Influencing the Formation of Biofilms on Medical Devices

The first step in medical device-related infection is the adherence of microorganisms to the surface of the device. Microorganisms may arrive near the surface of an indwelling device by direct contamination, or by either contiguous or haematogenous spread. Initial attachment relies upon non-specific, reversible interactions. Whether the microbes fully adhere and start to form biofilms on the device surface is dependent on several factors, namely the physical characteristics of the device and microorganism, the host, and the interactions between microbe and biomaterial. The following sections describe the roles these factors play in the establishment and maintenance of biofilms on medical devices.

1.12 Device-Related Factors

Studies of microbial adherence on different biomaterials have revealed several device-related factors that influence biofilm formation. Certain materials used in the design of medical devices are more favourable to microbial adherence than others. Generally, hydrophilic materials are considered less amenable to microbial attachment compared with hydrophobic surfaces, and negatively charged surfaces are less adhesive compared with positively charged.

The surface topography of a biomaterial is also an important consideration with smooth surfaces generally inhibiting biofilm formation compared with rougher ones. However, the extent of nano-roughness has to be at a level conducive to bacterial retention, and if the surface roughness exceeds a certain level then increased adhesion is no longer evident (Amoroso et al. 2006).

1.13 Role of Conditioning Films

The factors related to the material properties of a medical device are further complicated by the presence of a conditioning of its surface that rapidly occurs in vivo. Once a device has been implanted into a body, its biomaterial surface will rapidly become coated with a layer of proteins, platelets, and other components. This coating is referred to as the conditioning film, and depending on the molecules within the conditioning film, the hydrophobicity of the biomaterial surface can be greatly changed.

1.14 Treatment of Medical Device-Related Infections

Prevention of biofilm occurrence is a key component in the management and prevention of medical device infection. In the case of indwelling urinary catheters, minimising and targeted use can be achieved by identifying which animals are likely to experience repeated obstruction if catheterization is not performed (Lees 1996). Catheter removal should be implemented as soon as possible and verification of underlying infection determined by urine culture.

Many methods have been employed in efforts to remove biofilms from urinary catheters. Chemical treatments such as the use of triclosan, citric acid and ethylenediamine tetra-acetic acid (EDTA) have been utilised with varying degrees of success.

Novel tools in the armoury against biofilms could involve the targeted enzymic disruption of the biofilm's chemical matrix, which is now known to include extracellular DNA, polysaccharides and proteins (Flemming and Wingender 2010). Interfering with biofilm production through gene therapy or quorum-sensing inhibitors are also attractive therapeutic options. Coating biomaterial surfaces with organic molecules to prevent the development of a surface-conditioning film are also considered options (Arciola 2009).

References

Ahern BJ, Richardson DW, Boston RC, Schaer TP (2010) Orthopedic infections in equine long bone fractures and arthrodeses treated by internal fixation: 192 cases (1990–2006). Vet Surg 39(5):588–593

Amoroso PF, Adams RJ, Waters MG, Williams DW (2006) Titanium surface modification and its effect on the adherence of *Porphyromonas gingivalis*: an *in vitro* study. Clin Oral Implants Res 17:633–637

Anderson GG, O'Toole GA (2008) Innate and induced resistance mechanisms of bacterial biofilms. Curr Top Microbiol Immunol 322:85–105

Arber N, Pras E, Copperman Y et al (1994) Pacemaker endocarditis. Report of 44 cases and review of the literature. Medicine (Baltimore) 73:299–305

Arciola CR (2009) New concepts and new weapons in implant infections. Int J Artif Organs 32(9):533–536

Bach A, Just A, Berthold H, Ehmke H, Kirchheim H, Borneff-Lipp M, Sonntag HG (1998) Catheter-related infections in long-term catheterized dogs. Observations on pathogenesis, diagnostic methods, and antibiotic lock technique. Zentralbl Bakteriol 288(4):541–552

Baddour LM, Bettmann MA, Bolger AF et al (2003) Nonvalvular cardiovascular device-related infections. Circulation 108:2015–2031

Barsanti JA, Blue J, Edmunds J (1985) Urinary tract infection due to indwelling bladder catheters in dogs and cats. J Am Vet Med Assoc 187(4):384–388

Behlau I, Gilmore MS (2008) Microbial biofilms in ophthalmology and infectious disease. Arch Ophthalmol 126(11):1572–1581

Blomstedt B, Osterberg B, Bergstrand A (1977) Suture material and bacterial transport. An experimental study. Acta Chir Scand 143(2):71–73

Bluhm G (1985) Pacemaker infections. A clinical study with special reference to prophylactic use of some isoxazolyl penicillins. Acta Med Scand Suppl 699:1–62

Bonifait L, Grignon L, Grenier D (2008) Fibrinogen induces biofilm formation by *Streptococcus suis* and enhances its antibiotic resistance. Appl Environ Microbiol 74(15):4969–4972

Bubenik LJ, Hosgood GL, Waldron DR, Snow LA (2007) Frequency of urinary tract infection in catheterized dogs and comparison of bacterial culture and susceptibility testing results for catheterized and noncatheterized dogs with urinary tract infections. J Am Vet Med Assoc 231(6):893–899

Busscher HJ, Rinastiti M, Siswomihardjo W, van der Mei HC (2010) Biofilm formation on dental restorative and implant materials. J Dent Res 89(7):657–665

Chu CC, Williams DF (1984) Effects of physical configuration and chemical structure of suture materials on bacterial adhesion. A possible link to wound infection. Am J Surg 147(2):197–204

Clutterbuck AL, Woods EJ, Knottenbelt DC et al (2007) Biofilms and their relevance to veterinary medicine. Vet Microbiol 121:1–17

Coenye T, Peeters E, Nelis HJ (2007) Biofilm formation by *Propionibacterium acnes* is associated with increased resistance to antimicrobial agents and increased production of putative virulence factors. Res Microbiol 158(4):386–392, Epub 2007 Feb 21

Cohen MI, Bush DM, Gaynor JW et al (2002) Pediatric pacemaker infections: twenty years of experience. J Thorac Cardiovasc Surg 124:821–827

Costerton JW, Montanaro L, Arciola CR (2005) Biofilm in implant infections: its production and regulation. Int J Artif Organs 28(11):1062–1068

Darke PG, McAreavey D, Been M (1989) Transvenous cardiac pacing in 19 dogs and one cat. J Small Anim Pract 30:491–499

Darouiche RO (2004) Treatment of infections associated with surgical implants. N Engl J Med 350(14):1422–1429

Dasgupta MK, Costerton JW (1989) Significance of biofilm adherent bacterial microcolonies on Tenckhoff catheters of CAPD patients. Blood Purif 7:144–155

Dautle MP, Ulrich RL, Hughes TA (2002) Typing and subtyping of 83 clinical isolates purified from surgically implanted silicone feeding tubes by random amplified polymorphic DNA amplification. J Clin Microbiol 40:414–421

Dautle MP, Wilkinson TR, Gauderer MW (2003) Isolation and identification of biofilm microorganisms from silicon gastronomy devices. J Pediatr Surg 38(2):216–220

Davis LE, Cook G, Costerton JW (2002) Biofilm on ventriculo-peritoneal shunt tubing as a cause of treatment failure in coccidioidal meningitis. Emerg Infect Dis 8(4):376–379

del Rio A, Anguera I, Miro JM et al (2003) Surgical treatment of pacemaker and defibrillator lead endocarditis: the impact of electrode lead extraction on outcome. Chest 124:1451–1459

Domenech O, Santilli R, Pradelli D et al (2005) The implantation of a permanent transvenous endocardial pacemaker in 42 dogs: a retrospective study. Med Sci Monit 11:BR168–BR175

Donlan RM (2001a) Biofilms and device-associated infections. Emerg Infect Dis 7(2):277–281

Donlan RM (2001b) Biofilm formation: a clinically relevant microbiological process. Clin Infect Dis 33:1387–1392

Esteban J, Molina-Manso D, Spiliopoulou I, Cordero-Ampuero J, Fernández-Roblas R, Foka A, Gómez-Barrena E (2010) Biofilm development by clinical isolates of *Staphylococcus* spp. from retrieved orthopedic prostheses. Acta Orthop 81(6):674–679

Everett WG (1970) Suture materials in general surgery. Prog Surg 8:14–37

Fedtke I, Götz F, Peschel A (2004) Bacterial evasion of innate host defenses – the *Staphylococcus aureus* lesson. Int J Med Microbiol 294(2–3):189–194

Fine DM, Tobias AH (2007) Cardiovascular device infections in dogs: report of 8 cases and review of the literature. J Vet Intern Med 21(6):1265–1271

Flemming HC, Wingender J (2010) The biofilm matrix. Nat Rev Microbiol 8(9):623–633

Gorman SP, Mawhinney WM, Adair CG, Issouckis M (1993) Confocal laser scanning microscopy of peritoneal catheter surfaces. J Med Microbiol 38:411–417

Gustilo RB, Gruninger RP, Davis T (1987) Classification of type III (severe) open fractures relative to treatment and results. Orthopedics 10(12):1781–1788

Hetrick EM, Schoenfisch MH (2006) Reducing implant-related infections: active release strategies. Chem Soc Rev 35(9):780–789

Hill D, Rose B, Pajkos A, Robinson M, Bye P, Bell S, Elkins M, Thompson B, Macleod C, Aaron SD, Harbour C (2005) Antibiotic susceptabilities of *Pseudomonas aeruginosa* isolates derived from patients with cystic fibrosis under aerobic, anaerobic, and biofilm conditions. J Clin Microbiol 43(10):5085–5090

Jakubovics NS (2010) Talk of the town: interspecies communication in oral biofilms. Mol Oral Microbiol 25(1):4–14

Johnson DH (2010) Endoscopic intubation of exotic companion mammals. Vet Clin North Am Exot Anim Pract 13(2):273–289

Karchmer AW, Longworth DL (2003) Infections of intracardiac devices. Cardiol Clin 21:253–271

Kodjikian L, Burillon C, Rogues C et al (2003) Bacterial adherence of Staphylococcus epidermidis to intraocular lenses: a bioluminescence and scanning electron microscopy study. Invest Ophthalmol Vis Sci 44(10):4388–4394

Kodjikian L, Burillon C, Rogues C et al (2004) Intraocular lenses, bacterial adhesion and endophthalmitis prevention: a review. Biomed Mater Eng 14(4):395–409

Kuijer R, Jansen EJ, Emans PJ et al (2007) Assessing infection risk in implanted tissue-engineered devices. Biomaterials 28(34):5148–5154

Kunin CM (1987) Detection, prevention and management of urinary tract infections, 4th edn. Lea & Febiger, Philadelphia, PA, pp 245–288

Laguno M, Miro O, Font C et al (1998) Pacemaker-related endocarditis. Report of 7 cases and review of the literature. Cardiology 90:244–248

Lees GF (1996) Use and misuse of indwelling urethral catheters. Vet Clin North Am Small Anim Pract 26(3):499–505

Lemire GG, Morin JE, Dobell AR (1975) Pacemaker infections: a 12-year review. Can J Surg 18:181–184

Lin CT, Hu CK, Liu CH, Yeh LS (2007) Surgical outcome and ocular complications of evisceration and intraocular prosthesis implantation in dogs with end stage glaucoma: a review of 20 cases. J Vet Med Sci 69(8):847–850

Machado JD, Suen VM, Figueiredo JF, Marchini JS (2009) Biofilms, infection, and parenteral nutrition therapy. JPEN J Parenter Enteral Nutr 33(4):397–403

Marquette CH, Mensier E, Copin MC et al (1995) Experimental models of tracheobronchial stenoses: a useful tool for evaluating airway stents. Ann Thorac Surg 60:651–656

Marquette CH, Copin MC, Wallet F et al (1996) Relationship between microbiologic and histologic features in bacterial pneumonia. Am J Respir Crit Care Med 154:1784–1787

McDermid KP, Morck DW, Olson ME, Boyd ND, Khoury AF, Dasgupta MK et al (1993) A porcine model of *Staphylococcus epidermidis* catheter-associated infection. J Infect Dis 168:897–903

Osterberg B, Blomstedt B (1979) Effect of suture materials on bacterial survival in infected wounds. An experimental study. Acta Chir Scand 145(7):141–144

Oyama MA, Sisson DD, Lehmkuhl LB (2001) Practices and outcome of artificial cardiac pacing in 154 dogs. J Vet Intern Med 15:229–239

Pneumatikos IA, Dragoumanis CK, Bouros DE (2009) Ventilator-associated pneumonia or endotracheal tube-associated pneumonia? Anesthesiology 110:673–680

Provost PJ, Ortenburger AI, Caron JP (1989) Silicone ocular prosthesis in horses: 11 cases (1983–1987). J Am Vet Med Assoc 194(12):1764–1766
Ramirez P, Ferrer M, Torres A (2007) Prevention measures for ventilator-associated pneumonia: a new focus on the endotracheal tube. Curr Opin Infect Dis 20:190–197
Read RR, Eberwin P, Dasgupta MK, Grant SK, Lam K, Nickel C et al (1989) Peritonitis in peritoneal dialysis: bacterial colonization by biofilm spread along the catheter surface. Kidney Int 35:614–621
Safdar N, Kluger DM, Maki DG (2002) A review of risk factors for catheter-related bloodstream infection caused by percutaneously inserted, noncuffed central venous catheters: implications for preventive strategies. Medicine (Baltimore) 81(6):466–479
Schinabeck MK, Ghannoum MA (2006) Biofilm-related indwelling medical device infections. In: Pace JL, Rupp ME, Finch RG, editors. Biofilms, Infections, and Antimicrobial Therapy. CRC Press. pp 39–50
Selvig KA, Biagotti GR, Leknes et al (1998) Oral tissue reactions to suture materials. Int J Periodontics Restorative Dent 18(5):474–487
Sisson D, Thomas WP, Woodfield J et al (1991) Permanent transvenous pacemaker implantation in forty dogs. J Vet Intern Med 5:322–331
Stickler DJ (2008) Bacterial biofilms in patients with indwelling urinary catheters. Nat Clin Pract Urol 5(11):598–608
Tan RH, Dart AJ, Dowling BA (2003) Catheters: a review of the selection, utilisation and complications of catheters for peripheral venous access. Aust Vet J 81(3):136–139
Taylor WM, Grady AW (1998) Catheter-tract infections in rhesus macaques (*Macaca mulatta*) with indwelling intravenous catheters. Lab Anim Sci 48:448–454
Trimbos JB, Brohim R, van Rijssel EJ (1989) Factors relating to the volume of surgical knots. Int J Gynaecol Obstet 30(4):355–359
Venditti M (2009) Clinical aspects of invasive candidiasis: endocarditis and other localized infections. Drugs 69(Suppl 1):39–43
Wermert D, Marquette CH, Copin MC et al (1998) Influence of pulmonary bacteriology and histology on the yield of diagnostic procedures in ventilator-acquired pneumonia. Am J Respir Crit Care Med 158:139–147
Wess G, Thomas WP, Berger DM et al (2006) Applications, complications, and outcomes of transvenous pacemaker implantation in 105 dogs (1997–2002). J Vet Intern Med 20:877–884
Wortinger A (2006) Care and use of feeding tubes in dogs and cats. J Am Anim Hosp Assoc 42(5):401–406

Bovine Mastitis and Biofilms

Marielle B. Melchior

Abstract Biofilm formation in bovine mastitis *Staphylococcus aureus* isolates was studied since the beginnings of biofilm research, even before the name "biofilm" was actually invented. Compared to other major bovine mastitis pathogens, such as *E. coli* and *Streptococcus uberis* relatively much research information is available on *S. aureus* biofilm formation, biofilm antimicrobial susceptibility and the role of several biofilm related genes.

Recent research on biofilm formation from *E. coli* and *Str. uberis* shed an interesting light on the whole dynamic process of bacterial invasion, adherence, persistence and evasive strategies of these bacteria, and reveal parallels and differences between these bacteria and *S. aureus*. In this chapter we present the current knowledge on biofilm formation in the bovine udder as an holistic and dynamic process, with more or less strain specific adaptive strategies and mechanisms which all support perseverance and survival of these bacterial species under stressful circumstances.

1 Introduction

From all major pathogens in bovine mastitis biofilm behavior is best studied in *Staphylococcus aureus*. Recently more information became available regarding biofilm formation for *Escherichia coli* and *Streptococcus uberis* mastitis isolates. The reason for the preference of *S. aureus* is presumably related to the importance of this species both for humans and bovine mastitis, two research fields with social and economic importance, and this results in a synergetic exchange of strains and models.

It seems difficult to point out the beginnings of biofilm research in bovine mastitis since early results that still are of importance in the current developments,

M.B. Melchior (✉)
MBM Veterinary Consultancy, 8111 RA, Heeten, The Netherlands
e-mail: mbm@mbmelchior.nl

S.L. Percival et al. (eds.), *Biofilms and Veterinary Medicine*,
Springer Series on Biofilms 6, DOI 10.1007/978-3-642-21289-5_9,

such as the works of Baselga et al. (1993, 1994) do not use the word biofilm in their publications. However, the slime production that is considered an important feature for biofilm formation is thoroughly investigated in these papers.

During the 1990s, another virulence factor of mastitis pathogens was studied more intense than slime production; the adherence and invasion of host cells. Both for *Str. uberis* and for *E. coli,* much of the current knowledge is based on these results (for references, see further in this chapter), and at the same time these virulence factors were also studied for *S. aureus*.

An important reason to study biofilm formation in *S. aureus* mastitis isolates was the persistence of these infections, even though they were perfectly susceptible for the antimicrobials used (Owens et al. 1997; Sears et al. 1990; Sol et al. 1997). Given the availability of antimicrobials with good intracellular efficacy, it seemed not feasible that intracellular growth was the most important reason for persistence (Melchior et al. 2006b). Furthermore, there were several other epidemiological parameters, such as higher persistence for intramammary infections of longer duration, which seemed hard to explain by intracellular growth alone (Sol et al. 1997).

The presentation of biofilm formation of bacteria as a virulence factor for mastitis in general and *S. aureus* in particular (Melchior et al. 2006a, b), often led to questions regarding the importance and role of intracellular growth, since it was generally accepted that this is indeed an important virulence factor. Furthermore, besides studies regarding biofilm formation in *S. aureus* (Amorena et al. 1999), hardly any information was available regarding the biofilm behavior of other mastitis pathogens.

Interestingly, current research developments reveal a novel point of view; intracellular bacteria show biofilm behavior, as is described by Anderson (Anderson et al. 2003, 2010). Bringing these insights together with the current research results seems to bring an new perspective to the complex puzzle of bacterial infections of the bovine udder. The major pathogens *S. aureus*, *Str. uberis,* and *E. coli* each seem to be able to apply their own set of tools for host cell adherence and invasion. Based on the publications by Anderson researchers working in the field of bacterial biofilms now seem to claim that bacteria are also form biofilm-like polysaccharide-containing masses inside the host cell, called intracellular bacterial communities (IBCs). These findings ask for a clear definition of a bacterial biofilm. According to Costerton et al. (1999) biofilms are "*a structured community of bacterial cells enclosed in a self-produced polymeric matrix and adherent to an inert or living surface*", and this leads us to no clear clue about the intracellular biofilms. However, for research purpose we would like to investigate everything bacteria do in their natural lifestyle, and this broad view aims at revealing and understanding all evasive strategies of bacteria, which are applied to protect themselves against the host immune system and antimicrobial therapies.

Based on this idea, we would like to present biofilm formation in the bovine udder in a more holistic concept; that bacteria are equipped with a broad set of tools that they can apply in a dynamic way, whenever appropriate for sustainability or survival. Therefore, we support the idea that biofilm formation is an adaptation mechanism for a multitude of stress-circumstances and this would include

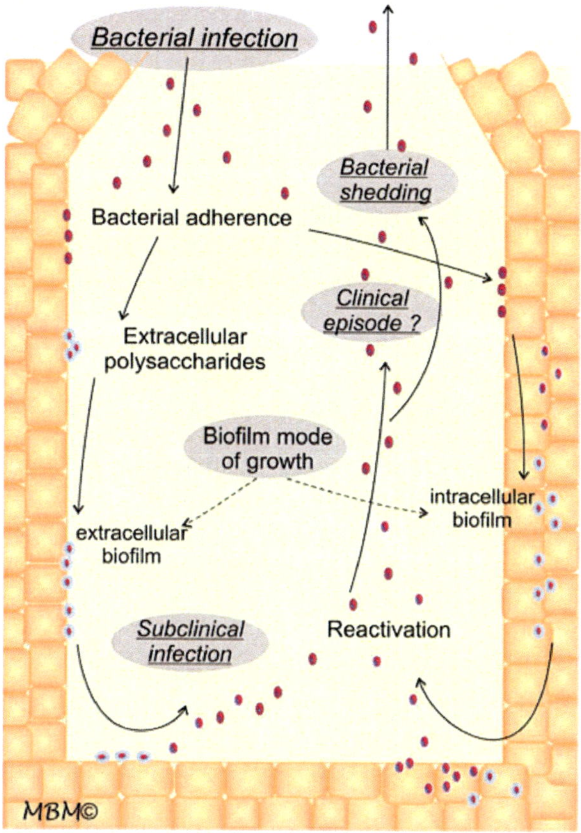

Fig. 1 Schematic overview for proposed model of bacterial infections in the udder. After bacterial infection the bacteria adhere to the epithelial surface. The bacteria protect themselves from the host defense systems by the formation of extracellular polysaccharides, or by invasion of the host tissue. If host defense systems are evaded successfully, this results in extracellular or intracellular biofilm growing bacteria. These subclinical infections can be reactivated resulting in increased shedding. Reactivation can also result in a clinical episode of the infection. The shedding results in further progress of the infection within the host, and between hosts

both extracellular and intracellular circumstances (see Fig. 1). Preferences for either extracellular or intracellular survival might be related to bacterial species and strain preferences. In this chapter, we aim to present an overview of the current knowledge on biofilm formation of the major bovine mastitis pathogens.

2 Staphylococcus aureus

2.1 Adherence, Invasion, and Biofilm Formation

In general, *S. aureus* can be seen primarily as an extracellular pathogen that infects the host by adherence to extracellular matrix components of the host cells (Foster and Hook 1998). Adherence is mediated by protein adhesions of the MSCRAMM (microbial surface components recognizing adhesive matrix molecules) family, which are anchored to the cell wall. We can distinguish three major families of

MSCRAMMs; the fibronectin binding-, the collagen binding-, and the fibrinogen binding MSCRAMMs. The evidence that these MSCRAMMs function in vivo as adhesions and virulence factors comes from studies that compare isogenic mutants with wild type strains in infection models.

The study by Lammers (Lammers et al. 1999) confirms the importance of the fibronectin binding proteins FnBPA and FnBPB in a *S. aureus* in vitro model of adherence and invasion on primary bovine mammary cells. Besides this, it was also shown that the levels of adhesion are dependent on the bacterial growth phase. The expression and synthesis of fibronectin-binding proteins is optimal during exponential growth phase, and downregulated during the stationary phase. The growth phases are controlled by the accessory gene regulator (*agr*) locus. This results in a 20-fold higher invasion in host cells for exponential growing bacteria compared to stationary growing bacteria.

The work of Hensen et al. shows that not all *S. aureus* strains are able to invade host cells, and that there are significant differences regarding the efficiency of invading strains (2000). The results show that invasion typically is a postadherence event, and that the efficiency of adherence and invasion are positively correlated. Culturing the bacteria in milk whey resulted in higher adherence compared to strains cultured in Tryptic Soy Broth (TSB). The adherence of milk whey grown bacteria was periodate sensitive, indicating a role for polysaccharides in this process; however, the TSB grown bacteria were not influenced by periodate indicating multiple mechanisms of adherence. This study also investigated the influence of capsule formation on adherence, and it was found that strong capsule formation hinders adherence. These results seem to indicate that polysaccharides are able to support adherence; however, abundant polysaccharide capsule formation reduces adherence. Further studies show that there are different genetic sources of extracellular polysaccharide production in *S. aureus*, and show that they have different roles in virulence behavior.

The *S. aureus* polysaccharide capsule is immunogenic and 11 different genotypes have been reported, but only four have been purified and chemically characterized (Tollersrud et al. 2000, 2001). The most prevalent capsular polysaccharide (CP) types from bovine mastitis isolates are CP5 and CP8 (Sordelli et al. 2000). The expression of the capsular polysaccharides is increased when bacteria are grown in milk (Sutra et al. 1990). These capsules have antiphagocytic properties, and antibodies against CP facilitate opsonophagocytosis (Sutra et al. 1990). It was shown by O'Brien et al. that immune sera against capsular serotypes 5 and 8 decrease adherence to bovine mammary epithelial cells (2000).

The second important source of extracellular polysaccharide production is the intercellular adhesion (*ica*) locus, which was found to play an important role in biofilm in *S. aureus* and *S. epidermidis* (Cramton et al. 1999; McKenney et al. 1998). The role for this polymeric *N*-actyl-glucosamine (PNAG) in biofilm development is based on the capability to support adherence to abiotic surfaces and support intercellular accumulation (O'Gara 2007). Initially it seemed that the presence of the *ica* genes was required for biofilm formation; however, later it was discovered that without the presence of the *ica* locus, biofilm accumulation could also be supported by the presence of several surface proteins (Cucarella et al. 2001; O'Gara 2007).

The *ica* locus is able to produce the capsular polysaccharide-adhesin (PS/A) or PNAG that mediates initial cell adherence to biomaterials and intercellular adherence. This adhesion polysaccharide protects the bacteria from opsonophagocytosis (Maira-Litran et al. 2002, McKenney et al. 2000). The early investigations for the role of the *ica* locus in virulence and disease indicated that this gene was significantly related to virulent strains and persistent infections in humans (Arciola et al. 2001, 2002). The first study investigating the prevalence of the *ica* locus in bovine mastitis isolates gave similar results; all mastitis isolates possessed this locus and nearly all produced biofilms in vitro (Vasudevan et al. 2003). As a result of this, the *ica* locus also appeared to be very important for bovine mastitis isolates. In a later study from a sample of Dutch bovine mastitis isolates it was found that not all strains from this sample contained the *ica* locus, and the presence was not related to the results of a quantitative biofilm assay (Melchior et al. 2009). Moreover, the amount of biofilm formation was influenced by the growth medium used, as growth in milk whey gave significantly different results than growth in TSB. Interestingly this study also revealed a strong correlation between the *agr*-type of the strains and the prevalence of the *ica* locus; *agr*-type 2 strains all contained this locus, whereas only 44% of the *agr*-type 1 strains contained the locus.

2.2 *Therapy of* S. aureus *Mastitis*

S. aureus is one of the most important causes of subclinical, clinical, recurrent, and chronic mastitis in dairy cattle (Barkema et al. 2006). Several studies have shown that the correlation between the antimicrobial susceptibility of *S. aureus* isolates determined in vitro and the actual bacteriological cure rate after antimicrobial treatment is only moderate (Pyorala and Pyorala 1998; Sol et al. 1997, 2000; Wilson et al. 1999). These epidemiological and clinical studies also revealed that treatment of older cows, or cows with high somatic cell counts (SCCs) in general is less successful and that treatment of cows with a high SCC due to a chronic infection often fails entirely (Apparao et al. 2009; Barkema et al. 2006; Owens et al. 1997; Taponen et al. 2003; Wilson et al. 1999).

It is shown in in vitro studies with a biofilm susceptibility model that the success of antimicrobial treatments might be improved by extending the duration of antimicrobial therapies (Amorena et al. 1999; Melchior et al. 2007). This was later confirmed in vivo within a study that tested the efficacy of extended treatment for chronic cases of *S. aureus* mastitis (Roy et al. 2009).

Reduced therapeutic efficacy of antimicrobial therapies that are related to infections of longer duration can also be modeled with an in vitro bacterial biofilm model (Amorena et al. 1999; Melchior 2007). Recently van den Borne showed, based on the results of a clinical trial, that early onset of antimicrobial treatments of

S. aureus bovine mastitis infections results in better therapy outcomes (van den Borne et al. 2010a).

Based on these research investigations we can conclude that bovine *S. aureus* is best treated in an early stage of the infection, with a therapy for an adequate time of duration, which is often longer than the standard therapy. The correct duration of these therapies should be long enough for the specific strain or strains present on the farm of interest, and must be evaluated per individual farm. Research outcomes have shown that strain differences exist, and that these result in significantly different cure rates when comparing farms and regions (Bradley and Green 2009; Melchior 2007; van den Borne et al. 2010b).

2.3 Vaccine Development

The findings, which showed that the CP are immunogenic and antibodies against CP facilitate opsonophagocytosis, led to the development of vaccines against CP serotypes 5, 8, and 336 (O'Brien et al. 2000; Tollersrud et al. 2000, 2001). The efficacy of vaccines based on CP antigen was developed with different conjugates, since CP antigen alone is not capable of inducing a protective immune response. Evaluation of the protective effect of several conjugates was tested in a mouse model, which led to promising results for further vaccine development (Calzolari et al. 1997; Giraudo et al. 1997; Han et al. 2000). This resulted in the registration of a *S. aureus* bacterin vaccine, based on CP antigens, in several countries and regions, however not in the European Union (EU). The efficacy of three *S. aureus* vaccines, among them the commercially available Lysigin©, was reviewed by Middleton (Middleton et al. 2009).

The second source of vaccine development is the presence of a surface polysaccharide poly-N-acetyl-beta-1,6-glucosamine (PNAG) antigen in *S. aureus* and *S. epidermidis* (McKenney et al. 1998, 2000), which is produced by the *ica* locus of these strains. PNAG proved to induce strong antibody responses in *S. aureus* infected mice and it was shown that PNAG is well-expressed during in vivo infections (McKenney et al. 1999, 2000). The efficacy of vaccines based on the production of antibodies against this slime associated antigenic complex (SAAC, which is also referred to as PNAG) was studied in cows, and the results indicated that the vaccines were not able to prevent infection; however, it did result in lower multiplication of the bacteria.(Prenafeta et al. 2010). Recently a bovine mastitis vaccine was registered in the EU named STARTVAC®, which contains PNAG or SAAC as one of the antigenic components. The possible differences in the presence of the *ica* locus for *S. aureus* strains on different farms and geographic regions might have some influence on the efficacy of this vaccine for *Staphylococcus* mastitis reduction (Melchior et al. 2009).

Besides the production of vaccines based on different antigenic polysaccharides, several vaccines are based on killed, avirulent, or mutant strains, which are able to reduce bacterial shedding or duration of infection (Calzolari et al. 1997; Giraudo et al. 1997; Leitner et al. 2003a, b; Pellegrino et al. 2010).

3 *Escherichia coli*

The formation of bacterial biofilms of *E. coli* in a host in general seems to be, based on current evidence, to a large extent an intracellular event. To subvert the host response, *E. coli* escapes into the cytoplasm of the infected cell and replicates into biomasses called IBCs (Anderson et al. 2010).

The majority of research on bacterial biofilm formation of *E. coli* is from human medicines, as the recognition that this type of behavior from *E. coli* causes different types of challenging infections; i.e. urinary tract infections, intestinal diseases or prostatitis, has led to a multitude of publications. In veterinary medicine research information is divided between recent studies; showing a change in the character of *E. coli* bovine mastitis infections to a more persistent and recurrent type of infection, and other studies showing the capability of *E. coli* bovine mastitis isolates to adhere and invade in vitro mammary epithelial cells.

Because of the spread of information about biofilm capabilities of *E. coli* between different kinds of research backgrounds, the overall picture will be drawn up based on the sequence of steps during infection, and the evidence that supports this from different backgrounds. The first important step for the formation of intracellular biofilms is the attachment to the host epithelial cells, followed by invasion, which in the meantime also implies successful evasion of the host immunity. Subsequently the bacteria find their niche for quiescent persistence in the epithelial cells, which includes protection for antimicrobials and the host defense system. Here they can stay for months without shedding, and the triggers for this process still has to be elucidated (Anderson et al. 2003, 2010).

3.1 *Curli Fibers Support Epithelial Adherence*

Noteworthy, the curli fibers were first discovered in 1989 on *E. coli* strains that caused bovine mastitis, and have since been studied for their key role in adherence and survival in bacterial biofilms (Arnqvist et al. 1992; Olsen et al. 1989). These proteinaceous extracellular fibers are involved in surface and cell–cell contacts that promote community behavior and host colonization.

The regulation of curli gene expression is rather complex and responsive to many environmental cues, of which several are of importance for their possible therapeutic consequences, which we will discuss later. The regulation of curli expression has been reviewed excellent by Barnhart and Chapman (2006), and the following has been summarized from this review.

The most important parts of the curli operon reading frame are; the regulatory part *CsgD*, the two curli subunit components *CsgA* and *CsgB*, and *CsgG* which forms the pore in the extracellular membrane, through which *CsgA* and *CsgB* can be brought outside the extracellular membrane.

It was first recognized that growth below 30°C promotes curli gene expression, and this still holds for most laboratory strains. However, it has been demonstrated that many clinical isolates of *E. coli* can express curli at 37°C. Furthermore, it has been shown on repeatedly that a mutation in the *CsgD* gene can be induced, resulting in good curli expression at 37°C.

There are three two component regulatory systems that regulate curli gene expression. The most import is the OmpR/EnvZ system that responds to changes in osmolarity and regulates the porins OmpF and OmpC. In general, we see that under circumstances of low osmolarity the OmpR/EnvZ system facilitates the opening of the porin gates, and in high osmolarity the gates are closed. At the same time, low osmolarity thus facilitates opening of the CsgG pore, and subsequently CsgA and CsgB can be brought efficiently outside the membrane. This system stimulates the expression and transportation of the curli outside the extracellular membrane where the actual action takes place.

The CpxA/R two-component system is activated in response to membrane stress and/or misfolded proteins. This system downregulates the curli expression.

The Rcs system responds specifically to outer membrane stress, and this results in stimulation of the capsule synthesis. The Rcs system also downregulates the csgG expression and this pathway is required for biofilm formation.

From this we can abstract that curli formation is important in the initial stages of biofilm formation, for initial adhesion, and is then turned off by Cpx and Rcs pathway for further maturation of the biofilm communities (Barnhart and Chapman 2006; Prigent-Combaret et al. 2001).

Outside the bacterial membrane the CsgA and CsgB components are able to form beta-sheet rich fibers, which are resistant to protease digestion, similar to amyloid fibers from eukaryotic cells. Curli bind many host proteins, among them plasminogen and tissue plasminogen activator, resulting in active plasmin. This might be an advantage for curli-producing bacteria, as plasmin degrades soft tissue and thus facilitates the access to deeper tissue.

3.2 Adherence and Invasion (Cellular and Abiotic)

The adherence and invasion of *E. coli* bovine mastitis isolates from single and recurrent infections was studied by Dopfer et al. in 2000 and Dogan et al. in 2006. Both used the MAC-T bovine mastitis epithelial cell line for their studies. In both studies the strains from persistent infection were more efficient in invasion, but not adherence, resulting in higher numbers of intracellular bacteria. These findings are supported by other publications (Barnhart and Chapman 2006), which show that curli synthesis supports epithelial invasion in different types of tissues (Kai-Larsen et al. 2010).

The adherence of *E. coli* to abiotic surfaces outside the host probably also supports the survival and spread of *E. coli* outside the udder of the cow, as has been shown to be of importance for other udder pathogens previously (Zadoks et al.

2001a, b). Because of the normal regulation and expression of curli genes, which is stimulated below 30°C, we can assume that environmental temperatures often facilitates curli expression. Furthermore, low osmolarity, induced by normal cleaning procedures with water in the milking parlor, and desiccation, from drying of surfaces between milking sessions, and lack of nutrients also stimulates curli expression and biofilm formation in general (Ferrieres and Clarke 2003; Prigent-Combaret et al. 1999; Vidal et al. 1998) . Unfortunately, once *E. coli* is living in the environment protected by a multi species biofilm community, it is able to withstand disinfectants like hypochlorus acid an monochloramine (Williams and Braun-Howland 2003), which are routinely used in dairy farms.

3.3 *Intracellular Persistence and Replication*

In a study from Dopfer et al. (1999), it was shown that almost 47% of recurrent *E. coli* infections within the same quarter were caused by a strain with the same DNA fingerprint. Further results from this study showed that *E. coli* with the same DNA fingerprint could also be found in another quarter of the same cow, or in a different cow, suggesting infections from a common source. Milk samples were taken during episodes of clinical mastitis, therefore possible shedding of bacteria between clinical periods was not evaluated.

The study from Bradley and Green showed that 86% of cases of recurrent *E. coli* mastitis within one quarter was implicated by the same DNA genotype, which is also suggestive of persistence between clinical episodes (2001). Also in this study the spread of the persistent genotype was suggested, as the same genotypes were found also in other quarters.

Bradley and Green also reported that typically none of the recurrent cases resulted in systemic signs of illness in the affected cows, and this is remarkable since the classic clinical *E. coli* symptoms normally include high fever and septicemia that could lead to death without the proper therapeutic interventions (2001).

Dogan was able to show that *E. coli* strains from persistent infections were better with respect to intracellular survival and replication in an in vitro model, resulting in higher number of bacteria after 48h of growth, as was previously been shown in other models of epithelial *E. coli* infections (Dogan et al. 2006). Finally Dogan demonstrated the presence of *E. coli* within the mammary epithelium from a cow with naturally occurring persistent *E. coli* mastitis.

3.4 *Immunity and Competition with Host Defenses*

The bacteria can outcompete the host defenses through several strategies. First efficient growth in milk means fermentation of lactose as the main source of energy. The results of Blum et al. (2008) show that lactose fermentation and growth in milk from *E. coli* isolates is positively correlated, and that mastitis isolates have better growth in milk compared to environmental isolates. Furthermore, this same study shows that phagocytosis by PMN was significantly lower compared to the environmental strains.

The synthesis of curli also seems to be of importance for protection against the immune defenses. In a study on uropathogenic *E. coli,* it was shown that production of curli is higher amongst these isolates and this facilitates adhesion, invasion, and biofilm production (Kai-Larsen et al. 2010). The epithelial cells of the urinary tract respond with an increase of antimicrobial peptide (LL-37) upon infection with uropathogenic *E. coli*. This peptide is able to cause bacterial lysis; however, curli binds this peptide and thus protect against this antimicrobial peptide. If curli have not been produced yet, LL-37 can inhibit curli formation by blocking the polymerization of the curli subunit CsgA; resulting in inhibition of biofilm formation in vitro. The outcome of an infection with curli producing *E. coli* depends on the timing and balance between curli production and antimicrobial peptide production; reduced curli expression by colonizing bacteria makes them more vulnerable for LL-37 (Kai-Larsen et al. 2010).

3.5 Prevention and Therapy of E. coli *Infections*

The more we learn about biofilm formation, and *E. coli* is not a exception to this, the more we come to realize that all strategies are aimed at survival under all kinds of stressful circumstances. The information about the regulation of curli fibers reveals however some interesting things for common practical knowledge.

It is common experience that housing and construction of stables and beds for the cows should prevent cows from getting "cold" udders, as this can increase the incidence of *E. coli* infections. Therefore, although stables should allow for enough air fresh for the cows, wind breakers sometimes have to be lowered to protect cows from draft. This might be in agreement with better curli expression at lower temperatures.

Biofilm formation is stimulation by several circumstances; i.e. osmolarity, membrane stress, and low nutrient levels. The latter is regulated through the Cpx system via σ^s, which is aimed at survival under starving conditions (Barnhart and Chapman 2006). One of the less well-understood therapies for *E. coli* infection is the infusion in the udder of isotonic and/or glucose solutions. The rationale for this therapy is that this dilutes and flushes out the toxic compound LPS, but it might also reduce the propensity of the bacteria to shift to a defensive biofilm mode of growth, which would leave them untouchable for the defense system or antimicrobials.

Once the bacteria are settled persistently in the udder, similar to the situation in the urinary tract, little seems to have any real effect on them. Repeatedly it is shown that intracellular *E. coli* are very resistant to antimicrobials in vitro, and in vivo results do not seem to contradict this. Furthermore, there is increasing evidence that so-called persister formation (Lewis 2000), resulting in untouchable bacteria is stimulated by the use of antimicrobials (Dorr et al. 2010; Hoffman et al. 2005).

Current knowledge seems to indicate that some bacterial species are better equipped to survive outside the host cells in the bovine udder, i.e. *S. aureus*, and others are better adapted for intracellular survival such as *E. coli*. However, because of the relations and correlations of bacterial characteristics important for adherence and invasion and for biofilm formation, these are presented as a whole process that can take place in the bovine udder.

Research investigations most often focus, for practical reasons, on single parts of the whole process. This shows which tools are used, and reveals the differences in abilities within bacterial species and between bacterial species. Furthermore, recent results show that *Str. uberis* is able to form biofilm communities in milk and that milk components, such as casein supports this (Varhimo et al. 2010). The evaluation of the antimicrobial susceptibility of these persistent intracellular infections, however, is difficult to capture in an in vitro model.

4 *Streptococcus uberis*

Epidemiological research of *Str. uberis* bovine mastitis infections have shown that these bacteria are often causing persistent and chronic infections, although infections of shorter duration occur, which end due to spontaneous cure or effective treatments (McDougall et al. 2004; Phuektes et al. 2001; Pullinger et al. 2007; Zadoks et al. 2003). Many of these infections are subclinical and when infections stay unnoticed and untreated can become chronic (Phuektes et al. 2001; Zadoks et al. 2003).

Several typing techniques have been used to study the spread of *Str. uberis* between farms and between cows on one farm (Douglas et al. 2000; McDougall et al. 2004; Pullinger et al. 2007; Zadoks et al. 2003). Although the use of different molecular techniques limits the possibility to compare these studies the outcomes agree on several results. The studies show that there is a multitude of different strains available of these environmental bacteria, and that cows become infected by different strains within one farm and between farms. When infections on one farm are followed for several lactations the type of *Str. uberis* isolates change over time, although sometimes a dominant strain can be found among a constantly changing group of isolates causing the new infections (Phuektes et al. 2001; Pullinger et al. 2007; Zadoks et al. 2003). These studies also show that persistent infections are in fact caused by one isolate, and are unlikely to be caused by reinfections, given the multitude of types available in the environment (Douglas et al. 2000; Pullinger et al. 2007; Zadoks et al. 2003).

The chronic and persistent character of *Str. uberis* infections seems to be related to the type of strain in some studies (Phuektes et al. 2001; Zadoks et al. 2003), although more recent outcomes did not reveal a relation between persistence and a particular type of isolates, which suggests that management or host factors are also of importance for the outcomes of the infection (Pullinger et al. 2007).

4.1 Adherence and Invasion

Based on the extensive work from Almeida and Oliver and coworkers, we are able to establish a picture of the evasive strategies of *Str. uberis*. First it was found that around 40% of the *Str. uberis* strains isolated from bovine mastitis is able to produce a capsule of hyaluronic acid (Almeida and Oliver 1993a, b). The hyaluronic acid polysaccharide capsule is protective for phagocytosis by the host defense mechanisms. The expression of the capsule is increased by lactose, skim milk, and casein although strain differences also seem to play a role (Almeida and Oliver 1993a; Matthews et al. 1994). The protective effect of the hyaluronic acid capsule is not surprising given the presence of host derived hyaluronic acid especially in the skin and connective tissue and the regulatory role of this in tissue repair and inflammation; the capsule is a smart decoy for the host immune system.

Strains that do not produce a protective capsule seem to be better equipped for adherence and invasion of host cells. *Str. uberis* is able to make use of several host factors during this process such as; collagen, glycosaminoglycans, fibrinogen, and lactoferrin (Almeida et al. 1996, 1999a, b, 2003, 2006; Almeida and Oliver 2001; Fang et al. 2000; Patel et al. 2009). Typically milk components such as beta-casein and lactoferrin, are also facilitating the adherence and invasion in host cells (Almeida et al. 2003; Fang et al. 2000).

Based on in vitro work it seems feasible that *Str. uberis* is able to survive intracellular for extended periods (Tamilselvam et al. 2006) and besides this, the data from epidemiological studies also provide strong evidence for persistent presence in the bovine udder (Phuektes et al. 2001; Pullinger et al. 2007; Zadoks et al. 2003). With the lack of histological evidence from ex vivo samples, revealing the presence of *Str. uberis* in the udder tissue, however, we are not able to elucidate the question of where the most important niche is for *Str. uberis*. Are they residing predominantly intracellular, quiescently hiding for the host immune system, or are they also able to survive in a bacterial community outside the host cells, in the connective tissue or the alveoli of the udder?

Recently it was shown that alpha- and beta-casein, import components of host milk, are able to support bacterial biofilm formation of *Str. uberis* in a classical biofilm assay (Varhimo et al. 2010). Interestingly it was previously shown that whole milk, skim milk, and casein are also able to stimulate capsule production (Almeida et al. 1999a; Matthews et al. 1994). These results suggest that *Str. uberis* biofilm can be composed of both proteinaceous and polysaccharide components, similar to Staphylococcal biofilms. Earlier it was found that beta-casein is also able to stimulate host cell adherence and invasion (Almeida et al. 2003). Besides the role of alpha- and beta-casein in biofilm accumulation Varhimo et al. also revealed that casein degradation by serine protease activity resulted in maximal biofilm production (2010). These results suggested that the extracellular proteolytic activity of *Str. uberis* contributes to an increased biofilm formation. Besides the bacterial proteolytic activity, we should take into account that milk also contains an endogenous proteolytic system, which can result, in response to bacterial presence, in increasing concentrations of plasmin from plasminogen and the polymorphonuclear proteolytics elastase, collagenase and several cathepsins. The endogenous milk protease plasmin, a serine protease, contributes to primary proteolyses of caseins (Le Roux et al. 2003). The presence of bacteria in the bovine udder will result to some extend to the recruitment of PMN, and therefore to an increase of proteolytic activity in milk. The results of Varhimo et al. might implicate that the host immune response influences the outcomes of the extracellular biofilm accumulation process.

References

Almeida RA, Oliver SP (1993a) Antiphagocytic effect of the capsule of *Streptococcus uberis*. Zentralbl Veterinärmed B 40(9–10):707–714

Almeida RA, Oliver SP (1993b) Growth curve, capsule expression and characterization of the capsular material of selected strains of *Streptococcus uberis*. Zentralbl Veterinärmed B 40(9–10): 697–706

Almeida RA, Oliver SP (2001) Role of collagen in adherence of *Streptococcus uberis* to bovine mammary epithelial cells. J Vet Med B Infect Dis Vet Public Health 48(10):759–763

Almeida RA, Luther DA, Kumar SJ, Calvinho LF, Bronze MS, Oliver SP (1996) Adherence of *Streptococcus uberis* to bovine mammary epithelial cells and to extracellular matrix proteins. Zentralbl Veterinärmed B 43(7):385–392

Almeida RA, Fang W, Oliver SP (1999a) Adherence and internalization of *Streptococcus uberis* to bovine mammary epithelial cells are mediated by host cell proteoglycans. FEMS Microbiol Lett 177(2):313–317

Almeida RA, Luther DA, Oliver SP (1999b) Incubation of *Streptococcus uberis* with extracellular matrix proteins enhances adherence to and internalization into bovine mammary epithelial cells. FEMS Microbiol Lett 178(1):81–85

Almeida RA, Luther DA, Nair R, Oliver SP (2003) Binding of host glycosaminoglycans and milk proteins: possible role in the pathogenesis of *Streptococcus uberis* mastitis. Vet Microbiol 94(2):131–141

Almeida RA, Luther DA, Park HM, Oliver SP (2006) Identification, isolation, and partial characterization of a novel *Streptococcus uberis* adhesion molecule (SUAM). Vet Microbiol 115(1–3): 183–191

Amorena B, Gracia E, Monzon M, Leiva J, Oteiza C, Perez M, Alabart JL, Hernandez-Yago J (1999) Antibiotic susceptibility assay for *Staphylococcus aureus* in biofilms developed in vitro. J Antimicrob Chemother 44(1):43–55

Anderson GG, Palermo JJ, Schilling JD, Roth R, Heuser J, Hultgren SJ (2003) Intracellular bacterial biofilm-like pods in urinary tract infections. Science 301(5629):105–107

Anderson GG, Goller CC, Justice S, Hultgren SJ, Seed PC (2010) Polysaccharide capsule and sialic acid-mediated regulation promote biofilm-like intracellular bacterial communities during cystitis. Infect Immun 78(3):963–975

Apparao MD, Ruegg PL, Lago A, Godden S, Bey R, Leslie K (2009) Relationship between in vitro susceptibility test results and treatment outcomes for gram-positive mastitis pathogens following treatment with cephapirin sodium. J Dairy Sci 92(6):2589–2597

Arciola CR, Baldassarri L, Montanaro L (2001) Presence of *icaA* and *icaD* genes and slime production in a collection of staphylococcal strains from catheter-associated infections. J Clin Microbiol 39(6):2151–2156

Arciola CR, Baldassarri L, Montanaro L (2002) In catheter infections by *Staphylococcus epidermidis* the intercellular adhesion (*ica*) locus is a molecular marker of the virulent slime-producing strains. J Biomed Mater Res 59(3):557–562

Arnqvist A, Olsen A, Pfeifer J, Russell DG, Normark S (1992) The Crl protein activates cryptic genes for curli formation and fibronectin binding in *Escherichia coli* HB101. Mol Microbiol 6(17):2443–2452

Barkema HW, Schukken YH, Zadoks RN (2006) The role of cow, pathogen, and treatment regimen in the therapeutic success of bovine *Staphylococcus aureus* mastitis. J Dairy Sci 89(6): 1877–1895

Barnhart MM, Chapman MR (2006) Curli biogenesis and function. Annu Rev Microbiol 60: 131–147

Baselga R, Albizu I, De La Cruz M, Del Cacho E, Barberan M, Amorena B (1993) Phase variation of slime production in *Staphylococcus aureus*: implications in colonization and virulence. Infect Immun 61(11):4857–4862

Baselga R, Albizu I, Amorena B (1994) *Staphylococcus aureus* capsule and slime as virulence factors in ruminant mastitis. A review. Vet Microbiol 39(3–4):195–204

Blum S, Heller ED, Krifucks O, Sela S, Hammer-Muntz O, Leitner G (2008) Identification of a bovine mastitis Escherichia coli subset. Vet Microbiol 132(1–2):135–148

Bradley AJ, Green MJ (2001) Adaptation of *Escherichia coli* to the bovine mammary gland. J Clin Microbiol 39(5):1845–1849

Bradley AJ, Green MJ (2009) Factors affecting cure when treating bovine clinical mastitis with cephalosporin-based intramammary preparations. J Dairy Sci 92(5):1941–1953

Calzolari A, Giraudo JA, Rampone H, Odierno L, Giraudo AT, Frigerio C, Bettera S, Raspanti C, Hernandez J, Wehbe M, Mattea M, Ferrari M, Larriestra A, Nagel R (1997) Field trials of a vaccine against bovine mastitis. 2. Evaluation in two commercial dairy herds. J Dairy Sci 80(5):854–858

Costerton JW, Stewart PS, Greenberg EP (1999) Bacterial biofilms: a common cause of persistent infections. Science 284(5418):1318–1322

Cramton SE, Gerke C, Schnell NF, Nichols WW, Gotz F (1999) The intercellular adhesion *(ica)* locus is present in *Staphylococcus aureus* and is required for biofilm formation. Infect Immun 67(10):5427–5433

Cucarella C, Solano C, Valle J, Amorena B, Lasa I, Penades JR (2001) *Bap*, a *Staphylococcus aureus* surface protein involved in biofilm formation. J Bacteriol 183(9):2888–2896

Dogan B, Klaessig S, Rishniw M, Almeida RA, Oliver SP, Simpson K, Schukken YH (2006) Adherent and invasive *Escherichia coli* are associated with persistent bovine mastitis. Vet Microbiol 116(4):270–282

Dopfer D, Barkema HW, Lam TJ, Schukken YH, Gaastra W (1999) Recurrent clinical mastitis caused by *Escherichia coli* in dairy cows. J Dairy Sci 82(1):80–85

Dopfer D, Almeida RA, Lam TJ, Nederbragt H, Oliver SP, Gaastra W (2000) Adhesion and invasion of *Escherichia coli* from single and recurrent clinical cases of bovine mastitis in vitro. Vet Microbiol 74(4):331–343

Dorr T, Vulic M, Lewis K (2010) Ciprofloxacin causes persister formation by inducing the TisB toxin in *Escherichia coli*. PLoS Biol 8(2):e1000317

Douglas VL, Fenwick SG, Pfeiffer DU, Williamson NB, Holmes CW (2000) Genomic typing of *Streptococcus uberis* isolates from cases of mastitis, in New Zealand dairy cows, using pulsed-field gel electrophoresis. Vet Microbiol 75(1):27–41

Fang W, Almeida RA, Oliver SP (2000) Effects of lactoferrin and milk on adherence of *Streptococcus uberis* to bovine mammary epithelial cells. Am J Vet Res 61(3):275–279

Ferrieres L, Clarke DJ (2003) The RcsC sensor kinase is required for normal biofilm formation in *Escherichia coli* K-12 and controls the expression of a regulon in response to growth on a solid surface. Mol Microbiol 50(5):1665–1682

Foster TJ, Hook M (1998) Surface protein adhesins of *Staphylococcus aureus*. Trends Microbiol 6(12):484–488

Giraudo JA, Calzolari A, Rampone H, Rampone A, Giraudo AT, Bogni C, Larriestra A, Nagel R (1997) Field trials of a vaccine against bovine mastitis. 1. Evaluation in heifers. J Dairy Sci 80(5): 845–853

Han HR, Pak S 2nd, Guidry A (2000) Prevalence of capsular polysaccharide (CP) types of *Staphylococcus aureus* isolated from bovine mastitic milk and protection of S. aureus infection in mice with CP vaccine. J Vet Med Sci 62(12):1331–1333

Hensen SM, Pavicic MJ, Lohuis JA, Poutrel B (2000) Use of bovine primary mammary epithelial cells for the comparison of adherence and invasion ability of *Staphylococcus aureus* strains. J Dairy Sci 83(3):418–429

Hoffman LR, D'Argenio DA, MacCoss MJ, Zhang Z, Jones RA, Miller SI (2005) Aminoglycoside antibiotics induce bacterial biofilm formation. Nature 436(7054):1171–1175

Kai-Larsen Y, Luthje P, Chromek M, Peters V, Wang X, Holm A, Kadas L, Hedlund KO, Johansson J, Chapman MR, Jacobson SH, Romling U, Agerberth B, Brauner A (2010) Uropathogenic Escherichia coli modulates immune responses and its curli fimbriae interact with the antimicrobial peptide LL-37. PLoS Pathog 6(7):e1001010

Lammers A, Nuijten PJ, Smith HE (1999) The fibronectin binding proteins of *Staphylococcus aureus* are required for adhesion to and invasion of bovine mammary gland cells. FEMS Microbiol Lett 180(1):103–109

Le Roux Y, Laurent F, Moussaoui F (2003) Polymorphonuclear proteolytic activity and milk composition change. Vet Res 34(5):629–645

Leitner G, Lubashevsky E, Glickman A, Winkler M, Saran A, Trainin Z (2003a) Development of a *Staphylococcus aureus* vaccine against mastitis in dairy cows. I. Challenge trials. Vet Immunol Immunopathol 93(1–2):31–38

Leitner G, Yadlin N, Lubashevsy E, Ezra E, Glickman A, Chaffer M, Winkler M, Saran A, Trainin Z (2003b) Development of a *Staphylococcus aureus* vaccine against mastitis in dairy cows. II. Field trial. Vet Immunol Immunopathol 93(3–4):153–158

Lewis K (2000) Programmed death in bacteria. Microbiol Mol Biol Rev 64(3):503–514

Maira-Litran T, Kropec A, Abeygunawardana C, Joyce J, Mark G III, Goldmann DA, Pier GB (2002) Immunochemical properties of the staphylococcal poly-*N*-acetylglucosamine surface polysaccharide. Infect Immun 70(8):4433–4440

Matthews KR, Almeida RA, Oliver SP (1994) Bovine mammary epithelial cell invasion by *Streptococcus uberis*. Infect Immun 62(12):5641–5646

McDougall S, Parkinson TJ, Leyland M, Anniss FM, Fenwick SG (2004) Duration of infection and strain variation in *Streptococcus uberis* isolated from cows' milk. J Dairy Sci 87(7):2062–2072

McKenney D, Hubner J, Muller E, Wang Y, Goldmann DA, Pier GB (1998) The *ica* locus of *Staphylococcus epidermidis* encodes production of the capsular polysaccharide/adhesin. Infect Immun 66(10):4711–4720

McKenney D, Pouliot KL, Wang Y, Murthy V, Ulrich M, Doring G, Lee JC, Goldmann DA, Pier GB (1999) Broadly protective vaccine for *Staphylococcus aureus* based on an in vivo-expressed antigen. Science 284(5419):1523–1527

McKenney D, Pouliot K, Wang Y, Murthy V, Ulrich M, Doring G, Lee JC, Goldmann DA, Pier GB (2000) Vaccine potential of poly-1-6 beta-D-*N*-succinylglucosamine, an immunoprotective surface polysaccharide of *Staphylococcus aureus* and *Staphylococcus epidermidis*. J Biotechnol 83(1–2):37–44

Melchior MB (2007) Biofilms: implications for the therapy of bovine *Staphylococcus aureus* mastitis. Thesis, University of Utrecht, Utrecht

Melchior MB, Fink-Gremmels J, Gaastra W (2006a) Comparative assessment of the antimicrobial susceptibility of *Staphylococcus aureus* isolates from bovine mastitis in biofilm versus planktonic culture. J Vet Med B Infect Dis Vet Public Health 53(7):326–332

Melchior MB, Vaarkamp H, Fink-Gremmels J (2006b) Biofilms: a role in recurrent mastitis infections? Vet J 171(3):398–407

Melchior MB, Fink-Gremmels J, Gaastra W (2007) Extended antimicrobial susceptibility assay for *Staphylococcus aureus* isolates from bovine mastitis growing in biofilms. Vet Microbiol 125(1–2):141–149

Melchior MB, van Osch MH, Graat RM, van Duijkeren E, Mevius DJ, Nielen M, Gaastra W, Fink-Gremmels J (2009) Biofilm formation and genotyping of *Staphylococcus aureus* bovine

mastitis isolates: evidence for lack of penicillin-resistance in Agr-type II strains. Vet Microbiol 137(1–2):83–89
Middleton JR, Luby CD, Adams DS (2009) Efficacy of vaccination against staphylococcal mastitis: a review and new data. Vet Microbiol 134(1–2):192–198
O'Brien CN, Guidry AJ, Fattom A, Shepherd S, Douglass LW, Westhoff DC (2000) Production of antibodies to *Staphylococcus aureus* serotypes 5, 8, and 336 using poly(DL-lactide-co-glycolide) microspheres. J Dairy Sci 83(8):1758–1766
O'Gara JP (2007) *ica* and beyond: biofilm mechanisms and regulation in *Staphylococcus epidermidis* and *Staphylococcus aureus*. FEMS Microbiol Lett 270(2):179–188
Olsen A, Jonsson A, Normark S (1989) Fibronectin binding mediated by a novel class of surface organelles on *Escherichia coli*. Nature 338(6217):652–655
Owens WE, Ray CH, Watts JL, Yancey RJ (1997) Comparison of success of antibiotic therapy during lactation and results of antimicrobial susceptibility tests for bovine mastitis. J Dairy Sci 80(2):313–317
Patel D, Almeida RA, Dunlap JR, Oliver SP (2009) Bovine lactoferrin serves as a molecular bridge for internalization of *Streptococcus uberis* into bovine mammary epithelial cells. Vet Microbiol 137(3–4):297–301
Pellegrino M, Giraudo J, Raspanti C, Odierno L, Bogni C (2010) Efficacy of immunization against bovine mastitis using a *Staphylococcus aureus* avirulent mutant vaccine. Vaccine 28(28): 4523–4528
Phuektes P, Mansell PD, Dyson RS, Hooper ND, Dick JS, Browning GF (2001) Molecular epidemiology of *Streptococcus uberis* isolates from dairy cows with mastitis. J Clin Microbiol 39(4):1460–1466
Prenafeta A, March R, Foix A, Casals I, Costa L (2010) Study of the humoral immunological response after vaccination with a *Staphylococcus aureus* biofilm-embedded bacterin in dairy cows: possible role of the exopolysaccharide specific antibody production in the protection from Staphylococcus aureus induced mastitis. Vet Immunol Immunopathol 134(3–4):208–217
Prigent-Combaret C, Vidal O, Dorel C, Lejeune P (1999) Abiotic surface sensing and biofilm-dependent regulation of gene expression in *Escherichia coli*. J Bacteriol 181(19):5993–6002
Prigent-Combaret C, Brombacher E, Vidal O, Ambert A, Lejeune P, Landini P, Dorel C (2001) Complex regulatory network controls initial adhesion and biofilm formation in *Escherichia coli* via regulation of the *csgD* gene. J Bacteriol 183(24):7213–7223
Pullinger GD, Coffey TJ, Maiden MC, Leigh JA (2007) Multilocus-sequence typing analysis reveals similar populations of *Streptococcus uberis* are responsible for bovine intramammary infections of short and long duration. Vet Microbiol 119(2–4):194–204
Pyorala SH, Pyorala EO (1998) Efficacy of parenteral administration of three antimicrobial agents in treatment of clinical mastitis in lactating cows: 487 cases (1989–1995). J Am Vet Med Assoc 212(3):407–412
Roy JP, DesCoteaux L, DuTremblay D, Beaudry F, Elsener J (2009) Efficacy of a 5-day extended therapy program during lactation with cephapirin sodium in dairy cows chronically infected with *Staphylococcus aureus*. Can Vet J 50(12):1257–1262
Sears PM, Smith BS, English PB, Herer PS, Gonzalez RN (1990) Shedding pattern of *Staphylococcus aureus* from bovine intramammary infections. J Dairy Sci 73(10):2785–2789
Sol J, Sampimon OC, Snoep JJ, Schukken YH (1997) Factors associated with bacteriological cure during lactation after therapy for subclinical mastitis caused by *Staphylococcus aureus*. J Dairy Sci 80(11):2803–2808
Sol J, Sampimon OC, Barkema HW, Schukken YH (2000) Factors associated with cure after therapy of clinical mastitis caused by *Staphylococcus aureus*. J Dairy Sci 83(2):278–284
Sordelli DO, Buzzola FR, Gomez MI, Steele-Moore L, Berg D, Gentilini E, Catalano M, Reitz AJ, Tollersrud T, Denamiel G, Jeric P, Lee JC (2000) Capsule expression by bovine isolates of *Staphylococcus aureus* from Argentina: genetic and epidemiologic analyses. J Clin Microbiol 38(2):846–850

Sutra L, Rainard P, Poutrel B (1990) Phagocytosis of mastitis isolates of *Staphylococcus aureus* and expression of type 5 capsular polysaccharide are influenced by growth in the presence of milk. J Clin Microbiol 28(10):2253–2258

Tamilselvam B, Almeida RA, Dunlap JR, Oliver SP (2006) *Streptococcus uberis* internalizes and persists in bovine mammary epithelial cells. Microb Pathog 40(6):279–285

Taponen S, Dredge K, Henriksson B, Pyyhtia AM, Suojala L, Junni R, Heinonen K, Pyorala S (2003) Efficacy of intramammary treatment with procaine penicillin G vs. procaine penicillin G plus neomycin in bovine clinical mastitis caused by penicillin-susceptible, gram-positive bacteria–a double blind field study. J Vet Pharmacol Ther 26(3):193–198

Tollersrud T, Kenny K, Reitz AJ Jr, Lee JC (2000) Genetic and serologic evaluation of capsule production by bovine mammary isolates of *Staphylococcus aureus* and other Staphylococcus spp. from Europe and the United States. J Clin Microbiol 38(8):2998–3003

Tollersrud T, Zernichow L, Andersen SR, Kenny K, Lund A (2001) *Staphylococcus aureus* capsular polysaccharide type 5 conjugate and whole cell vaccines stimulate antibody responses in cattle. Vaccine 19(28–29):3896–3903

van den Borne BH, Halasa T, van Schaik G, Hogeveen H, Nielen M (2010a) Bioeconomic modeling of lactational antimicrobial treatment of new bovine subclinical intramammary infections caused by contagious pathogens. J Dairy Sci 93(9):4034–4044

van den Borne BH, van Schaik G, Lam TJ, Nielen M (2010b) Therapeutic effects of antimicrobial treatment during lactation of recently acquired bovine subclinical mastitis: two linked randomized field trials. J Dairy Sci 93(1):218–233

Varhimo E, Varmanen P, Fallarero A, Skogman M, Pyorala S, Iivanainen A, Sukura A, Vuorela P, Savijoki K (2010) Alpha- and beta-casein components of host milk induce biofilm formation in the mastitis bacterium *Streptococcus uberis*. Vet Microbiol 149(3–4):381–389

Vasudevan P, Nair MK, Annamalai T, Venkitanarayanan KS (2003) Phenotypic and genotypic characterization of bovine mastitis isolates of *Staphylococcus aureus* for biofilm formation. Vet Microbiol 92(1–2):179–185

Vidal O, Longin R, Prigent-Combaret C, Dorel C, Hooreman M, Lejeune P (1998) Isolation of an *Escherichia coli* K-12 mutant strain able to form biofilms on inert surfaces: involvement of a new *ompR* allele that increases curli expression. J Bacteriol 180(9):2442–2449

Williams MM, Braun-Howland EB (2003) Growth of *Escherichia coli* in model distribution system biofilms exposed to hypochlorous acid or monochloramine. Appl Environ Microbiol 69(9):5463–5471

Wilson DJ, Gonzalez RN, Case KL, Garrison LL, Grohn YT (1999) Comparison of seven antibiotic treatments with no treatment for bacteriological efficacy against bovine mastitis pathogens. J Dairy Sci 82(8):1664–1670

Zadoks RN, Allore HG, Barkema HW, Sampimon OC, Grohn YT, Schukken YH (2001a) Analysis of an outbreak of *Streptococcus uberis* mastitis. J Dairy Sci 84(3):590–599

Zadoks RN, Allore HG, Barkema HW, Sampimon OC, Wellenberg GJ, Grohn YT, Schukkent YH (2001b) Cow- and quarter-level risk factors for *Streptococcus uberis* and *Staphylococcus aureus* mastitis. J Dairy Sci 84(12):2649–2663

Zadoks RN, Gillespie BE, Barkema HW, Sampimon OC, Oliver SP, Schukken YH (2003) Clinical, epidemiological and molecular characteristics of *Streptococcus uberis* infections in dairy herds. Epidemiol Infect 130(2):335–349

Biofilms and Antimicrobial Resistance in Companion Animals

Thomas W. Maddox

Abstract Bacterial resistance to antimicrobials is a complex interaction of bacterial populations, resistance mechanisms, resistance genes and antimicrobial agents. Although comparatively little research has focused on bacteria from companion animals, many of the mechanisms conferring resistance identified in bacteria originating from humans have also been recognised in bacterial isolates from dogs, cats and horses. In addition to these well documented resistance mechanisms, it has recently been acknowledged that biofilm formation can contribute to the resistance encountered in some bacterial populations. Biofilm-associated resistance appears to be multifactorial, with interaction of specific biofilm resistance mechanisms and potentially other classical antimicrobial resistance mechanisms. Currently, there is incomplete understanding of this complex situation, but work continues to better characterise the processes involved.

1 Introduction

Resistance to antimicrobials is a highly complex interaction of bacterial populations, resistance mechanisms, resistance genes and, most crucially, antimicrobial agents. When considering antimicrobial resistance arising in bacteria from companion animals, unravelling all of these interactions is challenging. However, the bacteria generally involved do not differ significantly for the different animal species, and many of the resistance mechanisms encountered are common to multiple bacterial species.

Antimicrobials are extensively used in human and veterinary medicine, and antimicrobial resistance amongst bacteria is recognised as an increasing problem

T.W. Maddox (✉)
Faculty of Health and Life Sciences, School of Veterinary Science, University of Liverpool, Leahurst Campus, Neston CH64 7TE, UK
e-mail: Thomas.Maddox@liverpool.ac.uk

S.L. Percival et al. (eds.), *Biofilms and Veterinary Medicine*,
Springer Series on Biofilms 6, DOI 10.1007/978-3-642-21289-5_10,

with significant economic implications, as well as increased patient morbidity and mortality (Ogeer-Gyles et al. 2006; Paladino et al. 2002). Inevitably much research has focused on bacteria from humans, especially pathogenic examples, and to a lesser extent on bacteria from food producing animals such as cattle and swine. Antimicrobial resistance in bacteria from companion animals has been less well characterised, but given these species' high level of interaction with humans, such resistance warrants further scrutiny.

2 Antimicrobial Development and Use

Since their initial development in the first half of the twentieth century, antimicrobials have become critical for the treatment of bacterial infectious disease, and their introduction is considered one of the most important of all medical interventions with regard to reducing human morbidity and mortality (Andersson and Hughes 2010). Many of early antimicrobials, such as penicillin and tetracycline, were true "antibiotics"; substances produced by microorganisms that were inhibitory to the growth of other microorganisms (Boothe et al. 1953; Fleming 1932). However, synthetic antimicrobial agents such as the sulphonamides were also developed in this early era (Domagk 1935), and the 20 years from 1945 saw the rapid development of a large number of classes of antimicrobials, including the cephalosprorins, aminoglycosides, marcolides and quinolones (Wright 2007). By the end of this period, the use of antimicrobial therapy had become integral for the control of infectious disease; however, bacteria resistant to the effects of antimicrobial agents were recognised shortly after widespread clinical use commenced (Waksman et al. 1945).

Although the uptake of antimicrobials into clinical use was slower for veterinary medicine, once established, antimicrobials rapidly became widely utilised. Currently in the UK, approximately 400 tonnes of antimicrobial drugs are used therapeutically for the treatment of food-producing and companion animals (VMD 2009). The emergence of resistance in bacteria from animals was noted in the early 1960s, with resistance to several classes of antimicrobial drugs identified in a number of bacterial species (McKay et al. 1965).

3 Antimicrobial Resistance

The breadth and diversity of the mechanisms through which bacteria can achieve resistance to antimicrobials is considerable. This should not be surprising given that such resistance mechanisms represent the culmination of an evolutionary response that significantly predates the clinical use of antimicrobial drugs. For example, genes encoding for resistance have existed in nature, in various forms, for at least as long as there have been microorganisms capable of producing antibacterial substances. Many of the most prevalent resistance genes have arisen from the need of antibiotic-producing organisms to be protected from the toxic effects of their products (D'Costa et al. 2007). Additionally, mutational events that can confer resistance to antimicrobials occur spontaneously within a bacterial population at an approximately constant rate (as with all mutations) (Lederberg and Lederberg 1952). Finally, certain other bacterial adaptations (such as biofilm formation) may fortuitously result in antimicrobial resistance.

4 Antimicrobial Resistance Mechanisms

In individual bacteria, resistance can be achieved by one of three key mechanisms, namely modification or protection of the antimicrobial target, production of antimicrobial inactivating enzymes or exclusion of the antimicrobial molecule from the cell (by decreased permeability or expulsion though efflux pumps). All of these mechanism, or interactions between multiple mechanisms, can result in an increase in the minimum inhibitory concentration (MIC) of an antimicrobial agent required to prevent bacterial growth. For most bacteria, antimicrobial resistance may be considered to be either intrinsic or acquired. Intrinsic resistance is the consequence of a structural or functional trait allowing tolerance of an antimicrobial class or drug for all members of a bacterial group. As such, intrinsic resistance is usually expressed by chromosomal genes and vertically inherited. Acquired resistance is a trait associated with only some strains of a bacterial species or genus and is due to a genetic change in the bacterial genome. This change may be via chromosomal mutation (endogenous resistance) but more commonly by horizontal acquisition of foreign genetic material (exogenous resistance).

Chromosomal mutations within bacteria arise at a relatively constant frequency, estimated at approximately 10^{-9} to 10^{-10} per gene, and the use of antimicrobial agents can allow selection and amplification of pre-existing resistant variants from this background population (Livermore 2003). Emergence and dissemination of resistance through such endogenous mechanisms can be slow, but bacteria can exist in a "hypermutable" state where the mutational rate is increased by up to 200 times (Hall and Henderson-Begg 2006; Livermore 2003). Once endogenous resistance has been achieved, its distribution throughout a bacterial population is accomplished through clonal expansion or proliferation of strains carrying the chromosomal mutations.

When considering exogenous resistance, examples of transferable genetic material involved include resistance plasmids, transposons, integrons and gene cassettes or other mobile genetic elements (Hall and Collis 1995; Roupas and Pitton 1974). These elements frequently encode inactivating enzymes, drug efflux pumps, alternate versions of the antimicrobial target or occasionally factors affording protection of the target molecule (Datta and Kontomichalou 1965; Dejonge et al. 1992; Heikkila et al. 1990; Roberts 2005; Tran and Jacoby 2002). The exchange of genetic material can take place via bacterial transformation, conjugation or transduction and may occur between members of the same species or even across genera.

The above mechanisms are crucial in explaining antimicrobial resistance when considering individual bacteria. However, more recently, it has also been recognised that resistance encountered in groups or populations of bacteria can also be achieved through the production of biofilms. This has led to the realisation that microbial biofilms may be responsible for the recalcitrance of many bacterial infections to standard antimicrobial therapies (Fux et al. 2003; Stewart and Costerton 2001). The resistance exhibited by bacteria in biofilms appears to be largely independent of the previously described mechanisms (Stewart and Costerton 2001), and even susceptible bacteria without a known genetic basis for resistance can become profoundly resistant when incorporated into a biofilm (Anderl et al. 2000).

5 Biofilm Formation

The exact nature of biofilms has been considered elsewhere, but in basic terms a biofilm can be defined as population of cells growing on a surface and enclosed in an exopolysaccharide matrix (Lewis 2001). Bacterial biofilms are often polymicrobial, frequently containing mixed populations of aerobic and anaerobic bacterial species (Hansen et al. 2007).

A number of types of infection have been associated with bacterial biofilms, including pneumonia, osteomyelitis, colitis, vaginitis, urethritis, conjunctivitis, otitis and gingivitis (Adair et al. 1999; Davies 2003; Gristina et al. 1985). Significantly, the placement of medical devices such as intravenous catheters, shunting and stenting devices, urinary catheters and surgical prostheses can result in biofilm formation at implant surfaces (Cormio et al. 1996; Donelli 2006; Ramsay et al. 1989; Sheehan et al. 2004). Whilst biofilm-associated infections have received comparatively little specific attention within veterinary medicine, there is little reason to suggest that the situation is likely to be substantially different; persistent implant-associated infections have long been recognised as a serious problem in companion animals (Ahern et al. 2010; Fine and Tobias 2007; Owen et al. 2004; Smith et al. 1989).

It is clear that infections involving biofilms can frequently be clinically significant and regularly necessitate medical treatment, often including antimicrobial therapy. Therefore, the development of any biofilm-associated antimicrobial resistance can have potentially serious implications for the patient involved. In addition, conventional agar antimicrobial sensitivity testing may underestimate such resistance, leading to false reports of bacterial susceptibility and subsequent treatment failure. The increase in resistance exhibited by bacteria in biofilms can be profound, rendering the cells 10–1,000 times less susceptible to specific antimicrobial agents compared with bacteria in planktonic cultures (Davies 2003; Gilbert et al. 1990).

6 Mechanisms of Antimicrobial Resistance

The majority of the research conducted to date does not indicate a large role for the so-called "classical" mechanisms of antimicrobial resistance in biofilm-associated resistance (Anderson and O'Toole 2008). However, many of the bacteria typically involved in biofilm formation (particularly the pathogenic examples) will carry one or more of such resistance mechanisms, and hence they warrant some further consideration. Additionally, there is some evidence suggesting that some of these mechanisms may have a limited part to play in biofilm-associated resistance. Resistance to the major classes of antimicrobial agents is summarised in Table 1 and detailed with respect to the more significant pathogenic bacteria encountered in companion animals in the following section.

7 β-Lactam Resistance

Resistance to penicillin and other early β-lactam antimicrobial agents (benzyl penicillin and ampicillin) was the earliest form of antimicrobial resistance to be recognised. In fact, the first β-lactamase enzyme (penicillinase) documented in *Escherichia coli* was identified prior to the release of penicillin for clinical use (Abraham and Chain 1988). All β-lactam antimicrobials exert their antibacterial effect through the same basic principle of binding to the bacterial penicillin binding protein (PBP) enzyme, preventing its normal biosynthetic function of assembling the bacterial cell wall. In turn, all β-lactamase enzymes confer bacterial resistance by the basic principle of cleaving the β-lactam's central structural ring, rendering the antimicrobial molecule inactive.

Over 470 novel β-lactamases have been identified and their corresponding encoding genes may be chromosomal, plasmid-borne, or found on transposable elements. The majority of β-lactamases are coded for by *bla* resistance genes, and these can be found on both the bacterial chromosome and plasmids. The predominant plasmid-mediated β-lactamases of Gram-negative rods are the TEM-1, TEM-2 and SHV-1 enzymes, the first of which (TEM-1) was identified in *E. coli* in 1965 by Datta and Kontomichalou. β-Lactamases can also be found in Gram-positive bacteria, with *blaZ* genes being prevalent in staphylococci from dogs and cats (Malik et al. 2007).

Separate from the bla_{TEM} and bla_{SHV} encoded enzymes, AmpC β-lactamases are encoded by *amp*C genes (Sanders et al. 1997; Sanders and Sanders 1988). These enzymes have a broader range of substrates than the TEM/SHV β-lactamases and, for the most part, are unaffected by β-lactamase inhibitors such as clavulanic acid or tazobactam. Initially, they were reported as a cause of β-lactam resistance in species such as *Enterobacter cloacae*, *Citrobacter freundii* and *Pseudomonas aeruginosa*, where mutations had allowed over-expression of their chromosomal *ampC* genes (Sanders 1987). However, these genes did not remain chromosomally

Table 1 Summary of antimicrobial classes used in veterinary medicine, their mechanisms of action, commonly encountered mechanisms of resistance, examples of encoding genes and examples of significant groups or species of resistant bacteria

Antimicrobial class	Antimicrobial mechanism	Antimicrobial Target	Resistance mechanism	Examples	Encoding genes	Example bacteria
β-lactams - penicillins - cephalosporins	Inhibition of bacterial cell wall synthesis	Penicillin binding protein (PBP)	Enzymatic inactivation	β-lactamases	bla_{TEM}, bla_{SHV}, *blaZ*,	*E. coli*, other *Enterobacteriaceae*
			Target modification	Altered PBP2a (reduced affinity)	*mecA*	*Staphylococcus aureus*
Glycopeptides - vancomycin - teicoplanin		Peptidoglycan in cell wall	Target modification	Altered peptidoglycan precursor	*vanA, vanB*	*Enterococcus faecium* and other enterococci
Trimethoprim	Inhibition of folic acid synthesis	Dihydrofolate reductase	Target modification	Altered dihydrofolate reductase	*dfr*	*Enterobacteriaceae*, Staphylococci
Quinolones/fluoroquinolones - enrofloxacin - marbofloxacin	Inhibition of DNA synthesis	DNA gyrase/ topoisomerase	Inactivating enzyme	Acetyltransferases	*AAC(6')-Ib-cr*	*Enterobacteriaceae*
			Target modification	Pentapeptides	*qnr*	
			Drug efflux pumps	Efflux pump	*qepA*	
Tetracyclines - chlortetracycline - oxytetracycline - doxycycline	Inhibition of protein synthesis	30S ribosomal subunit	Drug efflux	Efflux pump	*tetA, tetB*	*Enterobacteriacea*
			Target protection	Ribosomal protection	*tetM*	Staphylococci
Aminoglycosides - gentamicin - neomycin		30S ribosomal subunit	Inactivating enzymes	Acetyltransferases, adenylyltransferases	*AAC(6')-Ib*	*Enterobacteriaceae*
			Target modification	Methylases	*armA, rtmb*	Enterococci, staphylococci
Macrolides/lincosamides - erythromycin - clindamycin - lincomycin		50S ribosomal subunit	Target modification	Alteration of ribosomal binding site	*ermA, ermB*	Staphylococci
			Enzymatic inactivation	Esterases, acetyltransferases	*mph*(C)	Staphylococci
			Drug efflux pumps	Efflux pump	*mef*(A)	
Nitromidazoles - metronidazole	Disruption of DNA	DNA	Enzymatic inactivation	Reductase	*Nim*	Clostridia

bound for long, escaping onto plasmids and being acquired by other species such as *Klebsiella pneumoniae* and *E. coli* (Jacoby and Han 1996; Papanicolaou et al. 1990). *ampC* genes may be plasmid or chromosomal in location (Bradford et al. 1997; Hopkins et al. 2006). High-level resistance occasionally results from chromosomal mutations allowing hyperproduction of these enzymes, but the acquisition of plasmid-borne mechanisms appears more common in *E. coli* (Brinas et al. 2002). Plasmid-mediated AmpC β-lactamases have been identified in *E. coli* and *Enterobacter* species from dogs (Gibson et al. 2010b; Sidjabat et al. 2007) and *E. coli* and *Klebsiella pneumonia* originating from horses (Vo et al. 2007).

7.1 Extended Spectrum β-Lactamases

The increasing prevalence of resistance to the early β-lactam agents necessitated the development of further, related antimicrobials that were stable to hydrolysis by the β-lactamase enzymes. These were the extended-spectrum β-lactams (including the third and fourth generation cephalosporins), which rapidly became a critically important group of drugs in human medicine and to a lesser extent in veterinary medicine (WHO 2007). The group include antibiotics such as the cephalosporins; cefuroxime, cefotaxime and ceftazidime.

Resistance is conferred by extended-spectrum β-lactamase (ESBL) enzymes, with the first identified being SHV-2 (Kliebe et al. 1985). ESBL production confers resistance to most β-lactams, whilst usually retaining susceptibility to the β-lactamase inhibitors (Bradford 2001). The majority of ESBL enzymes are mutations derived from the TEM/SHV β-lactamases described previously, and these have been reported in canine and equine *E. coli* isolates (Briñas et al. 2003; Ewers et al. 2010; Vo et al. 2007). ESBL-producing bacteria of species other than *E. coli*, such as *Salmonell*a, have been identified frequently in food-production animals (Carattoli 2008; Liebana et al. 2004), and although rare in companion animals (Irina et al. 2007), have been reported in *Salmonella* from horses (Rankin et al. 2005).

In 1986, a novel ESBL-type with the ability to hydrolyse cefotaxime was identified in an *E. coli* isolated from a dog's faecal sample (Matsumoto et al. 1988). This enzyme appeared unrelated to previously identified SHV and TEM types, it was subsequently dubbed CTX-M (cefotaximase) and has since been reported in *E. coli* from humans and other animals (Bauernfeind et al. 1990; Hopkins et al. 2006; Liebana et al. 2006). There are now in excess of 60 types recognised, phylogenetically grouped into five clusters based on their amino acid identities: the CTX-M1 cluster (CTX-M1, 3, 10, 11, 12 and 15), the CTX-M2 cluster (CTX-M2, 4, 5, 6, 7 and 20, TOHO-2), the CTX-M8 cluster (CTX-M8), the CTX-M9 cluster (CTX-M9, 13, 14, 16, 17 and 19, TOHO-1) and the CTX-M25 cluster (CTX-M25 and 26) (Pitout and Laupland 2008). Of particular significance is CTX-M-15, which has become globally disseminated clonally through a close association with strains of *E. coli* belonging to sequence type (ST)131, allowing spread in a clonal fashion (Arpin et al. 2009; Naseer et al. 2009; Nicolas-Chanoine et al. 2008; Suzuki et al. 2009; Woodford et al. 2009). In conjunction with this, ESBL-producing bacteria, which were once considered

Table 2 Details and origins of extended spectrum β-lactamase enzymes encountered in Gram-negative bacteria

Enzyme type	Encoding genes	Origins	Number of variants	Common examples
TEM	*bla*$_{TEM}$	Mutations of TEM1,2	Over 100	TEM-10, -26
SHV	*bla*$_{SHV}$	Mutations of SHV 1	Over 50	SHV-2, -5, -12
CTX-M	*bla*$_{CTX-M}$	Mutations of chromosomal *Kluyvera* spp. enzyme	Over 40 (five sub-classes)	CTX-M-1, M-15, M-26
VEB	*bla*$_{VEB}$	Unknown	3	VEB-1
PER	*bla*$_{PER}$	Unknown	2	PER-1, -2
OXA	*bla*$_{OXA}$	Mutants of OXA-2	6	OXA-15, -11

largely nosocomial in origin, are now frequently being found in the wider community and away from healthcare settings (Naseer et al. 2009).

CTX-M enzymes do not appear closely related to SHV or TEM β-lactamases, showing approximately 40% or less identity to those enzymes (Tzouvelekis et al. 2000). However, they do show more similarities with chromosomally encoded β-lactamases of *Kluyvera ascorbata*, suggesting this bacterial species as the possible origin of the CTX-M enzyme (Oliver et al. 2001). Information on the prevalence CTX-M enzymes in bacteria from companion animals is lacking, as it is for most ESBL-producing bacteria, but *ctx-m* genes have been reported in equine bacterial isolates, including *E. coli* (Abraham and Chain 1988; Vo et al. 2007). ST131 *E. coli* have also been reported in companion animals (Pomba et al. 2009), including dogs and horses (Ewers et al. 2010), and in both cases the isolates concerned carried CTX-M-15.

Other variants of ESBL-producing enzymes exist, such as the OXA, PER and VEB types (detailed in Table 2), but these are rarely encountered in bacteria from companion animals, with only OXA having been reported in a collection of *Enterobacter* isolates from dogs (Sidjabat et al. 2007).

7.2 *Meticillin-Resistance in Staphylococci*

Meticillin (previously methicillin) constitutes a special case of β-lactam resistance principally encountered in the staphylococcal species. This β-lactam antimicrobial was first introduced into clinical use for medicine in 1959, largely as a response to extensive resistance to the early penicillins; however, resistance in *Staphylococcus aureus* was identified within a year (Jevons et al. 1961). Since then meticillin-resistance has become increasingly widespread, with studies variously reporting from 25% to over 60% of the *S. aureus* isolates identified in human hospitals to be meticillin-resistant (Fluit et al. 2001; Klevens et al. 2006). The clinical relevance of resistance to meticillin specifically is limited, as the drug is no longer in use. Crucially, however, the mechanism involved provides cross-resistance to most of the other β-lactams and related compounds.

Currently the best characterised meticillin-resistant staphylococcus is meticillin-resistant *S. aureus* (MRSA); however, meticillin-resistance is also encountered in other coagulase positive and coagulase negative staphylococci (MR-CPS and MR-CNS, respectively). Despite this, the presence of resistance in *S. aureus* predominates as a clinical concern, largely due to the pathogenic potential of this species. Low-level meticillin-resistance can be achieved by the hyper-production of classical penicillinases or related β-lactamase enzymes (Gal et al. 2001). However, of more significance is the "intrinsic" resistance primarily mediated by the production of an additional 78-kDa alternative penicillin-binding protein (PBP2a), which has a reduced affinity for the central ring of the β-lactam molecule. This alternative protein is capable of taking over the biosynthetic function of the other conventional penicillin binding proteins (PBP1, 2, 3 and 4) when these become saturated by a β-lactam antimicrobial and so peptidoglycan-layer synthesis is not disrupted (Brown and Reynolds 1980; Dejonge et al. 1992; Georgopapadakou et al. 1982).

Meticillin-resistance in staphylococcal species is principally conferred by the *mec*A gene, which encodes for PBP2a. The *mec*A gene is situated on a large (21–63 kb), mobile genetic element known as the staphylococcal cassette chromosome *mec* (SCC*mec*), which is located near the origin of replication of the *S. aureus* bacterial chromosome (Hanssen and Sollid 2006; Kuhl et al. 1978). Other regulatory and ancillary genes are situated on the cassette, and downstream of the *mec* region is a variable segment of DNA that terminates with an insertion-like element (IS431), which serves as a target for the recombination of other mobile genetic elements. Consequently, the SCC*mec* unit can integrate plasmids and transposons, acquiring genes for resistance to other antimicrobials.

Several differently sized variants of the SCC*mec* cassette have been identified and form the basis of a typing scheme for MRSA. Currently, seven main types of SCC*mec* (types I to VII) have been described, defined by their *mec* complex and the associated *ccr* genes (Kim et al. 2007). Whilst only SCC*mec* II and III harbour genes for resistance to antimicrobials other than β-lactams (due to incorporation of plasmid and transposon genes), other sites on the *S. aureus* chromosome, or additional plasmids, can be the location of further resistance genes. As such, MRSA is frequently resistant to other antimicrobial classes such as the fluoroquinolones, tetracyclines, aminoglycosides and potentiated sulphonamides (Blumberg et al. 1991; Saroglou et al. 1980; Skinner et al. 1988).

As previously indicated, meticillin-resistance amongst staphylococci is not solely restricted to *S. aureus* and the SCC*mec* cassette is frequently identified in coagulase-negative members of the genus, such as *S. epidermidis* and *S. sciuri*. Some meticillin resistant coagulase-negative staphylococci may be capable of causing disease under some circumstances or in immunocompromised or animals otherwise at-risk of opportunistic infections (Corrente et al. 2009; Pyorala and Taponen 2009). Perhaps more significantly, the potential for transfer of the SCC*mec* cassette may render such coagulase-negative species a reservoir of meticillin-resistance for *S. aureus*.

The carriage of MRSA in companion animals has been assessed; prevalence estimates for healthy dogs range from 0 to 4% (Baptiste et al. 2005; Boost et al. 2008; Weese and van Duijkeren 2010) and between 0 and 4.7% for horses (Baptiste et al. 2005; Vengust et al. 2006; Weese et al. 2005). Prevalence generally appears higher for hospitalised animals (Loeffler et al. 2005; van Duijkeren et al. 2010).

8 Tetracycline Resistance

The tetracyclines represent the original broad-spectrum antimicrobials. Tetracycline is bacteriostatic, reversibly inhibiting bacterial protein synthesis by preventing aminoacyl-tRNA from binding to the bacterial ribosome, with the other tetracyclines (oxytetracycline, chlortetracycline and doxycycline), functioning in a similar fashion (Roberts 1996). All tetracyclines are actively concentrated within bacterial cells, explaining their selective action against bacteria (Chopra and Roberts 2001).

The majority of tetracycline resistance mechanisms are acquired and conferred by one or more of over 30 currently described *tet* genes. These genes encode one of three mechanisms of resistance: a drug efflux pump, an enzyme for inactivation of the antimicrobial molecule or a ribosomal protection factor (Chopra and Roberts 2001). Efflux mechanisms appear most prevalent within Enterobacteriaceae, while ribosomal protection mechanisms are more common among Gram-positive organisms (Huys et al. 2004; Riesen and Perreten 2009; Tuckman et al. 2007). There have only been three genes identified that encode for inactivating enzymes and currently these appear to be neither clinically relevant nor widespread (Roberts 1996). The various *tet* genes can be located on plasmids, transposons and integrons (Chopra and Roberts 2001; Roberts 1996).

Tetracycline resistance is widely reported in Enterobacteriaceae from animals (Bryan et al. 2004; Harihara and Barnum 1973). A recent study of over 400 animal isolates showed a 14.5% prevalence of resistance (Grobbel et al. 2007), with previous studies showing a similar, or higher, prevalence (Bryan et al. 2004; Dunowska et al. 2006; Saenz et al. 2001). In a study of non-pathogenic *E. coli*, animal isolates a large number of *tet* genes were found, with the most prevalent being *tet*(B) (63% of isolates) and *tet*(A) (35%), and also *tet*(C), *tet*(D) and *tet*(M) (Bryan et al. 2004). Most of the isolates carried multiple genes and 97% of isolates carried at least one *tet* gene. A similar study of a smaller number of *E. coli* from various animals again showed high prevalence of *tet*A and *tet*B genes among these isolates (Saenz et al. 2004).

Various *tet* genes have also been less frequently identified in other bacterial species from companion animals; *tet*(K) and *tet*(M) in staphylococci (Schnellmann et al. 2006) and *tet*(L) in enterococci (Moura et al. 2010) from horses and *tet*(K) in staphylococci from dogs (Schwarz et al. 2008). However, a high prevalence of *tet*(A) and *tet*(B) genes was documented in canine *Clostridium perfringens* isolates (Kather et al. 2006).

9 Trimethoprim Resistance

Due to its low cost and the comparative ease of administration possible with oral formulations, trimethoprim (combined with sulfamethoxazole) has become amongst the most widely prescribed of antimicrobial drugs in the horse and to a lesser extent in other companion animals (van Duijkeren et al. 1994). It is a synthetic structural analogue of folic acid and exerts its effect by preventing the reduction of dihydrofolate to tetrafolate via competitive inhibition of the dihydrofolate reductase (DHFR) enzyme. Due to differences in the enzyme's structure between species, the affinity of trimethoprim for mammalian DHFR is low, and as a consequence it inhibits bacterial cells preferentially (Matthews et al. 1985).

High-level resistance to trimethoprim can be related to mutations in the chromosomal *dfr* gene, but this is relatively rarely reported in Enterobacteriaceae (Datta et al. 1979). The more frequently encountered mechanism is the production of an additional, alternative, DHFR enzyme which is less sensitive than the chromosomal enzyme to inhibition, with this being the most prevalent mechanism in *E. coli* (Heikkila et al. 1990). The alternative enzyme is present in addition to the native sensitive enzyme and provides an alternative synthesis route for tetrafolate. Such enzymes are encoded by *dfr* genes and often located on plasmids or transposons, allowing wide horizontal spread within many bacterial species (Blahna et al. 2006).

There are several types of transferable resistant DHFR enzymes, through phylogeny analysis these have been grouped into three families on the basis of amino acid sequence identities, with the first two families containing the majority of enzymes (Huovinen et al. 1995; White and Rawlinson 2001). Family 1 includes enzyme types DHFR I, V, VI, VII and Ib and are encoded by *dfrA* genes. Family 2 contains enzymes DHFR IIa, IIb and IIc, encoded by *dfrB* genes. It appears that *dfra1* is the most commonly acquired *dfr* gene (Blahna et al. 2006; Towner et al. 1994). Gene cassettes incorporated into integrons play a significant role in the trimethoprim resistance seen in Gram-negative bacteria; 15 of the 19 identified transferable DHFR enzymes are encoded by genes that form part of a cassette (Blahna et al. 2006; Leverstein-van Hall et al. 2003; White et al. 2000). Identification of trimethoprim resistance in an isolate appears predictive for the presence of integrons, and, as integrons often incorporate multiple resistance genes, trimethoprim resistance is often associated with multi-drug resistance (Leverstein-van Hall et al. 2003).

Trimethoprim resistance is well recognised in *E. coli* isolates from animals, including horses and dogs (Bucknell et al. 1997; Saenz et al. 2001; Vo et al. 2007). In a series of 104 pathogenic isolates causing metritis examined from horses, 15% were trimethoprim resistant, the highest prevalence of resistance found excluding streptomycin (Albihn et al. 2003). Several different *dfr* genes have been characterised in trimethoprim resistant companion animal *E. coli*, including *dfrA1*, *dfrA12* and *dfrA17* (Saenz et al. 2004; Vo et al. 2007). In Gram-positive bacteria, *dfrG* has been identified to have a high prevalence of over 90% in canine meticillin-resistant *Staphylococcus pseudintermedius* (Perreten et al. 2010), and *dfrA* and *dfrD* have been verified in equine staphylococcal isolates (Schnellmann et al. 2006).

10 Aminoglycoside Resistance

Aminoglycosides such as gentamicin and amikacin inhibit bacterial protein synthesis by irreversibly binding to the bacterial 30S ribosomal subunit at several domains (including the 16S rRNA subunit, a highly conserved region of ribosomal RNA), resulting in cell death (Kotra et al. 2000). Resistance to aminoglycosides is most frequently the result of inactivation of the antimicrobial molecule by aminoglycoside modifying enzymes such as acetyltransferases, phosphorylases and adenylyltransferases, of which over 50 have been identified (Kotra et al. 2000; Shaw et al. 1993). These enzymes, such as the 6′-*N*-acetyltransferase, AAC(6′)-Ib, are frequently located on plasmids and have become geographically widespread and disseminated into multiple bacterial species (Tran Van Nhieu et al. 1992).

However, resistance can also be achieved by modification of the bases of 16S rRNA subunit responsible for the binding between the ribosome and aminoglycoside molecule. Such modifications can include substitution or methylation, leading to a loss of affinity for the antimicrobial and hence high-level resistance. The substitution/methylation of the bases is carried out by methylating enzymes and genes encoding several such enzymes (16S rRNA methylases) have been identified on plasmids in the Enterobacteriaceae (Galimand et al. 2003; Yan et al. 2004). Two of the most widespread of these genes (*arm*A and *rtm*B) have been identified in *E. coli* isolates, but only at a quite low prevalence (Yan et al. 2004). Both *arm*A and *rtm*B have been identified in *E. coli* from pigs in Spain and China, where it is speculated that heavy agricultural use of aminoglycosides may be implicated in their emergence (Chen et al. 2007; Gonzalez-Zorn et al. 2005). Gentamicin is also quite widely used in equine species, but there are no reports of these enzymes in bacteria originating from horses or other companion animals.

11 Quinolone and Fluoroquinolone Resistance

The quinolone and fluoroquinolone antimicrobials exert their antibacterial action through the binding and inhibition of two bacterial topoisomerase enzymes: DNA gyrase (topoisomerase II) and topoisomerase IV. These enzymes are essential for bacterial cell replication; DNA gyrase catalyses the negative supercoiling of double-stranded DNA, altering its topology (Reece et al. 1991) and topoisomerase IV is involved in the segregation of replicated daughter chromosomes during DNA replication (Drlica and Zhao 1997). Both DNA gyrase and topoisomerase IV are tetrameric proteins composed of A and B subunits, encoded by the genes *gyrA* and *gyrB* (for DNA gyrase) and *parC* and *parE* (for topoisomerase IV).

High-level quinolone resistance in *E. coli* and other Enterobacteriaceae largely results from chromosomal mutations of the quinolone resistance-determining region (QRDR) of the *gyrA* and *parC* genes coding for the A subunits of both enzymes. Other mechanisms such as changes in outer membrane permeability and efflux proteins (or in their regulatory mechanisms) have also been identified, with all of these mechanisms being chromosomally mediated (Ruiz 2003). However, more recently plasmid-mediated resistance mechanisms have also been documented, beginning with the identification of the Qnr determinant in Enterobacteriaceae (Martinez-Martinez et al. 1998; Nordmann and Poirel 2005; Tran and Jacoby 2002).

11.1 Plasmid-Mediated Fluoroquinolone Resistance

Qnr proteins are pentapeptides that appear to bind to the bacterial DNA gyrase enzyme, protecting it from the actions of the quinolone antibiotics (Tran and Jacoby 2002; Tran et al. 2005). This plasmid-mediated resistance mechanism only confers relatively low-level quinolone resistance; however, this marginal increase may be sufficient to facilitate selection of higher level of resistance by allowing increased emergence of *chromosomal* QRDR mutations (Martínez-Martínez et al. 2003). Effectively Qnr proteins raise the mutant prevention concentration (MPC) of recipients with respect to the fluoroquinolones, meaning that least-susceptible single-step mutants can survive and potentially lead to the development of mutants with full clinical resistance. The original function of these Qnr proteins (i.e. in the absence of any antimicrobials) has yet to be fully elucidated, but chromosomal genes coding for related pentapeptides with sequence similarity have been found in Gram-positive and Gram-negative bacteria (Strahilevitz et al. 2009).

Several Qnr proteins have been identified since 1998: QnrA, QnrB and QnrS are encoded by the *qnr*A, *qnr*B and *qnr*S genes, respectively (Hata et al. 2005; Jacoby et al. 2006; Martinez-Martinez et al. 1998). More recently, two further types have been documented; QnrC in a clinical isolate of *Proteus mirabilis* (Wang et al. 2009) and QnrD in *Salmonella* (Cavaco et al. 2009), both of these originating from China. The Qnr proteins appear only distantly related to one another, with QnrB and QnrS showing merely 40 and 59% amino acid identity with QnrA, respectively (Nordmann and Poirel 2005). A number of sub-variants of the first three Qnr proteins are now recognised, and these tend to differ by only a small number of amino acid substitutions. All of the first three of the *qnr* genes groups have been reported in *E. coli* in humans and food-production animals (Cavaco et al. 2007; Corkill et al. 2005; Jiang et al. 2008; Kuo et al. 2009; Martínez-Martínez et al. 2003) but not from equine *E. coli* and only in *Enterobacter* isolates from cats and dogs (Gibson et al. 2010a). *qnrC* or *qnrD* have yet to be identified in *E. coli* from any animal species (Seo et al. 2010).

Until recently, it was believed that enzymatic modification of fluoroquinolones by bacteria was not a significant resistance mechanism, due to the fully synthetic nature of these antimicrobials. However, the identification of transconjugants obtained from a group of clinical *E. coli* carrying *qnrA* genes with a higher than expected ciprofloxacin MIC led to the detection of a further novel fluoroquinolone resistance mechanism. This mechanism was a variant of a previously identified aminoglycoside acetyltransferase enzyme [AAC(6′)-Ib]. Subsequently named AAC (6′)-Ib-cr (Robicsek et al. 2006), it conferred resistance to ciprofloxacin but not nalidixic acid. Similar to *qnr*-mediated resistance, although the degree of resistance conferred is small it enables selection of mutants with higher-level resistance. Furthermore, this mechanism appears to act additively with *qnr*-mediated resistance, allowing clinically significant fluoroquinolone resistance to be achieved through the acquisition of a single plasmid (Robicsek et al. 2006).

A third plasmid-mediated resistance mechanism has recently been documented, with the identification of a plasmid-borne quinolone efflux pump gene (*qepA*) carried by *E. coli* (Perichon et al. 2007). A high degree of similarity to efflux pumps from actinomycetes species has suggested an environmental bacterial origin for this mechanism (Perichon et al. 2007; Yamane et al. 2007). Again, this mechanism appears to only provide a low level of fluoroquinolone resistance. Whilst *qnr* and *aac(6′)-Ib-cr* determinants appear to widely distributed in human clinical isolates of Enterobacteriaceae, less information exists regarding the prevalence of this newly reported efflux mechanism.

Plasmid-mediated quinolone resistance seems to have an association with ESBL production. Many enterobacterial isolates with *qnr* genes also carry ESBL genes, which in some cases may be located on the same plasmid (Jacoby et al. 2006; Jiang et al. 2008; Lavigne et al. 2006). CTX-M-15 producing *E. coli* have been identified as carrying AAC(6′)-Ib-cr, encoded on the same plasmid, on multiple occasions (Karisik et al. 2006; Machado et al. 2006) and CTX-M-15 isolates from small animals have been shown to carry *aac(6′)-Ib-cr* and *qnrB* genes (Pomba et al. 2009). A high prevalence of all three plasmid-mediated resistances were identified in *E. coli* from some companion and food-producing animals in China (Aarestrup 2005) and in *Enterobacter* species from dogs (Gibson et al. 2010a), but these mechanisms have not been reported in bacteria from horses.

12 Glycopeptide Resistance

Glycopeptide antimicrobials, such as vancomycin and teicoplanin, are crucial drugs to human medicine for the treatment of severe, resistant Gram-positive infections and, partly because of this, they are uncommonly used for the treatment of animals. Rather than inhibiting enzymes involved in cell wall synthesis, these agents act by binding to the D-alanine–D-alanine termini of peptidoglycan precursors, preventing transglycosylation and transpeptidation of the bacterial cell wall (Arthur et al. 1996a, b).

Gram-negative bacteria are intrinsically resistant to glycopeptides, as the large molecules are unable to permeate the outer membrane (Woodford et al. 1995). Acquired resistance in Gram-positive species is most commonly achieved by the modification of the target site, with alteration of terminal dipeptide to D-alanine–D-lactate or D-alanine–D-serine reducing the affinity for the antimicrobial molecule. Vancomycin resistance is most commonly encountered in enterococcal species, where the *van* gene cluster is responsible for the production of the modified peptidoglycan precursor (Arthur et al. 1993; Leclercq et al. 1988). Several different *van* gene cluster named A–L have been identified, with *vanA* and *vanB* being the most prevalent in clinical isolates (Rossolini et al. 2010).

Glycopeptide resistance does not appear to be very prevalent in bacteria from companion animals (Poeta et al. 2005; Turkyilmaz et al. 2010), but *vanC* genes have been identified in isolates from animals (Rice et al. 2003), and *vanA* genes in enterococci from dogs and horses (Devriese et al. 1996; Moura et al. 2010). *van* genes can also be identified in other Gram-positive pathogens such as *S. aureus* (Weigel et al. 2007), but vancomycin resistant *S. aureus* have yet to be identified in companion animals.

13 Macrolide and Lincosamide Resistance

Several antimicrobials of this group are employed in veterinary medicine, including erythromycin, tylosin, clindamycin and lincomycin. Most Gram-negative bacilli are intrinsically resistant to these antimicrobials due to their low permeability across the outer membrane. In Gram-positive species, the agents work by reversibly inhibiting peptide bond formation by binding to the 50S ribosomal subunit during protein synthesis (Gaynor and Mankin 2003).

Resistance can result from target site modification, enzymatic inactivation and drug efflux. Target site modification was the first mechanism to be identified, with various plasmid- or transposon-borne *erm* genes encoding for rRNA methylases which dimethylate an adenine residue in a conserved region of the 23s rRNA (Leclercq and Courvalin 1991; Weisblum 1995). Genes of *erm* class A and B have been identified in enterococci from cats and dogs (Turkyilmaz et al. 2010) and *ermA*, *ermB* and *ermC* in staphylococci from the same species (Luethje and Schwarz 2007). This study identified *ermB* gene as the predominant macrolide resistance gene in *Staphylococcus intermedius* and streptococci. Erythromycin resistant staphylococci were not identified in a survey of 100 horses in Denmark (Bagcigil et al. 2007), but a study involving hospitalised horses identified *ermB* and *ermC* genes (Schnellmann et al. 2006). A very low prevalence of *ermB* and *ermQ* genes was identified in canine *Clostridium perfringens* isolates (Kather et al. 2006).

A variety of macrolide-inactivating enzymes have been documented, including esterases, acetyltransferases and nucleotidyltransferases, all of which are located on plasmids (Roberts et al. 1999). Several genes encoding for drug efflux pumps have also been identified. These mechanisms appear to be less common that target modification in bacteria from animals; efflux genes *mef*(A) and *msr*(D) and inactivating enzyme genes *mph*(C) and *lnu*(A) all had a low prevalence in streptococci and staphylococci from companion animals (Luethje and Schwarz 2007; Schnellmann et al. 2006).

14 Rifampicin Resistance

Rifampicin is the only antimicrobial of this group used occasionally in veterinary medicine, due to its effectiveness against mycobacterial species. It is also sporadically used it the treatment of MRSA infections. Their antimicrobial activity is due to inhibition of bacterial RNA polymerase; rifampicin binds to conserved amino acids in the active centre of the enzyme and blocks transcription initiation. Much of the encountered resistance is due to mutations of these amino acids (Tupin et al. 2010). These mutations often occur with high frequency, which compels the use of rifampicin almost exclusively in combination with other drugs.

Rifampicin resistance in staphylococci from companion animals appears to be rare (Kizerwetter-Swida et al. 2009; Vanni et al. 2009), but seems to be relatively common in enterococci (Ossiprandi et al. 2008).

14.1 Nitromidazole Resistance

Metronidazole is the only nitromidazole antimicrobial agent commonly employed in the veterinary treatment of animals. It is selectively absorbed by anaerobic bacteria and so has relatively little effect on aerobic cells. Once uptaken, the molecule is reduced to a toxic compound (a short-lived radical anion), which damages and destabilises bacterial DNA (Leiros et al. 2004). Resistance can be conferred by *Nim* genes that encode for reductase enzymes, which convert the antimicrobial agent to a non-bactericidal amine (Leiros et al. 2004).

Metronidazole resistance has been identified in some clostridial species from horses but resistance genes were not characterised (Baverud et al. 2003; Jang et al. 1997). *Nim* genes were not identified in metonidazole resistant *Clostridium perfringens* isolates from 124 dogs (Kather et al. 2006).

14.2 Resistance Arising in Biofilms

For the most part, the antimicrobial resistance demonstrated by bacteria in biofilms appears quite different from that exhibited by free-living or planktonic bacteria. The observation that a wide array of microorganisms in biofilms display increased resistance initially led to the suggestion that there might be a single, universal mechanism of biofilm-associated resistance. However, the exact means through which resistance arises has yet to be fully elucidated, and it seems likely that biofilm resistance is multifactorial, resulting from the integration of several different antimicrobial resistance mechanisms. Some of these mechanisms are physical factors directly related to the presence of the biofilm and others relate to the behaviour displayed by bacteria in biofilms. Firstly, the biofilm matrix itself

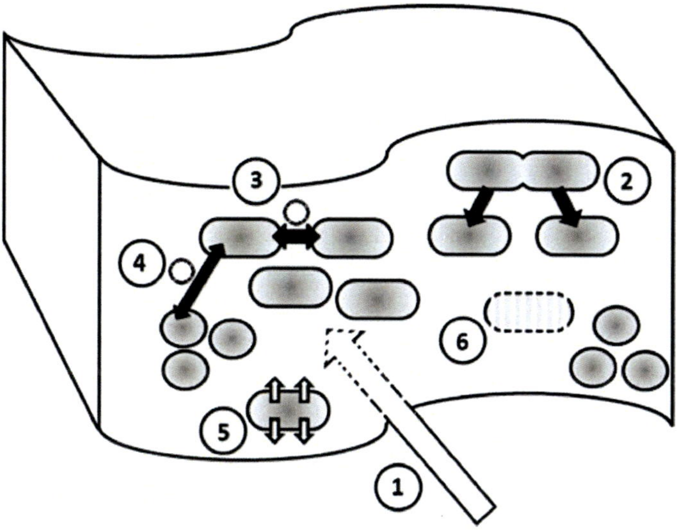

Fig. 1 Summary of the means through which antimicrobial resistance may arise in biofilms (1) decreased penetration of antimicrobial; (2) decreased bacterial growth rate; (3) exchange of antimicrobial resistance determinants between bacteria of same species; (4) exchange of antimicrobial resistance determinants between bacteria of different species; (5) induced expression or synergistic action of bacterial antimicrobial resistance mechanisms; (6) entry of bacteria into a protected "persister" state

may limit penetration of the antimicrobial agent. Secondly, the formation of microenvironments within the biofilm with reduced levels of oxygen or nutrients can slow the growth of bacteria and reduce their antimicrobial susceptibility. Thirdly, sub-populations of the bacteria within biofilms can differentiate into a "persistent" state with reduced antimicrobial sensitivity. Finally, there exist several resistance genes identified as being specifically regulated in biofilms. All of these mechanisms are summarised in Fig. 1 and considered in more detail below.

14.3 Reduced Antimicrobial Penetration

The biofilm can act as a physical barrier preventing passage of some antimicrobial molecules; for example, the negatively charged exopolysaccharide can effectively prevent positively charged aminoglycoside molecules from penetrating by binding the antimicrobial (Ishida et al. 1998; Shigeta et al. 1997). If the antimicrobial does penetrate further into the biofilm, then reduced diffusion through the matrix may mean that the agent fails to reach a sufficient concentration to affect bacteria in the deeper layers of the biofilm. The biofilm effectively acts as a diffusion barrier. However, this mechanism does not seem to be universal and does not apply to all biofilms or to all antimicrobial agents. Fluoroquinolones such as ciprofloxacin appear to diffuse freely across the biofilm matrix of *Klebsiella pneumoniae* and staphylococcal species (Anderl et al. 2000; Singh et al. 2010) and tetracycline through pathogenic *E. coli* biofilms (Stone et al. 2002). Additionally, structural analysis has shown that many biofilms contain multiple aqueous channels, which are unlikely to significantly impede antimicrobial molecule diffusion (Mangalappalli-Illathu et al. 2008; Wimpenny et al. 2000).

14.4 Reduced Growth

To be fully effective, many antimicrobial agents (such as some cephalosporins, aminoglycosides and fluoroquinolones) require that the target bacteria be undergoing growth, as they preferentially act on cells undergoing division and multiplication. The complex microenvironment created within a biofilm can result in specific regions having differing levels of oxygen and essential nutrients. Gradients in concentration of nutrients across biofilms have been documented, and anaerobic conditions have been identified in deep biofilm layers (de Beer et al. 1994). Bacteria present in these deprived regions of the biofilms will exhibit decreased growth rates and reduced metabolic rates have been verified in the central region of microcolonies within biofilms (Sternberg et al. 1999). Consequently, the bacteria in such regions will be less affected by these specific antimicrobials as they are undergoing reduced cell division. This mechanism may go some way to explaining why older biofilms typically demonstrate increasing levels of resistance, as more established biofilms have a more heterogeneous structure (Monzon et al. 2002).

14.5 Persistent Bacteria

A further mechanism speculated to play a role in the increased resistance seen in bacteria in biofilms is the development of so-called "persister" bacteria. This occurs when a sub-population of microorganisms enter a highly protected phenotypic state, similar to spore formation, which renders them more resistant to the actions of antimicrobial agents. Some supporting evidence for this is provided by the resistance seen in bacteria in newly formed biofilms (which are too thin to afford much physical protection and lack the varying microenvironments seen in more established biofilms) (Das et al. 1998). Additionally, it appears that in many biofilms most bacteria are rapidly killed on exposure to antimicrobials but a small number survive (Brooun et al. 2000; Goto et al. 1999). Persister formation may go some way to explaining the clinical problems encountered when treating infections involving biofilms; the majority of bacteria may be killed, but a nidus remains to re-establish infection when antimicrobial therapy ceases.

14.6 Resistance Genes

Although the conventional antimicrobial resistance mechanisms and genes seem unlikely to be primarily responsible for biofilm-associated resistance, in some instances such mechanism may act synergistically with biofilm formation. For example, in biofilms that contain β-lactamase producing bacteria, penetration of β-lactam antimicrobials such as ampicillin can be restricted as the enzyme will degrade the antimicrobial molecule (Anderl et al. 2000). It also appears that resistance mechanisms can be induced in biofilms; the *ampC* gene of *P. aeruginosa* can be strongly induced by the presence of the antimicrobial imipenem (Bagge et al. 2004). Further studies have identified the induction of a multi-drug resistant state in biofilm-forming *E. coli* through interplay between tetracycline resistance efflux pump *tetA(C)* and ampicillin resistance ($bla_{\text{TEM}-1}$) genes (May et al. 2009).

Conflicting evidence exists regarding the involvement of multi-drug efflux pumps to biofilm resistance; one study identified none of the four most well-characterised efflux pumps present in *P. aeruginosa* as contributing significantly to the resistance phenotype of the biofilm (De Kievit et al. 2001). However, others have identified a novel efflux pump that is more highly expressed in *P. aeruginosa* in biofilms than in planktonic cultures and its absence resulted in increased sensitivity to tobramycin, gentamicin and ciprofloxacin (Zhang and Mah 2008).

The simple extra survival time afforded bacteria within biofilms can allow for the development or acquisition of other, conventional resistance mechanisms. The de-repression of chromosomal β-lactamases results in the rapid development of resistance in biofilm-forming *P. aeruginosa* (Bagge et al. 2000). The biofilm environment also provides a suitable setting for the transfer of resistance determinants between bacterial strains (Ando et al. 2009; Kajiura et al. 2006). In particular, the polymicrobial nature of some biofilms may allow transfer from non-pathogenic to pathogenic species, with consequent significant clinical implications (Weigel et al. 2007). There is also in vitro and in vivo evidence that bacteria within biofilms can exist in a hypermutable state (Driffield et al. 2008; Oliver et al. 2000). Genes conferring protection against oxidative damage are down-regulated, increasing the mutability of the organism and creating conditions for the emergence of antimicrobial-resistant bacteria.

14.7 Further Considerations

The situation is further complicated by the fact that some antimicrobial agents can actually stimulate the production of the biofilm matrix. Low doses of tetracycline and erythromycin-activated expression of genes involved in exopolysaccharide formation in *S. epidermidis* (Rachid et al. 2000), and a similar effect has been noted with other biofilm-related genes in *E. coli* (Sailer et al. 2003). Furthermore, some classical resistance mechanisms may, in fact, contribute to the development of biofilms; inactivation of efflux pumps by pump inhibitors reduces biofilm formation, indicating that efflux systems may actually be required for biofilm formation (Kvist et al. 2008).

15 Conclusions

The various mechanisms through which bacteria achieve antimicrobial resistance are diverse. The resistance found in bacteria from companion animals is extensive, and mirrors that seen in bacteria from most animal species, including humans. Many of the significant resistant mechanisms that have been described in bacteria originating from numerous other species have also been identified in a range of bacterial species from dogs, cats and horses. In the clinical situation, the formation of biofilms undoubtedly contributes significantly to the resistance encountered in some circumstances. Biofilm-associated resistance would appear to be multifactorial, with interaction of specific biofilm mechanisms and perhaps other classical resistance mechanism. Although there is limited understanding of this complex state of affairs, clearly further work is required for a greater characterisation of all the processes involved.

References

Aarestrup FM (2005) Veterinary drug usage and antimicrobial resistance in bacteria of animal origin. Basic Clin Pharmacol Toxicol 96:271–281

Abraham EP, Chain E (1988) An enzyme from bacteria able to destroy Penicillin (Reprinted From Nature, Vol 146, Pg 837, 1940). Rev Infect Dis 10:677–678

Adair CG, Gorman SP, Feron BM, Byers LM, Jones DS, Goldsmith CE, Moore JE, Kerr JR, Curran MD, Hogg G, Webb CH, McCarthy GJ, Milligan KR (1999) Implications of endotracheal tube biofilm for ventilator-associated pneumonia. Intensive Care Med 25: 1072–1076

Ahern BJ, Richardson DW, Boston RC, Schaer TP (2010) Orthopedic infections in equine long bone fractures and arthrodeses treated by internal fixation: 192 cases (1990–2006). Vet Surg 39: 588–593

Albihn A, Baverud V, Magnusson U (2003) Uterine microbiology and antimicrobial susceptibility in isolated bacteria from mares with fertility problems. Acta Vet Scand 44:121–129

Anderl JN, Franklin MJ, Stewart PS (2000) Role of antibiotic penetration limitation in *Klebsiella pneumoniae* biofilm resistance to ampicillin and ciprofloxacin. Antimicrob Agents Chemother 44:1818–1824

Anderson GG, O'Toole GA (2008) Innate and induced resistance mechanisms of bacterial biofilms. Curr Top Microbiol Immunol 322:85–105

Andersson DI, Hughes D (2010) Antibiotic resistance and its cost: is it possible to reverse resistance? Nat Rev Microbiol 8:260–271

Ando T, Itakura S, Uchii K, Sobue R, Maeda S (2009) Horizontal transfer of non-conjugative plasmid in colony biofilm of *Escherichia coli* on food-based media. World J Microbiol Biotechnol 25:1865–1869

Arpin C, Quentin C, Grobost F, Cambau E, Robert J, Dubois V, Coulange L, Andre C, Sci Comm O (2009) Nationwide survey of extended-spectrum beta-lactamase-producing Enterobacteriaceae in the French community setting. J Antimicrob Chemother 63:1205–1214

Arthur M, Molinas C, Depardieu F, Courvalin P (1993) Characterization of Tn1546, a Tn3-related transposon conferring glycopeptide resistance by synthesis of depsipeptide peptidoglycan precursors in *Enterococcus faecium* BM4147. J Bacteriol 175:117–127

Arthur M, Reynolds P, Courvalin P (1996a) Glycopeptide resistance in enterococci. Trends Microbiol 4:401–407

Arthur M, Reynolds PE, Depardieu F, Evers S, DutkaMalen S, Quintiliani R, Courvalin P (1996b) Mechanisms of glycopeptide resistance in enterococci. J Infect 32:11–16

Bagcigil FA, Moodley A, Baptiste KE, Jensen VF, Guardabassi L (2007) Occurrence, species distribution, antimicrobial resistance and clonality of methicillin- and erythromycin-resistant staphylococci in the nasal cavity of domestic animals. Vet Microbiol 121:307–315

Bagge N, Ciofu O, Skovgaard LT, Hoiby N (2000) Rapid development in vitro and in vivo of resistance to ceftazidime in biofilm-growing *Pseudomonas aeruginosa* due to chromosomal beta-lactamase. APMIS 108:589–600

Bagge N, Schuster M, Hentzer M, Ciofu O, Givskov M, Greenberg EP, Hoiby N (2004) *Pseudomonas aeruginosa* biofilms exposed to imipenem exhibit changes in global gene expression and beta-lactamase and alginate production. Antimicrob Agents Chemother 48:1175–1187

Baptiste KE, Williams K, Willams NJ, Wattret A, Clegg PD, Dawson S, Corkill J, O'Neill T, Hart CA (2005) Methicillin resistant staphylococci in companion animals. Emerg Infect Dis 11: 1942–1944

Bauernfeind A, Grimm H, Schweighart S (1990) A new plasmidic cefotaximase in a clinical isolate of *Escherichia coli*. Infection 18:294–298

Baverud V, Gustafsson A, Franklin A, Aspan A, Gunnarsson A (2003) *Clostridium difficile*: prevalence in horses and environment, and antimicrobial susceptibility. Equine Vet J 35:465–471

Blahna MT, Zalewski CA, Reuer J, Kahlmeter G, Foxman B, Marrs CF (2006) The role of horizontal gene transfer in the spread of trimethoprim-sulfamethoxazole resistance among uropathogenic *Escherichia coli* in Europe and Canada. J Antimicrob Chemother 57:666–672

Blumberg HM, Rimland D, Carroll DJ, Terry P, Wachsmuth IK (1991) Rapid development of ciprofloxacin resistance in methicillin-susceptible and methicillin-resistant *Staphylococcus aureus*. J Infect Dis 163:1279–1285

Boost MV, O'Donoghue MM, James A (2008) Prevalence of *Staphylococcus aureus* carriage among dogs and their owners. Epidemiol Infect 136:953–964

Boothe JH, Morton J, Petisi JP, Wilkinson RG, Williams JH (1953) Tetracycline. J Am Chem Soc 75:4621

Bradford PA (2001) Extended-spectrum beta-lactamases in the 21st century: characterization, epidemiology, and detection of this important resistance threat. Clin Microbiol Rev 14:933–951

Bradford PA, Urban C, Mariano N, Projan SJ, Rahal JJ, Bush K (1997) Imipenem resistance in *Klebsiella pneumoniae* is associated with the combination of ACT-1, a plasmid-mediated AmpC beta-lactamase, and the loss of an outer membrane protein. Antimicrob Agents Chemother 41:563–569

Brinas L, Zarazaga M, Saenz Y, Ruiz-Larrea F, Torres C (2002) Beta-lactamases in ampicillin-resistant *Escherichia coli* isolates from foods, humans, and healthy animals. Antimicrob Agents Chemother 46:3156–3163

Briñas L, Moreno MA, Teshager T, Zarazaga M, Sáenz Y, Porrero C, Dominguez L, Torres T (2003) Beta-lactamase characterization in *Escherichia coli* isolates with diminished susceptibility or resistance to extended-spectrum cephalosporins recovered from sick animals in Spain. Microb Drug Resist 9:201–209

Brooun A, Liu SH, Lewis K (2000) A dose-response study of antibiotic resistance in *Pseudomonas aeruginosa* biofilms. Antimicrob Agents Chemother 44:640–646

Brown DFJ, Reynolds PE (1980) Intrinsic resistance to beta-lactam antibiotics in *Staphylococcus aureus*. FEBS Lett 122:275–278

Bryan A, Shapir N, Sadowsky MJ (2004) Frequency and distribution of tetracycline resistance genes in genetically diverse, nonselected, and nonclinical *Escherichia coli* strains, isolated from diverse human and animal sources. Appl Environ Microbiol 70:2503–2507

Bucknell DG, Gasser RB, Irving A, Whithear K (1997) Antimicrobial resistance in *Salmonella* and *Escherichia coli* isolated from horses. Aust Vet J 75:355–356

Carattoli A (2008) Animal reservoirs for extended spectrum beta-lactamase producers. Clin Microbiol Infect 14:117–123

Cavaco LM, Hansen DS, Friis-Moller A, Aarestrup FM, Hasman H, Frimodt-Moller N (2007) First detection of plasmid-mediated quinolone resistance (qnrA and qnrS) in *Escherichia coli* strains isolated from humans in Scandinavia. J Antimicrob Chemother 59:804–805

Cavaco LM, Hasman H, Xia S, Aarestrup FM (2009) qnrD, a novel gene conferring transferable quinolone resistance in *Salmonella enterica* Serovar Kentucky and Bovismorbificans strains of human origin. Antimicrob Agents Chemother 53:603–608

Chen L, Chen ZL, Liu JH, Zeng ZL, Ma JY, Jiang HX (2007) Emergence of RmtB methylase-producing *Escherichia coli* and *Enterobacter cloacae* isolates from pigs in China. J Antimicrob Chemother 59:880–885

Chopra I, Roberts M (2001) Tetracycline antibiotics: mode of action, applications, molecular biology, and epidemiology of bacterial resistance. Microbiol Mol Biol Rev 65:232

Corkill JE, Anson JJ, Hart CA (2005) High prevalence of the plasmid-mediated quinolone resistance determinant qnrA in multidrug-resistant Enterobacteriaceae from blood cultures in Liverpool, UK. J Antimicrob Chemother 56:1115–1117

Cormio L, VuopioVarkila J, Siitonen A, Talja M, Ruutu M (1996) Bacterial adhesion and biofilm formation on various double-J stents in vivo and in vitro. Scand J Urol Nephrol 30:19–24

Corrente M, D'Abramo M, Latronico F, Greco MF, Bellacicco AL, Greco G, Martella V, Buonavoglia D (2009) Methicillin-resistant coagulase negative staphylococci isolated from horses. New Microbiol 32:311–314

D'Costa VM, Griffiths E, Wright GD (2007) Expanding the soil antibiotic resistome: exploring environmental diversity. Curr Opin Microbiol 10:481–489

Das JR, Bhakoo M, Jones MV, Gilbert P (1998) Changes in the biocide susceptibility of *Staphylococcus epidermidis* and *Escherichia coli* cells associated with rapid attachment to plastic surfaces. J Appl Microbiol 84:852–858

Datta N, Kontomichalou P (1965) Penicillinase synthesis controlled by infectious R factors in Enterobacteriaceae. Nature 208:239

Datta N, Nugent M, Amyes SGB, McNeilly P (1979) Multiple mechanisms of trimethoprim resistance in strains of *Escherichia coli* from a patient treated with long-term co-trimoxazole. J Antimicrob Chemother 5:399–406

Davies D (2003) Understanding biofilm resistance to antibacterial agents. Nat Rev Drug Discov 2: 114–122

de Beer D, Stoodley P, Roe F, Lewandowski Z (1994) Effects of biofilm structures on oxygen distribution and mass transport. Biotechnol Bioeng 43:1131–1138

De Kievit TR, Parkins MD, Gillis RJ, Srikumar R, Ceri H, Poole K, Iglewski BH, Storey DG (2001) Multidrug efflux pumps: expression patterns and contribution to antibiotic resistance in *Pseudomonas aeruginosa* biofilms. Antimicrob Agents Chemother 45:1761–1770

Dejonge BLM, Chang YS, Gage D, Tomasz A (1992) Peptidoglycan composition of a highly methicillin-resistant *Staphylococcus aureus* strain – the role of penicillin binding protein-2a. J Biol Chem 267:11248–11254

Devriese LA, Ieven M, Goossens H, Vandamme P, Pot B, Hommez J, Haesebrouck F (1996) Presence of vancomycin-resistant enterococci in farm and pet animals. Antimicrob Agents Chemother 40:2285–2287

Domagk G (1935) A new class of disinfectant. Dtsch Med Wochenschr 61:829–832

Donelli G (2006) Vascular catheter-related infection and sepsis. Surg Infect (Larchmt) 7(Suppl 2): S25–S27

Driffield K, Miller K, Bostock JM, O'Neill AJ, Chopra I (2008) Increased mutability of *Pseudomonas aeruginosa* in biofilms. J Antimicrob Chemother 61:1053–1056
Drlica K, Zhao X (1997) DNA gyrase, topoisomerase IV, and the 4-quinolones. Microbiol Mol Biol Rev 61:377–392
Dunowska M, Morley PS, Traub-Dargatz JL, Hyatt DR, Dargatz DA (2006) Impact of hospitalization and antimicrobial drug administration on antimicrobial susceptibility patterns of commensal *Escherichia coli* isolated from the feces of horses. J Am Vet Med Assoc 228:1909–1917
Ewers C, Grobbel M, Stamm I, Kopp PA, Diehl I, Semmler T, Fruth A, Beutlich J, Guerra B, Wieler LH, Guenther S (2010) Emergence of human pandemic O25:H4-ST131 CTX-M-15 extended-spectrum-beta-lactamase-producing *Escherichia coli* among companion animals. J Antimicrob Chemother 65:651–660
Fine DM, Tobias AH (2007) Cardiovascular device infections in dogs: report of 8 cases and review of the literature. J Vet Intern Med 21:1265–1271
Fleming A (1932) On the specific antibacterial properties of penicillin and potassium tellurite – incorporating a method of demonstrating some bacterial antiagonisms. J Pathol Bacteriol 35: 831–842
Fluit AC, Wielders CLC, Verhoef J, Schmitz FJ (2001) Epidemiology and susceptibility of 3,051 *Staphylococcus aureus* isolates from 25 university hospitals participating in the European SENTRY study. J Clin Microbiol 39:3727–3732
Fux CA, Stoodley P, Hall-Stoodley L, Costerton JW (2003) Bacterial biofilms: a diagnostic and therapeutic challenge. Expert Rev Anti Infect Ther 1:667–683
Gal Z, Kovacs P, Hernadi F, Barabas G, Kiss L, Igloi A, Szabo I (2001) Investigation of oxacillin-hydrolyzing beta-lactamase in borderline methicillin-resistant clinical isolates of *Staphylococcus aureus*. Chemotherapy 47:233–238
Galimand M, Courvalin P, Lambert T (2003) Plasmid-mediated high-level resistance to aminoglycosides in Enterobacteriaceae due to 16S rRNA methylation. Antimicrob Agents Chemother 47:2565–2571
Gaynor M, Mankin AS (2003) Macrolide antibiotics: binding site, mechanism of action, resistance. Curr Top Med Chem 3:949–960
Georgopapadakou NH, Smith SA, Bonner DP (1982) Penicillin-binding proteins in a *Staphylococcus aureus* strain resistant to specific beta-lactam antibiotics. Antimicrob Agents Chemother 22:172–175
Gibson JS, Cobbold RN, Heisig P, Sidjabat HE, Kyaw-Tanner MT, Trott DJ (2010a) Identification of Qnr and AAC(6′)-1b-cr plasmid-mediated fluoroquinolone resistance determinants in multidrug-resistant *Enterobacter* spp. isolated from extraintestinal infections in companion animals. Vet Microbiol 143:329–336
Gibson JS, Cobbold RN, Trott DJ (2010b) Characterization of multidrug-resistant *Escherichia coli* isolated from extraintestinal clinical infections in animals. J Med Microbiol 59:592–598
Gilbert P, Collier PJ, Brown MR (1990) Influence of growth rate on susceptibility to antimicrobial agents: biofilms, cell cycle, dormancy, and stringent response. Antimicrob Agents Chemother 34:1865–1868
Gonzalez-Zorn B, Teshager T, Casas M, Porrero MC, Moreno MA, Courvalin P, Dominguez L (2005) armA and aminoglycoside resistance in *Escherichia coli*. Emerg Infect Dis 11:954–956
Goto T, Nakame Y, Nishida M, Ohi Y (1999) In vitro bactericidal activities of beta-lactamases, amikacin, and fluoroquinolones against *Pseudomonas aeruginosa* biofilm in artificial urine. Urology 53:1058–1062
Gristina AG, Oga M, Webb LX, Hobgood CD (1985) Adherent bacterial colonization in the pathogenesis of osteomyelitis. Science 228:990–993
Grobbel M, Lubke-Becker A, Alesik E, Schwarz S, Wallmann J, Werckenthin C, Wieler LH (2007) Antimicrobial susceptibility of *Escherichia coli* from swine, horses, dogs and cats as determined in the BfT-GermVet monitoring program 2004–2006. Berl Münch Tierärztl Wochenschr 120:391–401

Hall RM, Collis CM (1995) Mobile gene cassettes and integrons – capture and spread of genes by site-specific recombination. Mol Microbiol 15:593–600

Hall LMC, Henderson-Begg SK (2006) Hypermutable bacteria isolated from humans – a critical analysis. Microbiology 152:2505–2514

Hansen SK, Rainey PB, Haagensen JAJ, Molin S (2007) Evolution of species interactions in a biofilm community. Nature 445:533–536

Hanssen AM, Sollid JUE (2006) SCCmec in staphylococci: genes on the move. FEMS Immunol Med Microbiol 46:8–20

Harihara H, Barnum DA (1973) Drug resistance among pathogenic Enterobacteriaceae from animals in Ontario. Can J Public Health 64:69

Hata M, Suzuki M, Matsumoto M, Takahashi M, Sato K, Ibe S, Sakae K (2005) Cloning of a novel gene for quinolone resistance from a transferable plasmid in *Shigella flexneri* 2b. Antimicrob Agents Chemother 49:801–803

Heikkila E, Renkonen OV, Sunila R, Uurasmaa P, Huovinen P (1990) The emergence and mechanisms of trimethoprim resistance in *Escherichia coli* isolated from outpatients in Finland. J Antimicrob Chemother 25:275–283

Hopkins KL, Batchelor MJ, Liebana E, Deheer-Graham AP, Threlfalla EJ (2006) Characterisation of CTX-M and AmpC genes in human isolates of *Escherichia coli* identified between 1995 and 2003 in England and Wales. Int J Antimicrob Agents 28:180–192

Huovinen P, Sundstrom L, Swedberg G, Skold O (1995) Trimethoprim and sulfonamide resistance. Antimicrob Agents Chemother 39:279–289

Huys G, D'Haene K, Collard JM, Swings J (2004) Prevalence and molecular characterization of tetracycline resistance in *Enterococcus* isolates from food. Appl Environ Microbiol 70: 1555–1562

Irina F, Misic D, Ruzica A (2007) Investigation of the presence of extended spectrum beta-lactamases (ESBL) in multiresistant strains of *E. coli* and *Salmonella* species originated from domestic animals. Acta Vet Beograd 57:369–379

Ishida H, Ishida Y, Kurosaka Y, Otani T, Sato K, Kobayashi H (1998) In vitro and in vivo activities of levofloxacin against biofilm-producing *Pseudomonas aeruginosa*. Antimicrob Agents Chemother 42:1641–1645

Jacoby GA, Han P (1996) Detection of extended-spectrum beta-lactamases in clinical isolates of *Klebsiella pneumoniae* and *Escherichia coli*. J Clin Microbiol 34:908–911

Jacoby GA, Walsh KE, Mills DM, Walker VJ, Oh H, Robicsek A, Hooper DC (2006) qnrB, another plasmid-mediated gene for quinolone resistance. Antimicrob Agents Chemother 50: 1178–1182

Jang SS, Hansen LM, Breher JE, Riley DA, Magdesian KG, Madigan JE, Tang YJ, Silva J, Hirsh DC (1997) Antimicrobial susceptibilities of equine isolates of *Clostridium difficile* and molecular characterization of metronidazole-resistant strains. Clin Infect Dis 25:S266–S267

Jevons MP, Rolinson GN, Knox R (1961) Celbenin-resistant staphylococci. Br Med J 1:124

Jiang Y, Zhou ZH, Qian Y, Wei ZQ, Yu YS, Hu SN, Li LJ (2008) Plasmid-mediated quinolone resistance determinants qnr and aac(6′)-Ib-cr in extended-spectrum beta-lactamase-producing *Escherichia coli* and *Klebsiella pneumoniae* in China. J Antimicrob Chemother 61:1003–1006

Kajiura T, Wada H, Ito K, Anzai Y, Kato F (2006) Conjugative plasmid transfer in the biofilm formed by *Enterococcus faecalis*. J Health Sci 52:358–367

Karisik E, Ellington MJ, Pike R, Warren RE, Livermore DM, Woodford N (2006) Molecular characterization of plasmids encoding CTX-M-15 beta-lactamases from *Escherichia coli* strains in the United Kingdom. J Antimicrob Chemother 58:665–668

Kather EJ, Marks SL, Foley JE (2006) Determination of the prevalence of antimicrobial resistance genes in canine *Clostridium perfringens* isolates. Vet Microbiol 113:97–101

Kim J, Jeong JH, Cha HY, Jin JS, Lee JC, Lee YC, Seol SY, Cho DT (2007) Detection of diverse SCCmec variants in methicillin-resistant *Staphylococcus aureus* and comparison of SCCmec typing methods. Clin Microbiol Infect 13:1128–1130

Kizerwetter-Swida M, Chrobak D, Rzewuska M, Binek M (2009) Antibiotic resistance patterns and occurrence of *mecA* gene in Staphylococcus intermedius strains of canine origin. Pol J Vet Sci 12:9–13

Klevens RM, Edwards JR, Tenover FC, McDonald LC, Horan T, Gaynes R (2006) Changes in the epidemiology of methicillin-resistant *Staphylococcus aureus* in intensive care units in US hospitals, 1992–2003. Clin Infect Dis 42:389–391

Kliebe C, Nies BA, Meyer JF, Tolxdorffneutzling RM, Wiedemann B (1985) Evolution of plasmid-coded resistance to broad-spectrum cephalosporins. Antimicrob Agents Chemother 28:302–307

Kotra LP, Haddad J, Mobashery S (2000) Aminoglycosides: perspectives on mechanisms of action and resistance and strategies to counter resistance. Antimicrob Agents Chemother 44: 3249–3256

Kuhl SA, Pattee PA, Baldwin JN (1978) Chromosomal map location of methicillin resistance determinant in *Staphylococcus aureus*. J Bacteriol 135:460–465

Kuo HC, Chou CC, Tu C, Gong SR, Han CL, Liao JW, Chang SK (2009) Characterization of plasmid-mediated quinolone resistance by the qnrS gene in *Escherichia coli* isolated from healthy chickens and pigs. Vet Med 54:473–482

Kvist M, Hancock V, Klemm P (2008) Inactivation of efflux pumps abolishes bacterial biofilm formation. Appl Environ Microbiol 74:7376–7382

Lavigne JP, Marchandin H, Delmas J, Bouziges N, Lecaillon E, Cavalie L, Jean-Pierre H, Bonnet R, Sotto A (2006) qnrA in CTX-M-producing *Escherichia coli* isolates from France. Antimicrob Agents Chemother 50:4224–4228

Leclercq R, Courvalin P (1991) Bacterial resistance to macrolide, lincosamide, and streptogramin antibiotics by target modification. Antimicrob Agents Chemother 35:1267–1272

Leclercq R, Derlot E, Duval J, Courvalin P (1988) Plasmid-mediated resistance to vancomycin and teicoplanin in *Enterococcus faecium*. N Engl J Med 319:157–161

Lederberg J, Lederberg EM (1952) Replica plating and indirect selection of bacterial mutants. J Bacteriol 63:399–406

Leiros HKS, Kozielski-Stuhrmann S, Kapp U, Terradot L, Leonard GA, McSweeney SM (2004) Structural basis of 5-nitroimidazole antibiotic resistance – the crystal structure of NimA from *Deinococcus radiodurans*. J Biol Chem 279:55840–55849

Leverstein-van Hall MA, Blok HEM, Donders ART, Paauw A, Fluit AC, Verhoef J (2003) Multidrug resistance among Enterobacteriaceae is strongly associated with the presence of integrons and is independent of species or isolate origin. J Infect Dis 187:251–259

Lewis K (2001) Riddle of biofilm resistance. Antimicrob Agents Chemother 45:999–1007

Liebana E, Gibbs M, Clouting C, Barker L, Clifton-Hadley FA, Pleydell E, Abdalhamid B, Hanson ND, Martin L, Poppe C, Davies RH (2004) Characterization of beta-lactamases responsible for resistance to extended-spectrum cephalosporins in *Escherichia coli* and *Salmonella enterica* strains from food-producing animals in the United Kingdom. Microb Drug Resist 10:1–9

Liebana E, Batchelor M, Hopkins KL, Clifton-Hadley FA, Teale CJ, Foster A, Barker L, Threlfall EJ, Davies RH (2006) Longitudinal farm study of extended-spectrum beta-lactamase-mediated resistance. J Clin Microbiol 44:1630–1634

Livermore DM (2003) Bacterial resistance: origins, epidemiology, and impact. Clin Infect Dis 36: S11–S23

Loeffler A, Boag AK, Sung J, Lindsay JA, Guardabassi L, Dalsgaard A, Smith H, Stevens KB, Lloyd DH (2005) Prevalence of methicillin-resistant *Staphylococcus aureus* among staff and pets in a small animal referral hospital in the UK. J Antimicrob Chemother 56:692–697

Luethje P, Schwarz S (2007) Molecular basis of resistance to macrolides and lincosamides among staphylococci and streptococci from various animal sources collected in the resistance monitoring program Bff-GermVet. Int J Antimicrob Agents 29:528–535

Machado E, Coque TM, Canton R, Baquero F, Sousa JC, Peixe L, Portuguese Resistance Study Group (2006) Dissemination in Portugal of CTX-M-15-, OYA-1-, and TEM-1-producing

Enterobacteriaceae strains containing the aac(6′)-Ib-cr gene, which encodes an aminoglycoside- and fluoroquinolone-modifying enzyme. Antimicrob Agents Chemother 50:3220–3220
Malik S, Christensen H, Peng H, Barton MD (2007) Presence and diversity of the beta-lactamase gene in cat and dog staphylococci. Vet Microbiol 123:162–168
Mangalappalli-Illathu AK, Lawrence JR, Swerhone GDW, Korber DR (2008) Architectural adaptation and protein expression patterns of *Salmonella enterica* serovar Enteritidis biofilms under laminar flow conditions. Int J Food Microbiol 123:109–120
Martinez-Martinez L, Pascual A, Jacoby GA (1998) Quinolone resistance from a transferable plasmid. Lancet 351:797–799
Martínez-Martínez L, Pascual A, García I, Tran JH, Jacoby GA (2003) Interaction of plasmid and host quinolone resistance. J Antimicrob Chemother 51:1037–1039
Matsumoto Y, Ikeda F, Kamimura T, Yokota Y, Mine Y (1988) Novel plasmid-mediated 3-lactamase from *Escherichia coli* that inactivates oxyimino-cephalosporins. Antimicrob Agents Chemother 32:1243–1246
Matthews DA, Bolin JT, Burridge JM, Filman DJ, Volz KW, Kraut J (1985) Dihydrofolate reductase. The stereochemistry of inhibitor selectivity. J Biol Chem 260:392–399
May T, Ito A, Okabe S (2009) Induction of multidrug resistance mechanism in *Escherichia coli* biofilms by interplay between tetracycline and ampicillin resistance genes. Antimicrob Agents Chemother 53:4628–4639
McKay KA, Ruhnke HL, Barnum DA (1965) The results of sensitivity tests on animal pathogens conducted over the period 1956–1963. Can Vet J 6:103–111
Monzon M, Oteiza C, Leiva J, Lamata M, Amorena B (2002) Biofilm testing of *Staphylococcus epidermidis* clinical isolates: low performance of vancomycin in relation to other antibiotics. Diagn Microbiol Infect Dis 44:319–324
Moura I, Radhouani H, Torres C, Poeta P, Igrejas G (2010) Detection and genetic characterisation of vanA-containing *Enterococcus* strains in healthy Lusitano horses. Equine Vet J 42:181–183
Naseer U, Haldorsen B, Tofteland S, Hegstad K, Scheutz F, Simonsen GS, Sundsfjord A, Norwegian ESG (2009) Molecular characterization of CTX-M-15-producing clinical isolates of *Escherichia coli* reveals the spread of multidrug-resistant ST131 (O25:H4) and ST964 (O102:H6) strains in Norway. APMIS 117:526–536
Nicolas-Chanoine MH, Blanco J, Leflon-Guibout V, Demarty R, Alonso MP, Canica MM, Park YJ, Lavigne JP, Pitout J, Johnson JR (2008) Intercontinental emergence of *Escherichia coli* clone O25: H4-ST131 producing CTX-M-15. J Antimicrob Chemother 61:273–281
Nordmann P, Poirel L (2005) Emergence of plasmid-mediated resistance to quinolones in Enterobacteriaceae. J Antimicrob Chemother 56:463–469
Ogeer-Gyles JS, Mathews KA, Boerlin P (2006) Nosocomial infections and antimicrobial resistance in critical care medicine. J Vet Emerg Crit Care 16:1–18
Oliver A, Canton R, Campo P, Baquero F, Blazquez J (2000) High frequency of hypermutable *Pseudomonas aeruginosa* in cystic fibrosis lung infection. Science 288:1251–1253
Oliver A, Perez-Diaz JC, Coque TM, Baquero F, Canton R (2001) Nucleotide sequence and characterization of a novel cefotaxime-hydrolyzing beta-lactamase (CTX-M-10) isolated in Spain. Antimicrob Agents Chemother 45:616–620
Ossiprandi MC, Bottarelli E, Cattabiani F, Bianchi E (2008) Susceptibility to vancomycin and other antibiotics of 165 *Enterococcus* strains isolated from dogs in Italy. Comp Immunol Microbiol Infect Dis 31:1–9
Owen MR, Moores AP, Coe RJ (2004) Management of MRSA septic arthritis in a dog using a gentamicin-impregnated collagen sponge. J Small Anim Pract 45:609–612
Paladino JA, Sunderlin JL, Price CS, Schentag JJ (2002) Economic consequences of antimicrobial resistance. Surg Infect (Larchmt) 3:259–267
Papanicolaou GA, Medeiros AA, Jacoby GA (1990) Novel plasmid-mediated β-lactamase (MIR-1) conferring resistance to oxyimino- and a-methoxy β-lactams in clinical isolates of *Klebsiella pneumoniae*. Antimicrob Agents Chemother 34:2200–2209

Perichon B, Courvalin P, Galimand M (2007) Transferable resistance to aminoglycosides by methylation of G1405 in 16S rRNA and to hydrophilic fluoroquinolones by QepA-mediated efflux in *Escherichia coli*. Antimicrob Agents Chemother 51:2464–2469

Perreten V, Kadlec K, Schwarz S, Andersson UG, Finn M, Greko C, Moodley A, Kania SA, Frank LA, Bemis DA, Franco A, Iurescia M, Battisti A, Duim B, Wagenaar JA, van Duijkeren E, Weese JS, Fitzgerald JR, Rossano A, Guardabassi L (2010) Clonal spread of methicillin-resistant *Staphylococcus pseudintermedius* in Europe and North America: an international multicentre study. J Antimicrob Chemother 65:1145–1154

Pitout JDD, Laupland KB (2008) Extended-spectrum beta-lactamase-producing Enterobacteriaceae: an emerging public-health concern. Lancet Infect Dis 8:159–166

Poeta P, Costa D, Rodrigues J, Torres C (2005) Study of faecal colonization by vanA-containing *Enterococcus* strains in healthy humans, pets, poultry and wild animals in Portugal. J Antimicrob Chemother 55:278–280

Pomba C, da Fonseca JD, Baptista BC, Correia JD, Martinez-Martinez L (2009) Detection of the pandemic O25-ST131 human virulent *Escherichia coli* CTX-M-15-producing clone harboring the qnrB2 and aac(6′)-Ib-cr genes in a dog. Antimicrob Agents Chemother 53:327–328

Pyorala S, Taponen S (2009) Coagulase-negative staphylococci-emerging mastitis pathogens. Vet Microbiol 134:3–8

Rachid S, Ohlsen K, Witte W, Hacker J, Ziebuhr W (2000) Effect of subinhibitory antibiotic concentrations on polysaccharide intercellular adhesin expression in biofilm-forming *Staphylococcus epidermidis*. Antimicrob Agents Chemother 44:3357–3363

Ramsay JWA, Garnham AJ, Mulhall AB, Crow RA, Bryan JM, Eardley I, Vale JA, Whitfield HN (1989) Biofilms, bacteria and bladder catheters. A clinical study. Br J Urol 64:395–398

Rankin SC, Whichard JM, Joyce K, Stephens L, O'Shea K, Aceto H, Munro DS, Benson CE (2005) Detection of a bla(SHV) extended-spectrum beta-lactamase in *Salmonella enterica* serovar Newport MDR-AmpC. J Clin Microbiol 43:5792–5793

Reece RJ, Maxwell A, Wang JC (1991) DNA gyrase: structure and function. Crit Rev Biochem Mol Biol 26:335–375

Rice EW, Boczek LA, Johnson CH, Messer JW (2003) Detection of intrinsic vancomycin resistant enterococci in animal and human feces. Diagn Microbiol Infect Dis 46:155–158

Riesen A, Perreten V (2009) Antibiotic resistance and genetic diversity in *Staphylococcus aureus* from slaughter pigs in Switzerland. Schweiz Arch Tierheilkd 151:425–431

Roberts MC (1996) Tetracycline resistance determinants: mechanisms of action, regulation of expression, genetic mobility, and distribution. FEMS Microbiol Rev 19:1–24

Roberts MC (2005) Update on acquired tetracycline resistance genes. FEMS Microbiol Lett 245:195–203

Roberts MC, Sutcliffe J, Courvalin P, Jensen LB, Rood J, Seppala H (1999) Nomenclature for macrolide and macrolide-lincosamide-streptogramin B resistance determinants. Antimicrob Agents Chemother 43:2823–2830

Robicsek A, Strahilevitz J, Jacoby GA, Macielag M, Abbanat D, Park CH, Bush K, Hooper DC (2006) Fluoroquinolone-modifying enzyme: a new adaptation of a common aminoglycoside acetyltransferase. Nat Med 12:83–88

Rossolini GM, Mantengoli E, Montagnani F, Pollini S (2010) Epidemiology and clinical relevance of microbial resistance determinants versus anti-Gram-positive agents. Curr Opin Microbiol 13:582–588

Roupas A, Pitton JS (1974) R factor-mediated and chromosomal resistance to ampicillin in *Escherichia coli*. Antimicrob Agents Chemother 5:186–191

Ruiz J (2003) Mechanisms of resistance to quinolones: target alterations, decreased accumulation and DNA gyrase protection. J Antimicrob Chemother 51:1109–1117

Saenz Y, Zarazaga M, Brinas L, Lantero M, Ruiz-Larrea F, Torres C (2001) Antibiotic resistance in *Escherichia coli* isolates obtained from animals, foods and humans in Spain. Int J Antimicrob Agents 18:353–358

Saenz Y, Brinas L, Dominguez E, Ruiz J, Zarazaga M, Vila J, Torres C (2004) Mechanisms of resistance in multiple-antibiotic-resistant *Escherichia coli* strains of human, animal, and food origins. Antimicrob Agents Chemother 48:3996–4001

Sailer FC, Meberg BM, Young KD (2003) Beta-lactam induction of colanic acid gene expression in *Escherichia coli*. FEMS Microbiol Lett 226:245–249

Sanders CC (1987) Chromosomal cephalosporinases responsible for multiple resistance to newer β-lactam antibiotics. Annu Rev Microbiol 41:573–593

Sanders WE, Sanders CC (1988) Inducible beta-lactamases: clinical and epidemiologic implications for use of newer cephalosporins. Rev Infect Dis 10:830–838

Sanders CC, Bradford PA, Ehrhardt AF, Bush K, Young KD, Henderson TA, Sanders EW (1997) Penicillin-binding proteins and induction of AmpC beta-lactamase. Antimicrob Agents Chemother 41:2013–2015

Saroglou G, Cromer M, Bisno AL (1980) Methicillin-resistant *Staphylococcus aureus* – interstate spread of nosocomial infections with emergence of gentamicin–methicillin resistant strains. Infect Control Hosp Epidemiol 1:81–89

Schnellmann C, Gerber V, Rossano A, Jaquier V, Panchaud Y, Doherr MG, Thomann A, Straub R, Perreten V (2006) Presence of new mecA and mph(C) variants conferring antibiotic resistance in *Staphylococcus* spp. isolated from the skin of horses before and after clinic admission. J Clin Microbiol 44:4444–4454

Schwarz S, Kadlec K, Strommenger B (2008) Methicillin-resistant *Staphylococcus aureus* and *Staphylococcus pseudintermedius* detected in the BfT-GermVet monitoring programme 2004–2006 in Germany. J Antimicrob Chemother 61:282–285

Seo MR, Park YS, Pai H (2010) Characteristics of plasmid-mediated quinolone resistance genes in extended-spectrum cephalosporin-resistant isolates of *Klebsiella pneumoniae* and *Escherichia coli* in Korea. Chemotherapy 56:46–53

Shaw KJ, Rather PN, Hare RS, Miller GH (1993) Molecular-genetics of aminoglycoside resistance genes and familial relationships of the aminoglycoside-modifying enzymes. Microbiol Rev 57:138–163

Sheehan E, McKenna J, Mulhall KJ, Marks P, McCormack D (2004) Adhesion of *Staphylococcus* to orthopaedic metals, an in vivo study. J Orthop Res 22:39–43

Shigeta M, Tanaka G, Komatsuzawa H, Sugai M, Suginaka H, Usui T (1997) Permeation of antimicrobial agents through *Pseudomonas aeruginosa* biofilms: a simple method. Chemotherapy 43:340–345

Sidjabat HE, Hanson ND, Smith-Moland E, Bell JM, Gibson JS, Filippich LJ, Trott DJ (2007) Identification of plasmid-mediated extended-spectrum and AmpC beta-lactamases in *Enterobacter* spp. isolated from dogs. J Med Microbiol 56:426–434

Singh R, Ray P, Das A, Sharma M (2010) Penetration of antibiotics through *Staphylococcus aureus* and *Staphylococcus epidermidis* biofilms. J Antimicrob Chemother 65:1955–1958

Skinner S, Inglis B, Matthews PR, Stewart PR (1988) Mercury and tetracycline resistance genes and flanking repeats associated with methicillin resistance on the chromosome of *Staphylococcus aureus*. Mol Microbiol 2:289–292

Smith MM, Vasseur PB, Saunders HM (1989) Bacterial growth associated with metallic implants in dogs. J Am Vet Med Assoc 195:765–767

Sternberg C, Christensen BB, Johansen T, Toftgaard Nielsen A, Andersen JB, Givskov M, Molin S (1999) Distribution of bacterial growth activity in flow-chamber biofilms. Appl Environ Microbiol 65:4108–4117

Stewart PS, Costerton JW (2001) Antibiotic resistance of bacteria in biofilms. Lancet 358:135–138

Stone G, Wood P, Dixon L, Keyhan M, Matin A (2002) Tetracycline rapidly reaches all the constituent cells of uropathogenic *Escherichia coli* biofilms. Antimicrob Agents Chemother 46:2458–2461

Strahilevitz J, Jacoby GA, Hooper DC, Robicsek A (2009) Plasmid-mediated quinolone resistance: a multifaceted threat. Clin Microbiol Rev 22:664-

Suzuki S, Shibata N, Yamane K, Wachino J, Ito K, Arakawa Y (2009) Change in the prevalence of extended-spectrum-beta-lactamase-producing *Escherichia coli* in Japan by clonal spread. J Antimicrob Chemother 63:72–79

Towner KJ, Brennan A, Zhang Y, Holtham CA, Brough JL, Carter GI (1994) Genetic structures associated with spread of the type IA trimethoprim-resistant dihydrofolate-reductase gene amongst *Escherichia coli* strains isolated in the Nottingham area of the United-Kingdom. J Antimicrob Chemother 33:25–32

Tran Van Nhieu G, Bordon F, Collatz E (1992) Incidence of an aminoglycoside 6′-*N*-acetyltransferase, ACC(6′)-1b, in amikacin-resistant clinical isolates of Gram-negative bacilli, as determined by DNA–DNA hybridisation and immunoblotting. J Med Microbiol 36: 83–88

Tran JH, Jacoby GA (2002) Mechanism of plasmid-mediated quinolone resistance. Proc Natl Acad Sci USA 99:5638–5642

Tran JH, Jacoby GA, Hooper DC (2005) Interaction of the plasmid-encoded quinolone resistance protein Qnr with *Escherichia coli* DNA gyrase. Antimicrob Agents Chemother 49:118–125

Tuckman M, Petersen PJ, Howe AYM, Orlowski M, Mullen S, Chan K, Bradford PA, Jones CH (2007) Occurrence of tetracycline resistance genes among *Escherichia coli* isolates from the phase 3 clinical trials for tigecycline. Antimicrob Agents Chemother 51:3205–3211

Tupin A, Gualtieri M, Roquet-Baneres F, Morichaud Z, Brodolin K, Leonetti JP (2010) Resistance to rifampicin: at the crossroads between ecological, genomic and medical concerns. Int J Antimicrob Agents 35:519–523

Turkyilmaz S, Erdem V, Bozdogan B (2010) Investigation of antimicrobial susceptibility for enterococci isolated from cats and dogs and the determination of resistance genes by polymerase chain reaction. Turk J Vet Anim Sci 34:61–68

Tzouvelekis LS, Tzelepi E, Tassios PT, Legakis NJ (2000) CTX-M-type beta-lactamases: an emerging group of extended-spectrum enzymes. Int J Antimicrob Agents 14:137–142

van Duijkeren E, Vulto AG, Vanmiert A (1994) Trimethoprim sulfonamide combinations in the horse – a review. J Vet Pharmacol Ther 17:64–73

van Duijkeren E, Moleman M, van Oldruitenborgh-Oosterbaan MMS, Multem J, Troelstra A, Fluit AC, van Wamel WJB, Houwers DJ, de Neeling AJ, Wagenaar JA (2010) Methicillin-resistant *Staphylococcus aureus* in horses and horse personnel: an investigation of several outbreaks. Vet Microbiol 141:96–102

Vanni M, Tognetti R, Pretti C, Crema F, Soldani G, Meucci V, Intorre L (2009) Antimicrobial susceptibility of *Staphylococcus intermedius* and *Staphylococcus schleiferi* isolated from dogs. Res Vet Sci 87:192–195

Vengust M, Anderson MEC, Rousseau J, Weese JS (2006) Methicillin-resistant staphylococcal colonization in clinically normal dogs and horses in the community. Lett Appl Microbiol 43:602–606

VMD (2009) Sales of antimicrobial products used as veterinary medicines, growth promoters and coccidiostats in the UK in 2008. http://www.vmd.gov.uk/Publications/Antibiotic/salesanti08.pdf, Accessed November 2010

Vo ATT, van Duijkeren E, Fluit AC, Gaastra W (2007) Characteristics of extended-spectrum cephalosporin-resistant *Escherichia coli* and *Klebsiella pneumoniae* isolates from horses. Vet Microbiol 124:248–255

Waksman SA, Reilly HC, Schatz A (1945) Strain specificity and production of antibiotic substances: V. Strain resistance of bacteria to antibiotic substances, especially to streptomycin. Proc Natl Acad Sci USA 31:157–164

Wang MH, Guo QL, Xu XG, Wang XY, Ye XY, Wu S, Hooper DC, Wang MG (2009) New plasmid-mediated quinolone resistance gene, qnrC, found in a clinical isolate of *Proteus mirabilis*. Antimicrob Agents Chemother 53:1892–1897

Weese J, van Duijkeren E (2010) Methicillin-resistant *Staphylococcus aureus* and *Staphylococcus pseudintermedius* in veterinary medicine. Vet Microbiol 140:418–429

Weese JS, Rousseau J, Traub-Dargatz JL, Willey BM, McGeer AJ, Low DE (2005) Community-associated methicillin-resistant *Staphylococcus aureus* in horses and humans who work with horses. J Am Vet Med Assoc 226:580–583

Weigel LM, Donlan RM, Shin DH, Jensen B, Clark NC, McDougal LK, Zhu WM, Musser KA, Thompson J, Kohlerschinidt D, Dumas N, Limberger RJ, Patel JB (2007) High-level vancomycin-resistant *Staphylococcus aureus* isolates associated with a polymicrobial biofilm. Antimicrob Agents Chemother 51:231–238

Weisblum B (1995) Insights into erythromycin action from studies of its activity as inducer of resistance. Antimicrob Agents Chemother 39:797–805

White PA, Rawlinson WD (2001) Current status of the *aadA* and *dfr* gene cassette families. J Antimicrob Chemother 47:495–496

White PA, McIver CJ, Deng YM, Rawlinson WD (2000) Characterisation of two new gene cassettes, aadA5 and dfrA17. FEMS Microbiol Lett 182:265–269

WHO (2007) Critically important antimicrobials for human medicine. In: Report of the second WHO expert meeting, Copenhagen, 29–31 May 2007

Wimpenny J, Manz W, Szewzyk U (2000) Heterogeneity in biofilms. FEMS Microbiol Rev 24: 661–671

Woodford N, Johnson AP, Morrison D, Speller DCE (1995) Current perspectives on glycopeptide resistance. Clin Microbiol Rev 8:585–615

Woodford N, Carattoli A, Karisik E, Underwood A, Ellington MJ, Livermore DM (2009) Complete nucleotide sequences of plasmids pEK204, pEK499, and pEK516, encoding CTX-M enzymes in three major *Escherichia coli* lineages from the United Kingdom, all belonging to the international O25:H4-ST131 clone. Antimicrob Agents Chemother 53: 4472–4482

Wright GD (2007) The antibiotic resistome: the nexus of chemical and genetic diversity. Nat Rev Microbiol 5:175–186

Yamane K, Wachino JI, Suzuki S, Kimura K, Shibata N, Kato H, Shibayama K, Konda T, Arakawa Y (2007) New plasmid-mediated fluoroquinolone efflux pump, QepA, found in an *Escherichia coli* clinical isolate. Antimicrob Agents Chemother 51:3354–3360

Yan J, Wu J, Ko W, Tsai S, Chuang C, Wu H, Lu Y, Li J (2004) Plasmid-mediated 16S rRNA methylases conferring high-level aminoglycoside resistance in *Escherichia coli* and *Klebsiella pneumoniae* isolates from two Taiwanese hospitals. J Antimicrob Chemother 54:1007–1012

Zhang L, Mah TF (2008) Involvement of a novel efflux system in biofilm-specific resistance to antibiotics. J Bacteriol 190:4447–4452

Index

S.L. Percival et al. (eds.), *Biofilms and Veterinary Medicine*,
Springer Series on Biofilms 6, DOI 10.1007/978-3-642-21289-5,